Advances in III-V Semiconductor Nanowires and Nanodevices

Edited By

Jianye Li

University of Science & Technology Beijing
China

Deli Wang

University of California
USA

Ray R. LaPierre

McMaster University
Canada

CONTENT

CHAPTERS

FOREWORD

By Lars Samuelson, Solid State Physics/the Nanometer Structure Consortium, Lund University, Box 118, S-221 00 Lund, Sweden

Web: www.nano.lu.se/lars.samuelson/; Email: lars.samuelson@ftf.lth.se

III-V semiconductors constitute a wonderful materials system, allowing highly ideal materials with designable properties and, when combined into heterostructures, allow vastly varying structures and devices to be made. One of the limitations in the realization of advanced heterostructure systems using traditional planar epitaxy, is the requirement for a strict lattice-matching of adjacent layers, limiting what can be fabricated or making the realization of such heterostructure systems very demanding. This is when III-V nanowires came to the rescue: With the great progress in the late 1980's and early 1990's, in the laboratory of Dr. Hiruma at Hitachi Central Research Laboratories, the advent of III-V nanowires occurred based on metal-organic vapor phase epitaxy (MOVPE). Since then many different growth methods are being applied to nanowire growth, such as molecular beam epitaxy (MBE), chemical beam epitaxy (CBE), hydride vapor phase epitaxy (HVPE), each with advantages for special areas of applications. Already 20 years ago, Hiruma demonstrated high quality III-V nanowires, reported injection luminescence from nanowire pn-junctions and showed results for heterostructures between different materials in III-V nanowires.

It took until early 2000's before the realization of highly ideal and atomically abrupt heterostructures in III-V nanowires was reported by my group, including the demonstration of the ability to form such ideal heterostructures also for highly lattice mis-matched structures, first demonstrated for the InAs/InP materials system. In this case one benefits from the way indium alloys with the gold particle which induces growth, while the group V atoms, arsenic and phosphorous, do not alloy with the gold but are made to incorporate at the interface between the metal particle/droplet and the top-end of the growing semiconductor nanowire. This progress led to the realization of a large number of 1D heterostructure devices, such as resonant tunneling diodes, single-electron transistors, single/few-electron memory devices, single quantum-dot emitters as well as control of artificial atom-like structures revealing very clear shell-structures in level filling experiments.

Although this demonstration system InAs/InP appears straightforward and very promising, reality is often more complicated than the simple picture, as it was later shown. Frances Ross et al. investigated how the interface surface energies have to be proper in order for ideal nanowire heterostructures to form, often with one transition A->B being different from B->A, in a way that the formation of Stranski-Krastanow islands depends on the surface energies involved. I would also like to bring to the attention the very detailed study performed by Linus Fröberg et al., where even this most ideal materials combination and transition between InAs and InP is indeed quite complicated due to the fact that the equilibrium incorporation of indium in the gold, at conditions for growth, depends strongly on the As-to-P fraction, leading to transients during the transitions in-between these two binary materials InAs and InP. Apart from adding a complication to an otherwise simple picture, these insights allow the design of the most proper procedures for forming ideal heterostructures in nanowires.

In recent years, the control of the structural properties has developed into a very active field of research, for the materials properties in general and for the impact the structural properties have on optical and electrical materials properties. It is well known today that while bulk and thin film growth of the (regular) III-V materials lead to the cubic zinc blende (ZB) crystal structure, nanowire growth often selects the formation of the hexagonal, wurtzite (WZ) crystal structure, in which the consecutive planes in the direction perpendicular the 111B substrate surface, go down like ABCABC in ZB and as ABABAB in WZ. This field of intensive research has been pioneered by my own group as well as that of Erik Bakkers at Philips Research and, I believe it is fair to say that the state-of-the-art of the field has just been published in Nano Letters in an effort pushed forward primarily by Kimberly Dick and Philippe Caroff, where the ability was demonstrated to form a fixed number of monolayers of ZB and WZ segments that were repeated and formed into a perfect superlattice.

Apart from the nuisance of poor control of nanowires with mixed layer-by-layer stacking sequences, into ZB and WZ segments, this also leads to very novel properties, such as the spatially indirect nature of the band-to-band recombination, as shown first by Bao et al. in 2008, and later confirmed in studies of recombination of electrons confined in ultra-short segments of ZB nanowires, with holes at the valence band edge of surrounding WZ, in papers by Spirkovska et al. (2009), and also by Akopian et al. (2010) and by Jancu et al. (2010). In a just published study by Dick et al. 2010, it was also shown how a single-electron transistor could be formed with the source, drain and coulomb island being made from ZB NW material, and with two tunnel barriers surrounding the QD being made from ultra-thin WZ segments.

Among the most interesting III-V nanowire materials being studied today are those exhibiting very strong spin-orbit interaction and magnetically active properties. Quantum dots in InAs nanowires have been investigated from the perspective of their suitability for fabrication of quantum information devices and InSb nanowires have shown extremely high g-factors and spin-orbit interaction. Obviously, the ultra-thin cross-section of such spintronic NW-devices may suggest that they can also lead to a very high level of miniaturization.

Obviously much of the driving force to the field comes from the opportunities for device realization in III-V nanowires. Of special interest here is the ability to form these III-V materials on silicon wafers and to use top-down patterning to control formation of single NWs or arrays of nanowires, as pioneered by my former student Thomas Mårtensson (Thesis Lund University, 2008). Among the applications that have already been demonstrated, at least on a laboratory level or R&D, are wrap-gate field-effect transistors and different versions of solar cell structures, both very recently reviewed in two IEEE review articles, Wernersson et al., "Semiconductor nanowires: extending a narrowing road", and Borgström et al., "Nanowires with promise for photovoltaics". In the near future, it seems that nanowires may indeed enter the competition for GaN-based light-emitting diodes (LEDs) with the special advantages that they offer in the ability to be grown on silicon wafers and the much larger level of freedom of combining different materials, like the not really lattice-matched GaInN materials system. Considering the great cost and "productization" advantages with growth on large area and inexpensive silicon wafers, this may become the critical step for a general acceptance of LEDs for general illumination applications and a transition to Solid State Lighting with huge energy savings.

In this book a number of specialized chapters are included jointly giving a comprehensive and updated description of the field of III-V nanowires today.

Lars Samuelson
Solid State Physics, Materials Science
Quantum Physics, Nanoelectronics- & photonics
Nanoenergy, Lund University
Sweden

PREFACE

The field of semiconductor nanowires has become one of the most active research areas within the nanoscience community. This E-Book on "Advances in III-V Semiconductor Nanowires and Nanodevices" presents the latest knowledge in the highly topical research field of semiconductor nanowires. Semiconductor nanowires exhibit novel electronic and optical properties due to their unique one-dimensional structure and quantum confinement effects. In particular, III-V semiconductor nanowires have been of great scientific and technological interest for next generation optoelectronic devices including transistors, light emitting diodes, lasers, photodetectors, and solar cells. To fully exploit these properties and devices, current research has focused on rational synthetic control of these one-dimensional nanoscale building blocks, novel properties characterization, and device fabrication/integration. This special E-book is an account of recent progress in the synthesis, characterization, physical properties, device fabrication, and applications of binary compound and ternary alloy III-V semiconductor nanowires. The book begins with a description of general synthetic strategies, followed by a focus chapter on each of the major binary III-V material systems, a single chapter on the lesser studied but important ternary III-V material systems, and ends with three chapters on major application areas. Each chapter is prepared by renowned experts in the field, describing the current state of knowledge and key areas of research. The book is written at the expert level, although it may also serve as a guide for other researchers or graduate students wishing to enter the field.

Ray LaPierre, Associate Professor
Department of Engineering Physics
McMaster University
Canada

Jianye Li, Professor
Department of Physical Chemistry
University of Science & Technology Beijing
China

Deli Wang, Associate Professor
Department of Electrical & Computer Engineering
University of California San Diego
USA

LIST OF CONTRIBUTORS

Rienk E. Algra

Materials Innovation Institute, Delft, The Netherlands
IMM, Solid State Chemistry, Radboud University Nijmegen, The Netherlands

Erik P.A.M. Bakkers

Eindhoven University of Technology, The Netherlands
Delft University of Technology, The Netherlands

S. Crankshaw

Laboratoire des Matériaux Semiconducteurs, Ecole Polytechnique Fédérale de Lausanne, Switzerland

Dheeraj L. Dasa

Student, Department of Electronics and Telecommunications, Norwegian University of Science and Technology, Norway

Lou-Fé Feiner

Eindhoven University of Technology, The Netherlands

Takashi Fukui

Graduate School of Information Science and Technology, Hokkaido University, Japan
Research Center for Integrated Quantum Electronics, Hokkaido University, Japan

Frank Glas

Group Leader, CNRS-Laboratoire de Photonique et de Nanostructures, France

Shinjiroh Hara

Graduate School of Information Science and Technology, Hokkaido University, Japan
Research Center for Integrated Quantum Electronics, Hokkaido University, Japan

Jean-Christophe Harmand

Full-Time Researcher, CNRS-Laboratoire de Photonique et de Nanostructures, France

Kenji Hiruma

Graduate School of Information Science and Technology, Hokkaido University, Japan
Research Center for Integrated Quantum Electronics, Hokkaido University, Japan

Bin Hua

JST PRESTO Research Fellow, Japan Science and Technology Agency, Japan

Moira Hocevar

Delft University of Technology, The Netherlands

Fauzia Jabeen

Postdoctoral Research Fellow, CNRS-Laboratoire de Photonique et de Nanostructures, France

Ray LaPierre

Department of Engineering Physics, McMaster University, Canada

Ludovic Largeau

Full-Time Engineer, CNRS-Laboratoire de Photonique et de Nanostructures, France

Jianye Li

Department of Physical Chemistry, University of Science & Technology Beijing, China

Linsheng Liu

Postdoctoral Research Fellow, CNRS-Laboratoire de Photonique et de, Nanostructures, France
Presently at Beijing Institute of Auto-Testing Technology, China

Faustino Martelli

Institute for Microelectronics and Microsystems-Rome, CNR, Italy

Zetian Mi

Department of Electrical and Computer Engineering, McGill University, Canada

A. Fontcuberta i Morral

Laboratoire des Matériaux Semiconducteurs, Ecole Polytechnique Fédérale de Lausanne, Switzerland

Junichi Motohisa

Graduate School of Information Science and Technology, Hokkaido University, Japan

Gilles Patriarche

CNRS-Laboratoire de Photonique et de Nanostructures, France

Werner Prost

Scientist, Center for Semiconductor Technology and Optoelectronics, University Duisburg-Essen, Germany

Corinne Sartel

CNRS-Laboratoire de Photonique et de, Nanostructures, France
Presently Full-Time Engineer, Groupe d'Etudes de la Matière Condensée, Université Versailles- Saint-Quentin en Yvelines, CNRS-UMR 8635, France

Maria Tchernycheva

NRS-Laboratoire de Photonique et de Nanostructures, France
Presently Full-Time Researcher, Institut d'Electronique Fondamentale, Université Paris 11, CNRS-UMR 8622, France

Franz-Josef Tegude

Center for Semiconductor Technology and Optoelectronics, University Duisburg-Essen, Germany

Katsuhiro Tomioka

JST PRESTO Research Fellow, Japan Science and Technology Agency, Japan
Research Center for Integrated Quantum Electronics, Hokkaido University, Japan

Kumar S.K. Varadwaj

JST PRESTO Research Fellow, Japan Science and Technology Agency, Japan

Marcel A. Verheijen

Senior Scientist, Philips Research Laboratories Eindhoven, The Netherlands

Deli Wang

Department of Electrical & Computer Engineering, University of California, San Diego, USA

Helge Weman

Department of Electronics and Telecommunications, Norwegian University of Science and Technology, Norway

Hongmin Zhu

Department of Physical Chemistry, University of Science & Technology Beijing, China

CHAPTER 1

General Synthetic Strategies for III-V Nanowires

Jianye Li* and Hongmin Zhu

Department of Physical Chemistry, University of Science & Technology Beijing, China

Abstract: III-V semiconductor nanowires are expected to play a significant role in future nanoscale electronic and optoelectronic devices. In this chapter, we attempt to review the general synthetic strategies for III-V compound nanowires. We first summarize various III-V nanowire growth techniques such as chemical vapor deposition, laser ablation, metal-organic chemical vapor deposition, molecular beam epitaxy, chemical beam epitaxy, hydride vapour phase epitaxy, wafer annealing, and low-temperature solution methods. Subsequently, we discuss mechanisms involved to generate III-V nanowires from different synthetic schemes and conditions, including vapor-liquid-solid, vapor-solid-solid, solution-liquid-solid, vapor-solid (self-catalytic, oxide-assisted, and axial screw dislocation), ligand-aided solution-solid, and reactive Si-assisted growth.

Keywords: III-V compounds, semiconductor, nanowires, synthetic strategy, growth technique, growth mechanism.

INTRODUCTION

Semiconductor nanowires have novel electronic and optical properties owing to their unique one-dimensional structure and possible quantum confinement effects in two dimensions [1-3]. III-V compound semiconductor nanowires are expected to play a critical role in future nanoscale electronic and optoelectronic devices with excellent performance, such as laser [4], light emitting diodes (LED) [5], solar cells [6], and piezoelectric nanogenerators [7]. III-V semiconductor nanowires have been of great scientific and technological interest because by synthesizing them, the level of functionality in future nanosystems may be greatly enhanced [1-7]. In the past decade, an overwhelming number of articles on the synthesis of III-V semiconductor nanowires have been published [1-7].

Usually, growth mechanisms are used to refer to the general phenomenon whereby a nanowire is formed, and growth methods denote the experimentally employed chemical processes that incorporate a mechanism to obtain these nanowires [2].

III-V semiconductor nanowires can be synthesized through a variety of growth methods. In the past years, various techniques have been developed to realize III-V nanowires, such as chemical vapor deposition [8-13], laser ablation [14-16], metal-organic chemical vapor deposition or metal-organic vapour phase epitaxy [17-20], molecular beam epitaxy [21-23], chemical beam epitaxy [24], hydride vapour phase epitaxy [25], wafer annealing [26], and low-temperature solution methods [27].

However, the detailed growth mechanisms for III-V nanowires that apply for different synthetic schemes and conditions remain an active subject of investigation and debate [28]. A growth mechanism must explain how nanowire growth occurs, provide a kinetic and thermodynamic rationale, and be predictable and applicable [2]. In the presence of catalytic metal particles, the growth of III-V nanowires is frequently interpreted *via* the vapor-liquid-solid (VLS) [10, 12, 15, 21, 28] and vapor-solid-solid (VSS) [28-30] mechanisms. In the absence of metal particles, III-V nanowire growth evolution is often interpreted *via* the vapor-solid (VS) growth (including self-catalytic or group-III catalyzed VLS growth [20], oxide-assisted growth [31], and axial screw dislocation mechanism [32]), ligand-aided solution-solid growth [33], and reactive Si-assisted growth [34] mechanisms.

Herein, we review the general growth methods and mechanisms of III-V compound semiconductor nanowires.

III-V NANOWIRE GROWTH METHODS

Chemical Vapor Deposition

Chemical vapor deposition (CVD) is a chemical process often used to deposit solid materials. Synthesizing nanowires *via* a commonly used thermal CVD method in a conventional tube furnace is facile, simple and cheap. In

*Address correspondence to Jianye Li: Department of Physical Chemistry, University of Science & Technology Beijing, Beijing, 100083, China; E-mail: jyli@ustb.edu.cn

Jianye Li, Deli Wang and Ray R. LaPierre (Eds)

the past decade, the common thermal CVD method has been the most popular technique and widely used for the growth of III-V compound nanowires.

Han *et al.* first proposed an approach to the template-based CVD growth of gallium nitride nanowires (nanorods) *via* a carbon nanotube-confined reaction [8]. In this report, the carbon nanotubes served as a scaffold against which GaN nanowires with similar morphologies were synthesized. That is, the in situ generated GaN nanowires were shaped into nanostructures with morphology complementary to that of the carbon nanotubes (NTs). The reaction of the synthesis of GaN nanowires (NWs) was expressed as

$$2Ga_2O_{(g)} + C \,(NTs) + 4NH_3 \rightarrow 4GaN \,(NWs) + H_2O + CO + 5H_2$$

The reaction was carried out at high temperature in a conventional furnace with a horizontal tube. A 4:1 molar mixture of Ga-Ga_2O_3 powders was used as the precursor of Ga_2O and the synthesized GaN nanowires had a diameter similar to that of the original carbon nanotubes [8].

Cheng *et al.* demonstrated an approach to large-scale template-assisted CVD growth of single crystalline GaN nanowires in confined spaces provided by nanosized-pore Al_2O_3 membranes. The growth was carried out at 1273 K and the reaction for the GaN nanowire synthesis was

$$Ga_2O_{(g)} + 2NH_3 \rightarrow 2GaN \,(NWs) + H_2O + 2H_2$$

As reported in Ref 8, the Ga_2O gas was generated from the same starting material, a 4:1 molar mixture of Ga-Ga_2O_3 powders. Instead of in the carbon nanotubes, the GaN nanowires were synthesized in anodic alumina templates [9].

With the reaction of gallium and ammonia, GaN nanowires were synthesized by a template-independent CVD method at 920-940 °C [10]. In this report, the GaN nanowires were deposited on $LaAlO_3$ single crystal substrate without using a template. The nanowires grew *via* a vapor-liquid-solid growth mechanism and the used catalyst was from the decomposition of nickel (II) nitrate [10].

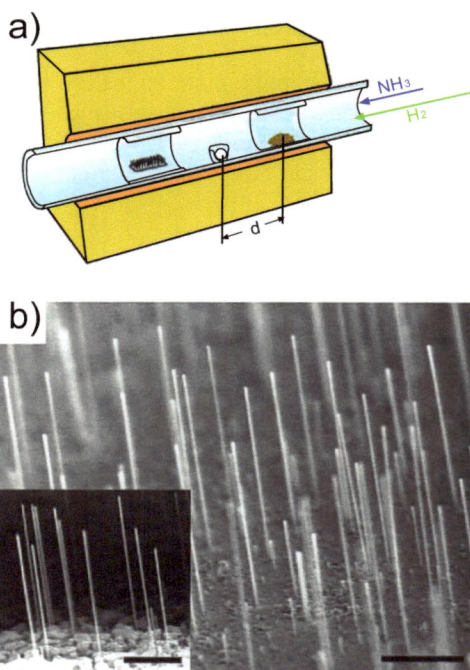

Figure 1: (**A**) Schematic of growth apparatus, where the separation between Ga and Mg_3N_2 starting materials is *d*. The Ga precursor was fixed at a distance of 0.5 in. upstream from the substrate. (**B**) Field emission scanning electron microscopy (FE-SEM) images of as grown GaN nanowires on sapphire substrate. Scale bars are 10 microns. Reprinted in part with permission from Ref. [5]. Copyright 2003 American Chemical Society.

Zhong and coworkers reported the preparation of p-type GaN nanowires *via* a metal-catalyzed CVD technique as shown in Fig. **1** [5]. The metal catalyst they used was nickel and nickel (II) nitrate was the catalyst precursor. Ammonia, gallium metal and magnesium nitride were used as the N, Ga, and Mg sources, respectively. The substrate was *c*-plane sapphire and the nanowires grew normal to the sapphire surface with lengths from 10 to 40 microns. They found that the growth direction and orientation of the GaN nanowire was consistent with epitaxial growth from the *c*-plane sapphire surface, and suggested that substrate-nanowire epitaxy was important in defining the nanowire growth direction [5].

Cheng *et al.* described an approach to the catalyst-free, vapor-solid growth of indium nitride nanowires *via* a CVD method [13]. In this report, a 1:1 mixture of indium/indium oxide powders was used as the starting material. At 700 °C, the starting material reacted with ammonia to synthesize the InN nanowires. The synthesized InN nanowires were high-purity, single-crystal hexagonal wurtzite and intrinsic *n* type. The nanowires had uniform diameters that ranged from 70 to 150 nm and lengths that varied between 3 and 30 microns [13].

He *et al.* proposed an extraordinarily simple CVD system for the successful synthesis of thin, long, uniform, and single-crystal InAs nanowires. The nanowires were grown employing a reaction of liquid/vapor indium with GaAs substrates in a horizontal oven with a quartz tube [35].

Using the same versatile template-independent CVD technique, He *et al.* realized controllable growth of good-quality III-V ternary nanowires such as InGaN and InGaAs nanowires. The InGaN nanowires were obtained *via* reaction of Ga and As with ammonia on SiC or GaAs substrate, and the InGaAs nanowires were synthesized by reaction of indium with GaAs [36]. Moreover, InGaAsN nanowires were also grown by this simple CVD system and it was the first controlled realization of III-V quaternary nanowires [36].

Laser Ablation

Laser ablation is a laser-assisted method and the key feature of this technique is that the catalyst used to define nanowire VLS growth can be selected from phase diagram data and/or knowledge of chemical reactivity [14]. Through laser ablation, the source material is ablated into a vapor phase, which may have the same composition as the precursor, and then the vapor phase is transferred to a substrate where nanowires nucleate and grow [37-39]. Nanowires can be readily obtained directly from solid starting materials by this method. Laser ablation is particularly useful in the synthesis of nanowires with a high-melting temperature because a high-energy laser can ablate solid materials in a very short time [39].

The famous Lieber group at Harvard University made significant progress toward the development of a general synthetic approach to single-crystalline compound semiconductor nanowires *via* the laser catalytic growth [14-16, 40,41].

Duan and coworkers reported the bulk synthesis of single crystalline GaN nanowires by using a pulsed Nd:YAG laser (1064 nm, 8 ns pulse width, 10 Hz repetition, 2.5 W average power) to vaporize a solid target containing GaN and catalyst Fe with an atomic ratio of 0.95:0.05. The target was positioned with a quartz tube at the center of a furnace and the resulting liquid Fe nanoclusters formed at elevated temperature direct the growth and define the diameter of GaN nanowires through a VLS growth mechanism. The grown GaN nanowires deposited on the inner quartz tube wall and had diameters on the orders of 10 nm and lengths greatly exceeding 1micron [14].

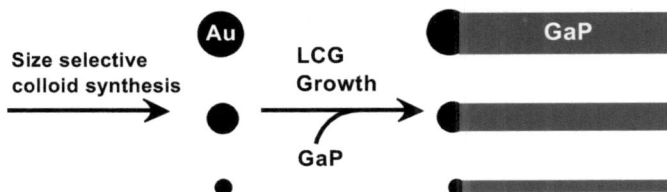

Figure 2: Schematic of laser ablation grown well-defined GaP nanowires with monodisperse gold colloids as catalysts. Reprinted with permission from Ref. [40]. Copyright 2000 American Chemical Society.

Duan *et al.* also reported the synthesis of GaAs nanowires in bulk quantities and high purity by pulsed Nd:YAG laser ablation growth and the targets used in the synthesis was a mixture of $(GaAs)_{0.95}M_{0.05}$, where M=Au, Ag, or Cu [15].

Using GaP as the target, Gudiksen *et al.* obtained GaP nanowires with diameters of 10, 20, and 30 nm and lengths greater than 10 microns on well-defined gold colloids coated substrates by the laser ablation growth [40]. Fig. **2** is a schematic of laser ablation growth of well-defined GaP nanowires using monodisperse gold colloids as catalysts.

Metal-Organic Chemical Vapor Deposition

Metal-organic chemical vapor deposition (MOCVD) is a chemical vapor deposition technique of epitaxial growth of materials, especially compound semiconductors from the surface reaction of organic compounds/metalorganics and metal hydrides containing the required chemical elements. Alternative names for this method include metal-organic vapour phase epitaxy (MOVPE), organometallic vapour phase epitaxy (OMVPE) and organometallic chemical vapour deposition (OMCVD) [42]. In the past few years, commercial MOCVD or MOVPE reactors have become common tools for the growth of III-V compound nanowires.

The notable Yang group at University of California Berkeley synthesized high-quality GaN nanowires *via* metal-initiated MOCVD method for the first time [17]. In their report, trimethylgallium (TMG) and ammonia source materials were used as Ga and N precursors. The substrate was silicon or *c*-plane and *a*-plane sapphire with thermally evaporated thin film of Ni, Fe, or Au. The GaN nanowire growth occurred *via* VLS mechanism at a substrate temperature of 800-1000 °C. The reaction was carried out in an oxygen-free environment at atmospheric pressure. Nitrogen was used as a carrier gas and percolated through the TMG precursor and coupled with a second nitrogen line. The GaN nanowires grown on a gold-coated *c*-plane sapphire substrate had diameters of 15-100 nm and lengths of 1-5 microns. It was observed that there were preferred nanowire orientations on this substrate, indicating the epitaxial growth of nanowires on a sapphire substrate [17].

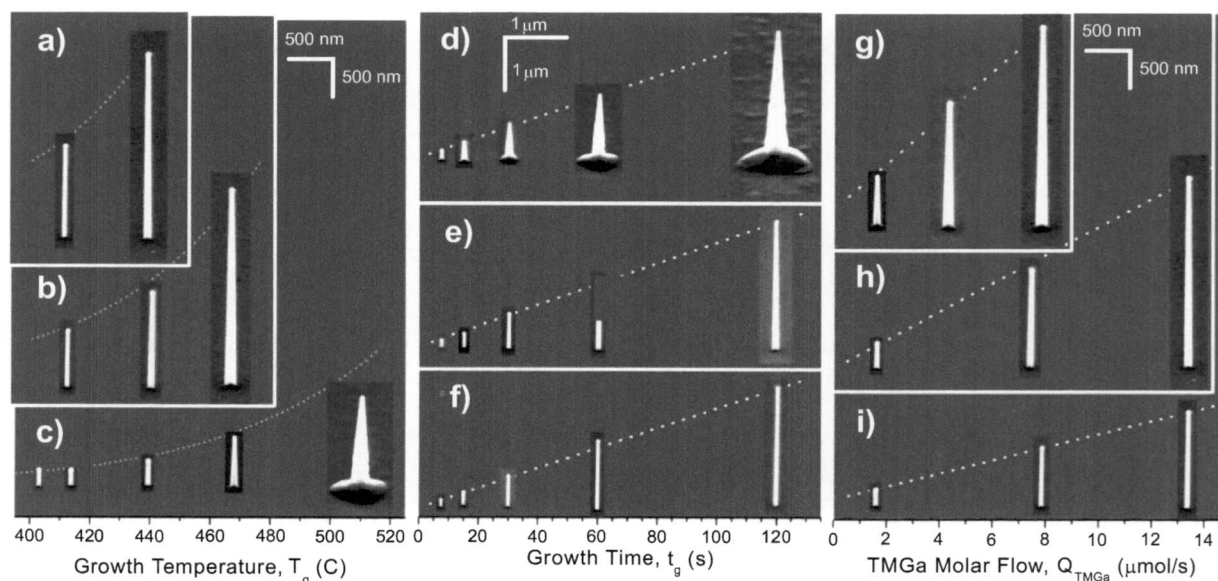

Figure 3: (a-c) SEM images of GaAs nanowire morphology as a function of growth temperature (T_g) for various TMGa molar flows of 13.4 **(a)**, 7.8 **(b)**, and 1.6 µmol/s **(c)**. **(d-f)** SEM of GaAs nanowires as a function of growth time (t_g) for various growth temperatures of 510 **(d)**, 440 **(e)**, and 410 °C **(f)**. **(g-i)** SEM of GaAs nanowires as a function of TMGa molar flow for various growth temperatures of 468 **(g)**, 440 **(h)**, and 410 °C **(i)**. Adapted with permission from Ref. [19]. Copyright 2008 American Chemical Society.

MOCVD/MOVPE grown GaAs nanowires were reported by several groups [19, 43-46]. Soci and coworkers demonstrated an approach to MOCVD growth of GaAs nanowires with trimethyl-gallium (TMGa) and arsine (AsH$_3$) as the gas precursors and hydrogen as the carrier gas. Using Au nanoparticles as the seed catalyst, epitaxial GaAs nanowires were achieved on GaAs (111) B substrates [19]. As shown in Fig. **3**, the epitaxial growth of GaAs nanowires was systematically investigated as a function of relevant growth parameters, including temperature, growth time, arsine (AsH$_3$) and trimethyl-gallium (TMGa) flow rates, and gold nanoparticle catalyst size.

Novotny *et al.* reported the MOCVD growth of vertically aligned InP nanowires without use of a deposited metal catalyst. The nanowires were grown on InP(111)B substrates in a horizontal MOCVD reactor. Trimethylindium (TMIn) and phosphine (PH$_3$) were used as the In and P sources respectively, and hydrogen as the carrier gas. Vertical growth from the InP(111)B substrate together with transmission electron microscopy analysis indicated epitaxial growth from the substrate in the [111]B direction. They concluded that a surface reconstruction induced indium droplets to form on the surface and thus acted as nucleation sites for nanowire growth [20].

Very recently, self-catalyzed epitaxial growth of vertical InP nanowires on silicon by MOVPE method was reported [47]. The experiments were carried out in a Veeco D125 MOVPE reactor with trimethylindium (TMIn) and tertiarybutylphosphine (TBP) as the starting materials. Liquid indium droplets were formed in situ and used to catalyze the nanowire deposition. The vertical fraction, growth rate, and tapering of the nanowires increased with temperature and V/III ratio. The nanowires were single-crystal zincblende InP, and contained a high density of rotational twins perpendicular to the [111] growth direction [47].

The Wang group at University of California San Diego demonstrated the MOCVD growth of InAs nanowires on thermally grown SiO$_2$/Si substrates with catalyst nanoparticles on top [48-50]. The growth was performed in a H$_2$ atmosphere with Arsine (AsH$_3$) and Trimethylindium (TMIn) precursors and the synthesized InAs nanowires were n-type with lengths of ~10 microns and diameters of ~60–120 nm [50]. Recently, they reported the catalyst-free, direct heteroepitaxial growth of vertical InAs nanowires on Si(111) substrates by MOCVD method. The grown nanowires were over a large area, and showed very uniform diameters and a zincblende crystal structure [51].

Molecular Beam Epitaxy

Molecular beam epitaxy (MBE) is a method to deposit single crystals. In the late 1960s, J. R. Arthur and Alfred Y. Cho invented MBE at Bell Telephone Laboratories [52]. The MBE method is particularly useful in the synthesis of III-V compound nanowires.

GaN nanowires were grown by radio frequency plasma-assisted MBE on Si(111) in N-rich conditions [22,53]. Calarco and coworkers investigated the nucleation process of GaN nanowires in terms of nucleation density and nanowire evolution with time for a given set of growth parameters. They found that the nanowire density increased rapidly with time and then saturates [22]. Bertness *et al.* analyzed the spontaneous growth of c-axis GaN nanowires in MBE under a number of different growth parameters, including relative flux of N species to Ga, type of N species present, AlN buffer layer thickness, and substrate orientation. They concluded that the nucleation mechanism for nanowires included formation of nanocolumns in the AlN buffer layer, and the propagation of the nanowires in GaN growth appeared to be driven by the differences in growth rates among crystallographic planes under N-rich conditions [53].

Figure 4: (a) Top view SEM image of InP/InAsP/InP nanowires grown at 390 °C. Cross-sectional SEM images of the nanowires grown at **(b)** 390 °C, **(c)** 420 °C, and **(d)** 420 °C overgrown with InP at 390 °C. The scale bar is the same for (b-d). Reprinted with permission from Ref. [59]. Copyright 2007 American Chemical Society.

Plasma-assisted MBE was successfully applied to produce InN nanowires [54-58]. Stoica *et al.* reported the plasma-assisted MBE growth of InN nanowires on Si(111) at a deposition temperature which varied within a range of 440-525 °C [56,57]. The growth parameters for uniform nanowires with a high crystalline quality were optimized [56], and the influence of the growth parameters on the photoluminescence spectra was investigated [57]. Chen *et al.* demonstrated that vertically aligned hexagonal InN nanowires could be densely grown on Si(111) substrates by plasma-assisted MBE at low and high growth temperatures. The InN nanowires were single crystalline with growth direction along the c-axis. The size distribution of the low temperature grown InN nanowires was quite uniform, while the high temperature grown nanowires exhibited a broad, bimodal distribution [58].

Tchernycheva and coworkers reported the Au-assisted MBE growth of InP nanowires with embedded InAsP insertions as shown in Figs. **4** & **5** [59]. The InP/InAsP/InP heterostructure nanowires were grown on InP(111)B substrates in an MBE system equipped with solid sources supplying In atoms and cracker As and P sources to produce dimers. They found that the growth temperature affected the nucleation on the nanowire lateral surface and the nanowires grew in two steps: to fabricate an axial heterostructure first, and then cover it by a shell.

The LaPierre group at McMaster University grew GaAs nanowires on the Au-functionalized single-walled carbon nanotube composite films *via* the VLS process in a gas source MBE system. The as-grown GaAs nanowires were oriented in a variety of angles on the single-walled carbon nanotube surface and grew along the <0001>direction with wurtzite crystal structure [60]. Shtrikman *et al.* obtained stacking-faults-free zincblende GaAs nanowires in a Riber 32 solid source MBE system through a VLS gold assisted growth process [61].

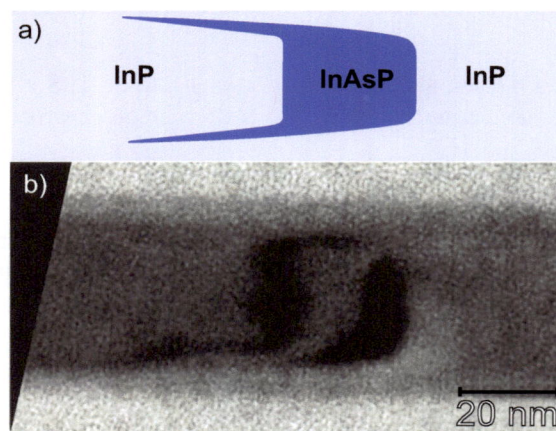

Figure 5: (a) Schematic of the insertion shape. **(b)** Transmission electron microscopy (TEM) image of InAsP insertion and the nanowire growth direction goes from left to right. Reprinted with permission from Ref. [59]. Copyright 2007 American Chemical Society.

Ihn *et al.* realized Au-activated InAs nanowires on Si substrates by a solid source MBE technique. The grown nanowires possessed a high crystalline quality, a high aspect ratio, and [111] growth direction. They found that the epitaxial growth of InAs nanowires was very sensitive to the surface condition of the Si substrates [62].

Cornet and coworkers reported the gas source MBE growth of InGaAs/InP core-shell and axial heterostructure nanowires on InP (111) B substrates using self-assembled Au particles as catalyst [63]. They analyzed the nanowire heterostructures and found significant transition regions of the material interfaces and a distinct bulging morphology attributed to transient group-III material in the Au catalyst. They also observed a sidewall growth of InGaAs on InP and a deficiency of Ga in the InGaAs section [63].

Chemical Beam Epitaxy

Chemical Beam Epitaxy (CBE) is a powerful growth technique that was first demonstrated by W.T. Tsang in 1984 [64]. It provides a capability for the epitaxial growth of semiconductor layer systems, especially III-V semiconductor materials with control at the atomic limit. The epitaxial growth is performed in an ultrahigh vacuum system and the reactants are in the form of molecular beams of reactive gases [65].

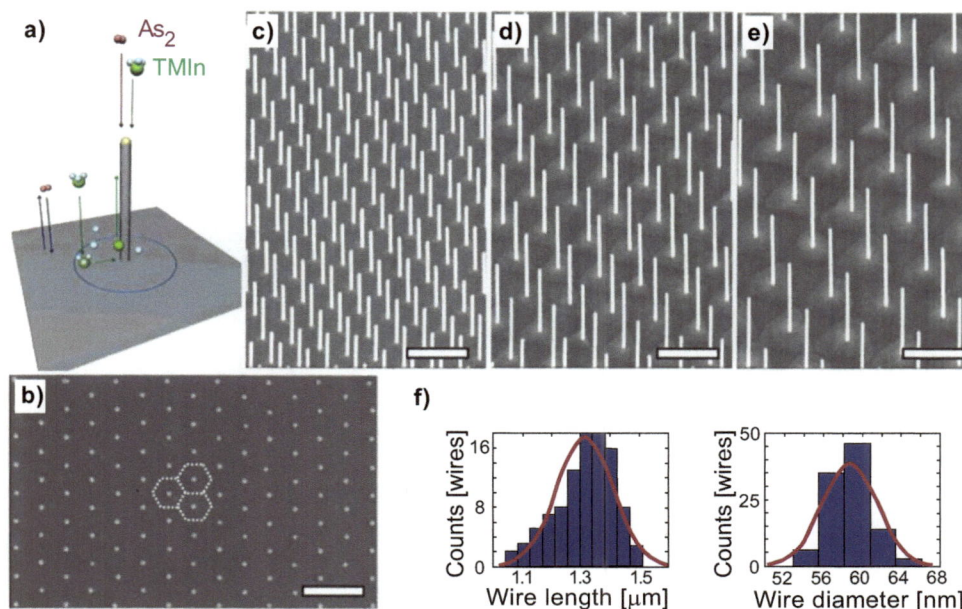

Figure 6: InAs nanowires grown from Au disk arrays. **(A)** Schematics of precursor transporting to the growth point. **(B)** Top view of Au particles defined by EBL. The dotted areas indicating the available surface collection area. **(C-E)** Tilted views of nanowire arrays with interwire distances of 0.5, 0.75, and 1 micron, but with an identical diameter. The nanowire's growth rate increases when the interwire distance increases. **(F)** Histograms of nanowire diameter and length. Red curves fit the data. All scale bars are 1 micron. Reprinted with permission from Ref. [66]. Copyright 2004 American Chemical Society.

The Samuelson group at Lund University developed the CBE technique for the synthesis of size-selected GaAs epitaxial nanowires grown on a crystalline substrate [24]. In this report, size-selected gold aerosol particles were used as catalysts and the as-grown nanowires were rod shaped, with a uniform diameter between 10 and 50 nm, correlated to the size of the catalytic seed. Individual nanowires could be nucleated in a controlled manner at specific positions on a substrate with accuracy on the nanometer level [24]. Using TMIn (trimethylindium) and TBAs (tertiarybutylarsine) as precursors for the group III and V compounds respectively, they also realized the CBE growth of InAs nanowires and investigated the role of surface diffusion in the nanowire growth [66]. They presented growth studies of InAs nanowires nucleated from lithographically positioned Au seeds on InAs(111)B substrates and found that the nanowires grown in a CBE system exhibited high aspect ratios and high homogeneity in length and width. They performed investigations of wire growth rate as a function of diameter, density, and time. The results indicated that 80% of the growth was due to In species diffusing from the (111)B substrate surface [66]. Fig. **6** displays the InAs nanowires grown from Au disk arrays.

Poole *et al.* reported the spatially controlled, nanoparticle-free growth of InP nanowires by CBE method [67]. The nanowires were grown in a Riber 32P chemical beam epitaxy system. Trimethylindium (TMI), and precracked PH_3 and AsH_3 were used as the sources. The growth was performed on (001) InP substrates at a temperature of 505 °C. The InP nanowires were precisely positioned on the InP substrate and always aligned along the available substrate <111>*A* directions. The nanowires had diameters as small as 40 nm, and typical lengths of 600 nm.

Chao *et al.* demonstrated catalyst-free growth of indium nitride nanowires on sapphire substrate by a homemade CBE system [68]. In this report, the growth chamber was vacuumed to reach a base pressure of 10^{-10} Torr by a turbo molecular pump. Trimethylindium and trimethylaluminum were used as group III precursors and delivered to the growth chamber by heating the metal-organic sources without carrier gases. The active nitrogen radicals were supplied by a radio-frequency plasma source.

The *C*-plane sapphire substrate was heated to 850 °C for thermal cleaning, then cooled down to 800 °C for nitridation, followed by the growth of a thin amorphous AlN nucleation layer at 700 °C. InN nanowires were grown on the nucleation layer and commenced at 500 °C. The synthesized InN nanowires were single-crystalline wurtzite, nearly unidirectional along the <001> direction, and with the diameters varying in the range of 20–40 nm with In/N flow ratio [68].

Epitaxial nanowire arrays of InAs/InP longitudinal heterostructure grown on an InAs(111)B substrate *via* CBE technique was realized by the Samuelson group [69,70]. The nanowire growth was Au-assisted and took place in a CBE system using trimethylindium, precracked *tert*-butylarsine, and precracked *tert*-butylphosphine as precursors. Prior to growth, size-selected Au aerosol particles were deposited on a InAs(111)B substrate and the sample was deoxidized in the growth chamber under As pressure at 520 °C. The growth was started at 425 °C for 530 nm of InAs and then further lowered to 390 °C for the superlattice growth with 20 segments of 20 nm (InAs) and 10 nm (InP). The structure was ended with an InAs segment of 65 nm [70].

Hydride Vapour Phase Epitaxy

Hydride vapour phase epitaxy (HVPE) is an epitaxial growth technique often employed to produce III-V compound semiconductors and the commonly used carrier gasses include ammonia, hydrogen and various chlorides. J.J. Tietjen and J.A. Amick developed the HVPE technology in 1966 at RCA Laboratories for the preparation of single crystal $GaAs_{1-x}P_x$ layers, in which gaseous hydrides of arsine and phosphine served as sources of arsenic and phosphorus [71]. HVPE provides a better process control than the commonly used CVD techniques and is of particular interest due to its high growth rates and simplicity of basic reactor design. [25].

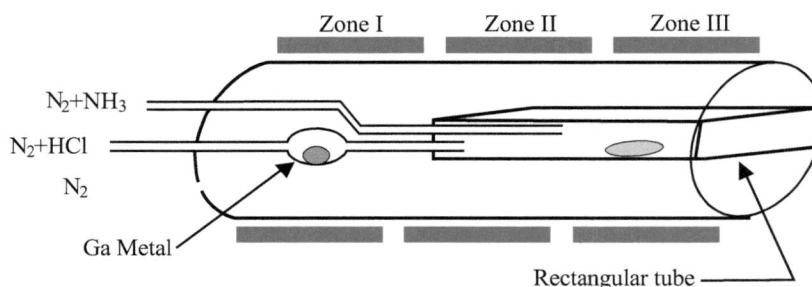

Figure 7: Schematic of horizontal HVPE system for GaN nanowire growth. Reprinted with permission from Ref. [25], *Nanotechnology*, 2005, Vol. 16, 2342–2345. Copyright 2005 IOP Publishing.

Seryogin *et al.* demonstrated the catalytic growth of GaN nanowires by the HVPE method [25]. In this work, nickel–gold was used as a catalyst and the nanowire growth was limited to catalyst-patterned areas. The growth of GaN nanowires was carried out in a custom designed HVPE rector at atmospheric pressure. The reactor, as shown in Fig. **7**, consisted of 75 mm quartz tube and placed in a three-zone Mellen horizontal furnace. Two heating zones were used for hydride reactions of column-III metals with HCl gas. Only gallium was used as the source metal to form GaCl in a direct reaction between molten Ga and HCl gas diluted with nitrogen:

$$Ga_s + HCl_g \leftrightarrow GaCl_g + 1/2H_2$$

The third zone was the growth zone and ammonia was delivered to this zone to react with GaCl gas. GaN nanowires were formed *via* the HVPE process on catalyst-patterned silicon (111) or *c*-plane sapphire substrates [25]:

$$GaCl_g + NH_{3g} \rightarrow GaN + HCl_g + H_2$$

Kim *et al.* reported the growth of InGaN nanowires through a horizontal HVPE system. By a reaction of HCl gas with Ga and In metal, the Ga and In precursor were synthesized first. Then, the precursors were transported to the substrate area to react with NH_3 and form InGaN nanowires on (1 1 1) silicon substrate [72].

Recently, Ramdani *et al.* realized the first growth of GaAs nanowires *via* Au-assisted VLS mechanism in a novel HVPE environment. The growth was carried out on 4° misoriented vicinal (001) GaAs substrates in a hot wall horizontal reactor kept at atmospheric pressure (AP-HVPE). Forty micrometer long rodlike <111> single-crystalline GaAs nanowires with a cubic zincblende structure were formed in 15 min with a mean density of 10^6 cm^{-2}. They explained the synthesis of such long nanowires in such a short time by the growth physics of near-equilibrium HVPE [73].

Wafer Annealing

Zhi *et al.* reported a wafer annealing method for the growth of III-V semiconductor nanowires, in which InAs, InP, and GaP nanowires were formed by annealing Au film coated semiconductor wafers in N_2 atmosphere at an appropriate temperature in a region of 550 – 650 °C. This method was based on a VLS growth mechanism and the composition of the resulted III-V nanowires was determined by both the substrates and the chemical conditions of growth [26].

Fig. **8** is a schematic of the experimental setup. To synthesize InAs nanowires, a polished InAs wafer was coated with Au film by magnetron sputtering. The Au coated wafer with a Mo sample holder was placed into a growth chamber. InAs nanowires were obtained by heating at 600 °C with nitrogen flow for 1 hour. Using a Au film covered InP wafer as a substrate, InP nanowires were formed at 650 °C by this technique. Putting an InP wafer by an Au film coated GaSb with a distance of less than 5 mm, GaP nanowires were easily fabricated at 550 °C though this process [26].

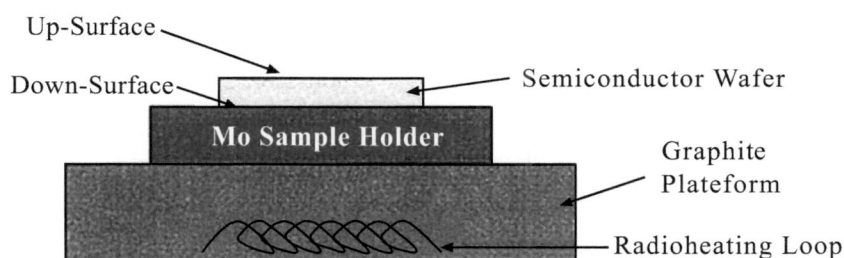

Figure 8: Schematic of the wafer annealing experimental setup. Reprinted with permission from *Applied Physics Letters*, 2004, Vol. 85, 1802-1804. Copyright 2004 American Institute of Physics.

Low-Temperature Solution

The Buhro group at Washington University developed a low-temperature (below 203 °C) solution method for synthesizing III-V compounds InP, InAs, and GaAs nanowires (nanowhiskers) [27]. In this technique, a metal with a low melting point such as indium *etc.* was used as a catalyst, and the desired III-V compound generated through the decomposition of organometallic precursors. The nanowires, having widths of 10 to 150 nanometers and lengths of up to several micrometers, were produced through well-studied simple organometallic reactions at low temperatures in hydrocarbon solvents. The growth was *via* a solution-liquid-solid process, which is analogous to the VLS growth mechanism [27].

Fanfair *et al.* realized solution-phase metal nanocrystal-seeded synthesis of III-V compound nanowires, in which bismuth (Bi) nanocrystals were used for solution-liquid-solid growth of crystalline InAs, GaP, GaAs, and InP nanowires at temperatures 300~340 °C in trioctylphosphine (TOP) and trioctylphosphine oxide (TOPO), and trioctylamine (TOA). It provided a scalable, rational, and general approach for growing high aspect ratio crystalline III-V semiconductor nanowires [74].

Xu *et al.* demonstrated a modified (120~180 °C) solvothermal route for the growth of InAs nanowires. The synthesized nanowires had diameters of 7~70 nm and lengths of up to several microns. The reaction could be as follows [33]:

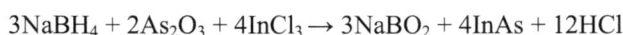

$$3NaBH_4 + 2As_2O_3 + 4InCl_3 \rightarrow 3NaBO_2 + 4InAs + 12HCl$$

The different techniques for making semiconductor nanowires are frequently placed into two categories, *i.e.* the bottom-up and top-down approaches. The bottom-up approaches, such as the III-V nanowire growth methods reviewed above, start with individual atoms and molecules and builds up the desired nanowires [75]. The top-down approaches rely on dimensional reduction *via* selective etching and various nanoimprinting techniques. Similarly, there are two such general approaches for the integration of nanowire building blocks: chemical assembly such as Langmuir-Blodgett (LB) techniques, and lithographical processes [75]. The top-down approaches for making III-V semiconductor nanowires go beyond the scope of this chapter.

III-V NANOWIRE GROWTH MECHANISMS

Vapor-Liquid-Solid Growth

Wagner and Ellis first described the vapor-liquid-solid (VLS) growth mechanism in 1964, during their studies of growth of semiconductor whiskers from vapor sources using Au particles as catalysts [76]. Later, Givargizov thermodynamically and kinetically justified the VLS process [77]. The VLS mechanism is common for the growth of one-dimensional (1D) materials, in which a liquid catalyst acts as the energetically favored site for absorption of gas-phase reactants to direct the growth of the 1D materials. The catalyst particles are typically detected at tips of the 1D materials grown through a VLS mechanism [37]. In the past years, generating nanowires from a rich variety of inorganic materials *via* the VLS process has been widely reexamined. Among all vapor-based methods, those involving the VLS mechanism seem to be the most successful in generating III-V nanowires with single crystalline structures.

Duan and coworkers exploited the predictable growth of single crystalline GaN nanowires by using laser-assisted catalytic method. In this report, a pulsed laser was used to vaporize a solid target containing GaN and catalyst Fe, and the resulting liquid nanoclusters produced at elevated temperature direct the growth and define the diameter of the nanowires through a VLS growth mechanism. Fig. **9** (A) is a schematic of the VLS growth of a GaN nanowire [14]. Fig. **9** (B) shows a diffraction contrast TEM image of a higher (darker) contrast faceted nanoparticle terminated GaN nanowire and the inset is the convergent beam electron diffraction pattern recorded along the <001> zone axis over the area indicated by the white circle [14]. In the report of catalytic HVPE growth of GaN nanowires by Seryogin *et al.*, a VLS growth mechanism of the nanowires was also demonstrated [25].

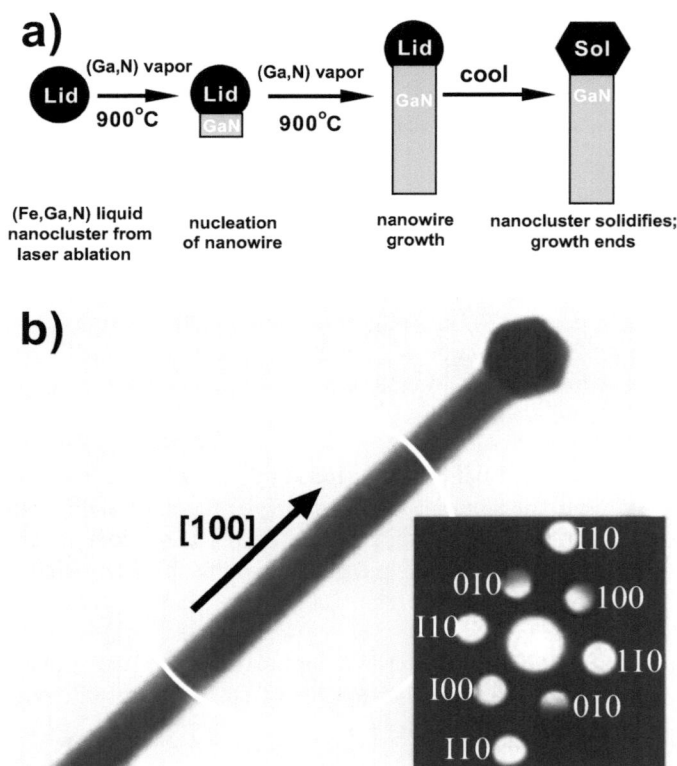

Figure 9: (A) Schematic of VLS growth of a GaN nanowire. **(B)** Diffraction contrast TEM image of a higher (darker) contrast faceted nanoparticle terminated GaN nanowire. (Inset) Convergent beam electron diffraction pattern recorded along the <001> zone axis over the area indicated by the white circle. Adapted with permission from Ref. [14]. Copyright 2000 American Chemical Society.

Using In and NH_3 as the starting materials and Au as the catalyst, Cai *et al.* reported the CVD synthesis of InN nanowires. They systematically investigated the nanowire growth and proposed that the growth was governed by a VLS mechanism [78].

Gudiksen *et al.* realized the diameter-selective synthesis of GaP nanowires by a laser-assisted catalytic method. In this experiment, gold catalyst clusters were coated on a SiO_2 substrate and pulsed laser ablation was used to generate the Ga and P reactants from a solid target of GaP. *Via* a VLS mechanism, the GaP nanowires were grown on the gold cluster coated substrates. The VLS growth mechanism of the GaP nanowires was verified by the presence of the gold nanocluster catalyst at the nanowire end, and the nanowires only grown with catalyst [40].

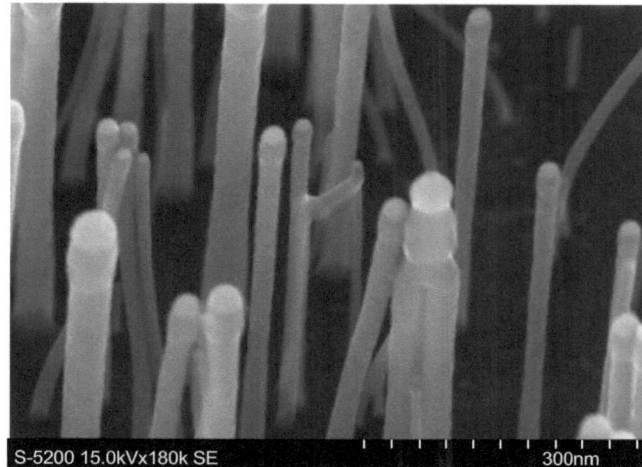

Figure 10: SEM image of VLS grown InP nanowires with Au nanoparticles at tips. Reprinted from *Physica* E, Vol. 24, S. Bhunia *et al.*, Systematic investigation of growth of InP nanowires by metalorganic vapor-phase epitaxy, 138-142, Copyright (2004), with permission from Elsevier.

Bhunia *et al.* revealed the VLS growth mechanism of vertically aligned InP nanowires synthesized by a MOVPE technique [79]. They used colloidal Au nanoparticles as the seed to control the nanowire growth and the Au nanoparticles were clearly observed at the tip of each nanowire (Fig. **10**).

Many groups demonstrated that GaAs nanowires could be grown *via* a VLS mechanism. Soci *et al.* reported MOCVD growth of GaAs nanowires though a VLS mechanism with Au nanoparticles as the catalyst [19]. Mohseni *et al.* realized GaAs nanowires grown *via* a VLS process in a gas source MBE system, in which gold was used as the catalyst particles were typically observed at tips of the GaAs nanowires [60]. Very recently, Ramdani *et al.* revealed the growth of GaAs nanowires *via* Au-assisted VLS mechanism by a HVPE method [73].

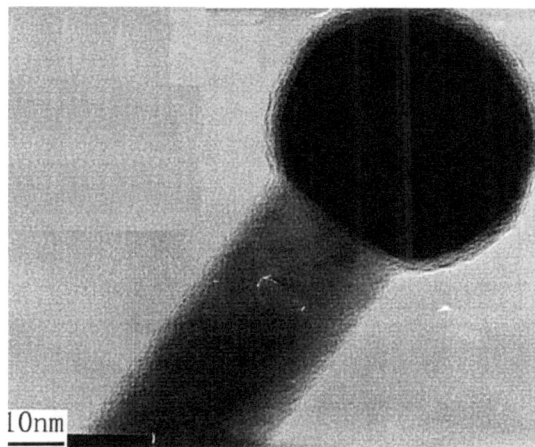

Figure 11: TEM image of a VLS grown InAs nanowire with a gold catalyst particle at the tip. Reprinted in part with permission from *Applied Physics Letters*, 2004, Vol. 85, 1802-1804. Copyright 2004 American Institute of Physics.

Park *et al.* reported the observation of InAs nanowires growth using a simple closed CVD system, in which the growth of the InAs nanowires proceeded through the VLS mechanism with Au-In alloy acting as the liquid catalyst

[80]. Dayeh *et al.* demonstrated MOCVD growth of InAs nanowires *via* a VLS mechanism with catalyst nanoparticles on top [48-50].

The VLS growth mechanism of III-V semiconductor nanowires was also reexamined in the wafer annealing growth of InAs, InP, and GaP nanowires reported by Zhi *et al.* Annealing only metal (gold) film coated semiconductor wafers formed the nanowires and the detected gold catalyst particle at the ends of the nanowires indicated the VLS mechanism of the nanowires [26]. Fig. **11** is a TEM image of the VLS grown InAs nanowire by a wafer annealing method.

Vapor-Solid-Solid Growth

Vapor-solid-solid (VSS) growth from a solid particle was first proposed for Ag-catalyzed silicon whiskers in 1979 [28]. VSS growth mechanism is a solid-phase diffusion mechanism, in which a solid catalyst acts as the energetically favored site for absorption of gas-phase reactants to direct the nanowire growth [28-30, 81-83]. VSS mechanism has been reported for the growth of III-V semiconductor nanowires such as GaAs [29] and InAs nanowires [30, 81].

Persson and coworkers discovered that the CBE grown GaAs nanowires could proceed *via* a VSS mechanism [29]. Revealed by *in situ* heating experiments of the GaAs nanowires in a TEM, the growth relied on a solid-phase diffusion mechanism of a single component through a solid gold seed particle. It was also supported by highly resolved chemical analysis and finite element calculations of the mass transport and composition profiles.

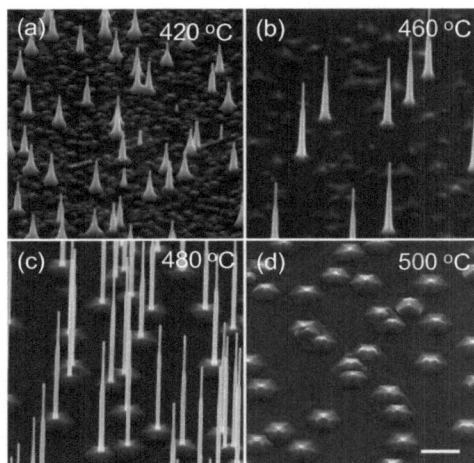

Figure 12: SEM images of Au-assisted InAs nanowires grown at different temperatures, viewed 45 degrees from normal to the surface: **(a)** 420 °C, **(b)** 460 °C, **(c)** 480 °C, and **(d)** 500 °C. The scale bar is 1 µm. No high-temperature annealing step is performed before growth; for all samples the growth time was 8 min. Reprinted with permission from Ref. [30]. Copyright 2005 American Chemical Society.

Dick *et al.* discovered a failure of the VLS mechanism in Au-assisted growth of InAs nanowires. They showed that InAs nanowires could be grown by pure Au-assisted MOVPE only in the temperature range where the particle formed a solid alloy with supplied In. They asserted that the InAs nanowire growth was assisted by a solid particle *via* a VSS mechanism, and their conclusion was supported by a compared growth of InAs nanowires in the same system assisted by a layer of SiO$_x$ [30]. Fig. **12** shows SEM images of the Au-assisted InAs nanowires grown at different temperatures, and Fig. **13** is high-resolution images of the top of a single nanowire.

Vapor-Solid Growth

Vapor-solid (VS) mechanism is generally referred to as a process of one-dimensional material growth from vapor phase precursors in the absence of a catalyst or obvious VLS evidence. There are many plausible growth mechanisms to consider, and a synthetic experiment might grow nanowires through a combination of these mechanisms. In consideration of thermodynamics and kinetics, the VS growth of nanowires could be possible *via* (*a*) a self-catalytic VLS growth, *(b)* an oxide-assisted growth, *(c)* Frank's screw dislocation mechanism, (*d*) an anisotropic growth mechanism, or (*e*) a different defect-induced growth model [2, 31, 32, 84-86]. Although the exact mechanisms responsible for VS growth are not completely elucidated and many of these proposed VS growth mechanisms lack compelling thermodynamic and kinetic validation of nanowire growth [2], many III-V semiconductor nanowires have been made using these processes.

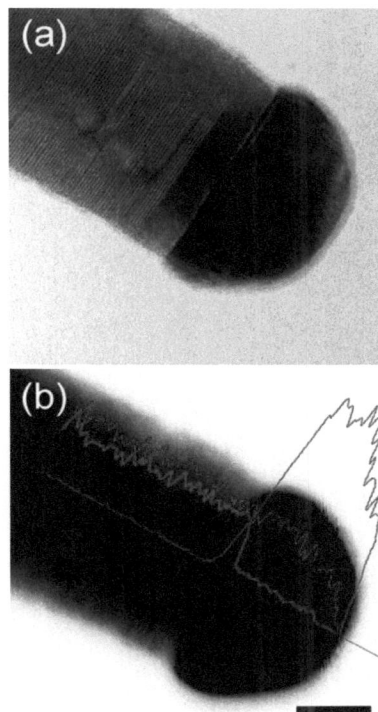

Figure 13: High-resolution images of the top of a single nanowire. **(a)** TEM image, showing the crystalline nature of the wire and the seed particle. Stacking faults in the wire are visible as lines perpendicular to the wire growth direction. **(b)** Scanning transmission electron microscopy (STEM) image of the same nanowire as in (a), with the corresponding energy dispersive X-ray spectroscopy (EDS) line scan data superimposed. For this illustration green = Au, red = In, and blue =As. The nanowire is grown at 480 °C. The scale bar is 10 nm. Reprinted with permission from Ref. [30]. Copyright 2005 American Chemical Society.

Self-catalytic VLS growth is a process for nanowires of binary and more complex stoichiometries, in which one of the nanowire elements serves as the VLS catalyst to direct the nanowire growth. The major advantage of a self-catalytic VLS mechanism is that it avoids undesired contamination from foreign elements typically used as VLS catalysts [2]. The generally reported self-catalytic VLS mechanism for III-V semiconductor nanowires is group-III catalyzed VLS process [20, 85, 86].

Stach *et al.* demonstrated the self-catalytic growth of GaN nanowires. They used *in situ* TEM to observe directly self-catalytic growth of the nanowires by heating GaN thin film in 10^{-7} torr vacuum. At the start, decomposition of the GaN film directs the formation of isolated liquid Ga nanoparticles, and then the resultant vapor species re-dissolves into the Ga droplets and initiates VLS nanowires after supersaturating the metal and creating a liquid-Ga/solid-GaN interface [85]. Fig. **14** shows the steps in the VLS process observed in the TEM study. These experimental findings provide significant new insight into the III-V nanowire growth and aid the understanding of the growth mechanisms.

Novotny *et al.* reported the growth of vertically aligned InP nanowires by MOCVD method without the use of a deposited metal catalyst. They concluded that a surface reconstruction induced indium droplets to form on the surface and thus act as nucleation sites for nanowire VLS growth [20]. Dayeh and coworkers examined nucleation and growth of InAs nanowires from excess In with and without Au nanoparticles during MOCVD growth, and found that excess liquid In nucleated InAs nanowires *via* the self-catalytic VLS mechanism on an InAs(111)B substrate [86].

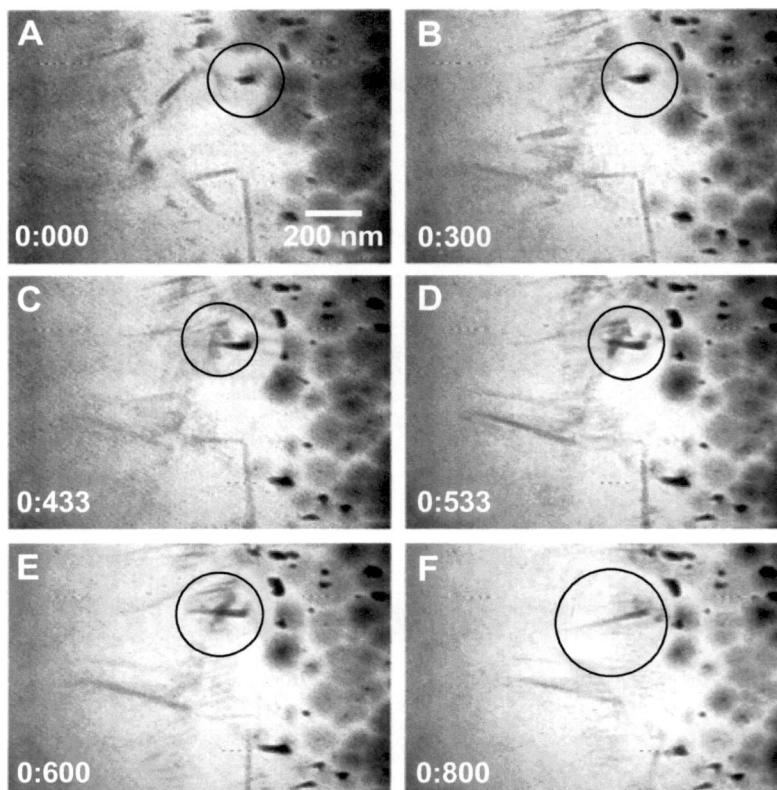

Figure 14: A series of video frames grabbed from observations of GaN decomposition at about 1050 °C, showing the real-time growth process of a GaN nanowire. The number on the bottom left corner of each image is the time (second: millisecond). Reprinted with permission from Ref. [85]. Copyright 2003 American Chemical Society.

Lee and co-workers found that the growth of Si nanowires was greatly enhanced when SiO_2-containing Si powder was used as target. Based on their experimental observations, they proposed an oxide-assisted growth mechanism. By this process, no metal catalyst is required for the growth of nanowires [87, 88]. By oxide-assisted laser ablation of a mixture of GaAs and Ga_2O_3, they realized the oxide-assisted growth of GaAs nanowires [31].

Frank explained the formation of a hollow core screw dislocation by a total energy minimization approach. At present, Frank-Eshelby's theory for nanopipes in thin cylinders takes on new significance as these conditions can be physically realized in nanowires [32, 84, 89, 90]. Using high-resolution TEM, Jacobs *et al.* analyzed the GaN nanowires synthesized by a catalyst-free VS growth process [32]. The cross section TEM studies revealed hollow core screw dislocations/nanopipes in the nanowires. They found that the hollow cores were located at or near the center of the nanowires along the axis of a screw dislocation. Their observations, carried out in cross section with the electron beam parallel to the Burgers vector, identified the axial screw dislocation mechanism in GaN nanowires [32].

Anisotropic growth of crystals induced by different surface energies is the basis for the formation of most elongated nanocrystals [39]. In an anisotropic mechanism, nanowire growth can be accomplished by the preferential reactivity and binding of gas phase reactants along specific crystal facets and also the desire for a system to minimize surface energies [2]. Mayers *et al.* reported that trigonal tellurium nanowires were easily grown due to anisotropic bonding [91].

In dislocation and defect-induced growth models, specific defects have larger sticking coefficients for gas phase species, thus allowing enhanced reactivity and deposition of gas phase reactants at these defects/dislocations [2]. For example, high-resolution TEM investigations revealed that the defects and silicon oxide outer layers existing at the nanowire tips played important roles for the Si nanowire growth [39].

Solution-Liquid-Solid Growth

Buhro and coworkers developed a low-temperature solution method for the synthesis of III-V semiconductor nanowires (nanowhiskers) [27]. In this report, III-V compounds InP, InAs, and GaAs nanowires were grown *via* a solution-liquid-solid (SLS) process.

The SLS growth mechanism is analogous to the VLS mechanism. In a typical procedure, a metal with a low melting point such as indium *etc.* was used as a catalyst, the desired III-V compound generated through the decomposition of organometallic precursors, and the III-V nanowires were synthesized by low-temperature solution phase reactions *via* a SLS growth mechanism [27].

Fanfair *et al.* reported solution-phase metal nanocrystal-seeded growth of III-V semiconductor nanowires [74]. Using Bi nanocrystals as the seeds, crystalline InAs, GaP, GaAs, and InP nanowires were synthesized *via* an SLS growth mechanism at temperatures 300~340 °C in TOP and TOPO, and TOA. The Bi nanocrystals were observed at the nanowire tips (Fig. **15**), confirming their function as the crystallization seeds [74].

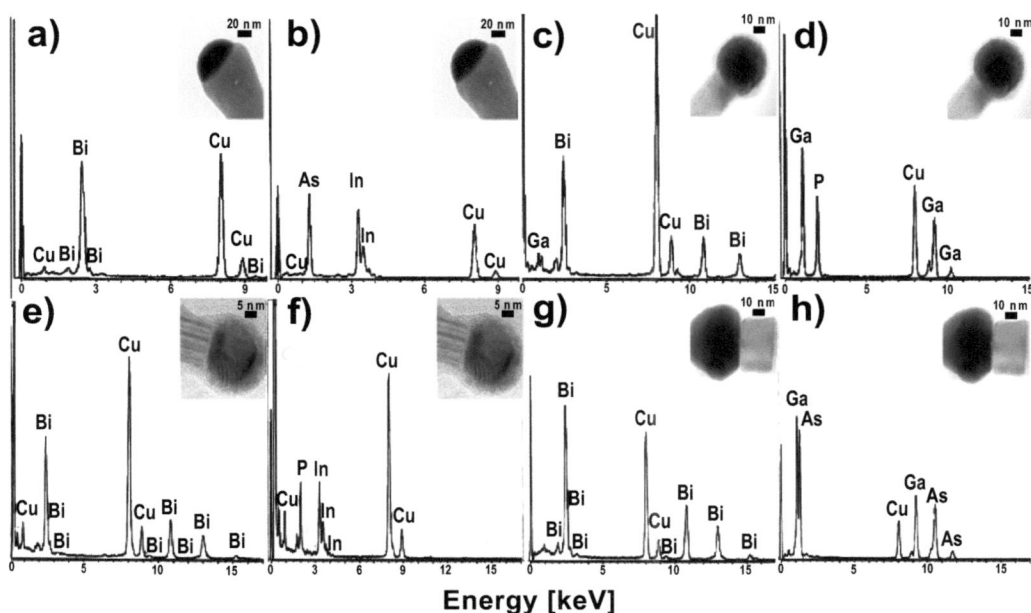

Energy [keV]

Figure 15: (A, C, E, G) EDS data obtained from nanocrystals at the tip of **(A)** InAs, **(C)** GaP, **(E)** InP, and **(G)** GaAs nanowires as compared to EDS from the **(B)** InAs, **(D)** GaP, **(F)** InP, and **(H)** GaAs nanowires. InAs, InP, and GaP nanowires were produced in TOP/TOPO. GaAs nanowires were produced in TOA. Reprinted with permission from Ref. [74]. Copyright 2005 American Chemical Society.

Other Growth Mechanisms

Other proposed III-V nanowire growth mechanisms have been reported, for example the ligand-aided solution-solid growth [33], and the reactive Si-assisted growth [34].

Xu *et al.* used a combination of concentration-driven and ligand-aided solution–solid (LSS) growth mechanisms to explain the morphology evolution of InAs nanowires [33]. In their experiment, the InAs nanowires were grown at 120 °C and no evidence was found that the SLS mechanism should be responsible for the nanowire formation. Considering the fact that a liquid indium droplet inducing VLS or SLS growth should be above the indium melting point, and a similar reaction route had confirmed that the SLS mechanism would not work at 120 °C for indium droplet directed InAs nanowire growth [92], they proposed a LSS growth mechanism to explain the formation of InAs nanowires at such a low temperature: With the ligands or capping reagents preferentially adsorbing or bonding to a specific InAs surface, the growth rates of different surfaces were redefined. As a result, the nanowires were generated [33].

Park and coworkers demonstrated the reactive Si-assisted growth of InAs nanowires using nanosized Si clusters in a closed system without any metal catalyst [34]. They found that the presence of oxygen inhibited the VLS mechanism during the nanowire growth. More significantly, in the case of SiO_x, a phase separation formed a stable SiO_2 and reactive, nanosized Si clusters at high temperature. They proposed that these clusters consequently served as the catalytic sites for the growth of InAs nanowires with a process different from VLS, VSS, and oxide assisted growth mechanism [34].

CONCLUSIONS

The area of III-V semiconductor nanowires has been constantly gaining interest and momentum among science and engineering communities since the late 1990s. In this chapter, we attempted to review the general synthetic strategies for III-V semiconductor nanowires. We first summarized various growth techniques to realize the synthesis of the III-V nanowires such as CVD, laser ablation, MOCVD or MOVPE, MBE, CBE, HVPE, wafer annealing, and low-temperature solution methods. Subsequently, we described the general mechanisms whereby III-V nanowires were obtained from different synthetic schemes and conditions, including VLS, VSS, SLS, VS (self-catalytic, oxide-assisted, and axial screw dislocation), ligand-aided solution-solid, and reactive Si-assisted growth.

ACKNOWLEDGEMENTS

Jianye Li thanks the financial supports from the Fundamental Research Funds for the Central Universities of China, Beijing Natural Science Foundation (2093038), National Natural Science Foundation of China (21071016), and Scientific Research Foundation for the Returned Overseas Chinese Scholars, Ministry of Education of China.

REFERENCES

[1]   Hu, J.T.; Odom, T.W.; Lieber, C.M. Chemistry and physics in one dimension: synthesis and properties of nanowires and nanotubes. *Acc. Chem. Res.* **1999**, *32*, 435.

[2]   Law, M.; Goldberger, T.; Yang, P.D. Semiconductor nanowires and nanotubes. *Annu. Rev. Mater. Res.* **2004**, *34*, 83.

[3]   Li, Y.; Qian, F.; Xiang, J.; Lieber, C.M. Nanowire electronic and optoelectronic devices. *Mater. Today*, **2006**, *9*, 18.

[4]   Qian, F.; Li, Y.; Gradecak, S.; Park, H. G.; Dong, Y.; Ding, Y.; Wang, Z. L.; Lieber, C. M. Multi-quantum-well nanowire heterostructures for wavelength-controlled lasers. *Nat. Mater.* **2008**, *7*, 701.

[5]   Zhong, Z.H.; Qian, F.; Wang, D.L.; Lieber, C.M. Synthesis of p-type gallium ntride nanowires for electronic and photonic nanodevices. *Nano Lett.* **2003**, *3*, 343.

[6]   Tang, Y. B.; Chen, Z. H.; Song, H. S.; Lee, C. S.; Cong, H. T.; Cheng, H. M.; Zhang, W. J.; Bello, I.; Lee, S. T. Vertically aligned p-type single-crystalline GaN nanorod arrays on n-type Si for heterojunction photovoltaic cells. *Nano Lett.* **2008**, *8*, 4191.

[7]   Huang, C.; Song, J.; Lee, W.; Ding, Y.; Cao, Z.Y.; Hao, Y.; Chen, L.J.; Wang, Z. L. GaN nanowire arrays for high-output nanogenerators. *J. Am. Chem. Soc.* **2010**, *132*, 4766.

[8]   Han, W.; Fan, S. H.; Li, Q. Q.; Hu, Y. D. Synthesis of gallium nitride nanorods through a carbon nanotube-confined reaction. *Science* **1997**, *277*, 1287.

[9]   Cheng, G.S.; Zhang, L.D.; Zhu, Y.; Fei, G.T.; Li, L.; Mo, C.M.; Mao, Y.Q. Large-scale synthesis of single crystalline gallium nitride nanowires. *Appl. Phys. Lett.* **1999**, *75*, 2455.

[10]   Chen, X. L.; Li, J. Y.; Cao, Y. G.; Lan, Y. C.; Li, H.; He, M.; Wang, C. Y.; Zhang, Z.; Qiao, Z. Y. Straight and smooth GaN nanowires. *Adv. Mater.* **2000**, *12*, 1432.

[11]   Liu, J.; Zhang, X.; Zhang, Y.; He, R.; Zhu, J. Novel synthesis of AlN nanowires with controlled diameters. *J. Mater. Res.* **2001**, *16*, 3133.

[12]   Li, J. Y.; Lu, C. G.; Maynor, B.; Huang, S. M.; Liu, J. Controlled growth of long GaN nanowires from catalyst Patterns fabricated by "dip-pen" nanolithographic techniques. *Chem. Mater.* **2004**, *16*, 1633.

[13]   Cheng, G. S.; Stern, E.; Turner-Evans, D.; Reed, M. A. Electronic properties of InN nanowires. *Appl. Phys. Lett.* **2005**, *87*, 253103.

[14]   Duan, X. F.; Lieber, C. M. Laser-assisted catalytic growth of single crystal GaN nanowires. *J. Am. Chem. Soc.* **2000**, *122*, 188.

[15]   Duan, X. F.; Wang, J.; Lieber, C. M. Synthesis and optical properties of gallium arsenide nanowires. *Appl. Phys. Lett.* **2000**, *76*, 1116.

[16] Huang, Y.; Duan, X. F.; Cui, Y.; Lieber, C. M. Gallium nitride nanowire nanodevices. *Nano Lett.* **2002**, *2*, 101.

[17] Kuykendall, T.; Pauzauskie, P.; Lee, S.; Zhang, Y. F.; Goldberger, J.; Yang, P. D. Metalorganic chemical vapor deposition route to GaN nanowires with triangular cross sections. *Nano Lett.* **2003**, *3*, 1063.

[18] Kuykendall, T.; Pauzauskie, P.; Zhang, Y. F.; Goldberger, J.; Sirbuly, D.; Denlinger, J.; Yang, P. D. Crystallographic alignment of high density gallium nitride nanowire arrays. *Nat. Mater.* **2004**, *3*, 524.

[19] Soci, C.; Bao, X. Y.; Aplin, D. P. R.; Wang, D. L. A systematic study on the growth of GaAs nanowires by metal-organic chemical vapor deposition. *Nano Lett.* **2008**, *8*, 4275.

[20] Novotny, C. J.; Yu, P. K. L. Vertically aligned, catalyst-free InP nanowires grown by metalorganic chemical vapor deposition. *Appl. Phys. Lett.* **2005**, *87*, 203111.

[21] Dubrovskii, V. G.; Cirlin, G. E.; Soshnikov, I. P.; Tonkikh, A. A.; Sibirev, N. V.; Samsonenko, Y. B.; Ustinov, V. M. Diffusion-induced growth of GaAs nanowhiskers during molecular beam epitaxy: Theory and experiment. *Phys. Rev. B* **2005**, *71*, 205325.

[22] Calarco, R.; Meijers, R. J.; Debnath, R. K.; Stoica,T.; Sutter, E.; Luth, H. Nucleation and growth of GaN nanowires on Si(111) performed by molecular beam epitaxy. *Nano Lett.* **2007**, *7*, 2248.

[23] Sadowski, J.; Dłuzewski, P.; Kret, S.; Janik, E.; Łusakowska, E.; Kanski, J.; Presz, A.; Terki,F.; Charar,S.; Tang, D. GaAs:Mn Nanowires grown by molecular beam epitaxy of (Ga,Mn)As at MnAs segregation conditions. *Nano Lett.* **2007**, *7*, 2724.

[24] Ohlsson, B. J.; Bjork, M. T.; Magnusson, M. H.; Deppert, K.; Samuelson, L.; Wallenberg, L. Size-, shape-, and position-controlled GaAs nano-whiskers. *Appl. Phys. Lett.* **2001**, *79*, 3335.

[25] Seryogin, G.; Shalish1, I.; Moberlychan, W.; Narayanamurti, V. Catalytic hydride vapour phase epitaxy growth of GaN nanowires. *Nanotechnology* **2005**, *16*, 2342.

[26] Zhi, C. Y.; Bai, X. D.; Wang, E. G. Synthesis of semiconductor nanowires by annealing. *Appl. Phys. Lett.* **2004**, *85*, 1802.

[27] Trentler, T. J.; Hickman, K. M.; Goel, S. C.; Vian, A. M.; Gibbons; Buhro, W. E. Solution-liquid-solid growth of crystalline III-V semiconductors: An analogy to vapor-liquid-solid growth. *Science,* **1995**, *270,* 1791.

[28] Dayeh, S. A.; Yu, E. T.; Wang, D. L. III-V nanowire growth mechanism: V/III ratio and temperature effects. *Nano Lett.* **2007**, *7*, 2486.

[29] Persson, A. I.; Larsson, M. W.; Stenstro¨m, S.; Ohlsson, B. J.; Samuelson, L.; Wallenberg, L. R. Solid-phase diffusion mechanism for GaAs nanowire growth. *Nat. Mater.* **2004**, *3*, 677.

[30] Dick, K. A.; Deppert, K.; Martensson, T.; Mandl, B.; Samuelson, L.; Seifert, W. Failure of the vapor–liquid–solid mechanism in Au-assisted MOVPE growth of InAs nanowires. *Nano Lett.* **2005**, *5*, 761.

[31] Shi, W. S.; Zheng, Y. F.; Wang, N.; Lee, C. S.; Lee, S. T. Oxide-assisted growth and optical characterization of gallium-arsenide nanowires. *Appl. Phys. Lett.* **2001**, *78*, 3304.

[32] Jacobs, B. W.; Crimp, M. A.; McElroy, K.; Ayres, V. M. Electronic and structural characteristics of zinc-blende wurtzite biphasic homostructure GaN nanowires. *Nano Lett.* **2008**, *8*, 4353.

[33] Xu, X.; Wei, W.; Qiu, X.; Yu, K.; Yu, R.; Si, S.; Xu, G.; Huang, W.; Peng, B. Synthesis of InAs nanowires *via* a low-temperature solvothermal route. *Nanotechnology* **2006**, *17*, 3416.

[34] Park, H. D.; Prokes, S. M.; Twig, M. E.; Cammarata, R. C.; Gaillot, A. C. Si-assisted growth of InAs nanowires. *Appl. Phys. Lett.* **2006**, *89*, 223125.

[35] He, M.; Fahmi, M. M. E.; Mohammad, S. N. InAs nanowires and whiskers grown by reaction of indium with GaAs. *Appl. Phys. Lett.* **2003**, *82*, 3749.

[36] He, M.; Mohammad, S. N. Novel chemical-vapor deposition technique for the synthesis of high-quality single-crystal nanowires and nanotubes. *J. Chem. Phys.* **2006**, *124*, 064714.

[37] Morales, A. M.; Lieber, C. M. A Laser Ablation method for the synthesis of crystalline semiconductor nanowires. *Science* **1998**, *279*, 208.

[38] Yang, P.D. Chemistry and physics of semiconductor nanowires. *MRS Bull.* **2005**, *30*, 85.

[39] Wang, N.; Cai, Y.; Zhang, R. Q. Growth of nanowires. *Mat. Sci. Eng. R* **2008**, *60*, 1.

[40] Gudiksen, M.S.; Lieber, C. M. Diameter-selective synthesis of semiconductor nanowires. *J. Am. Chem. Soc.* **2000**, *122*, 8801.

[41] Duan, X. F.; Lieber, C. M. General synthesis of compound semiconductor nanowires. *Adv. Mater.* **2000**, *12*, 298.

[42] http://en.wikipedia.org/wiki/MOVPE.

[43] Borgstrom, M.; Deppert, K.; Samuelson, L.; Seifert, W. Size- and shape-controlled GaAs nano-whiskers grown by MOVPE: a growth study. *J. Cryst. Growth* **2004**, *260* (1-2), 18.

[44] Paiano, P.; Prete, P.; Speiser, E.; Lovergine, N.; Richter, W.; Tapfer, L.; Mancini, A. M. GaAs nanowires grown by Au-catalyst-assisted MOVPE using tertiarybutylarsine as group-V precursor. *J. Cryst. Growth* **2007**, *298*, 620.

[45] Joyce, H. J.; Gao, Q.; Tan, H. H.; Jagadish, C.; Kim, Y.; Zhang, X.; Guo, Y. N.; Zou, J. Twin-free uniform epitaxial GaAs nanowires grown by a two-temperature process. *Nano Lett.* **2007**, *7* (4), 921.

[46] Bauer, J.; Gottschalch, V.; Paetzelt, H.; Wagner, G.; Fuhrmann, B.; Leipner, H. S. MOVPE growth and real structure of vertical-aligned GaAs nanowires. *J. Cryst. Growth* 2007, *298*, 625.

[47] Gao, L.; Woo, R. L.; Liang, B.; Pozuelo, M.; Prikhodko, S.; Jackson, M.; Goel, N.; Hudait, M. K.; Huffaker, D. L.; Goorsky, M. S.; Kodambaka, S.; Hicks, R. F. Self-catalyzed epitaxial growth of vertical indiumpPhosphide nanowires on silicon. *Nano Lett.* **2009**, *9*, 2223.

[48] Dayeh, S. A.; Aplin, D. P. R.; Zhou, X.; Yu, P. K. L.; Yu, E. T.; Wang, D. L. High electron mobility InAs nanowire field-effect transistors. *Small* **2007**, *3*, 326.

[49] Dayeh, S. A.; Yu, E. T.; Wang, D. L. Growth of InAs nanowires on SiO_2 substrates: nucleation, evolution, and the role of Au nanoparticles. *J. Phys. Chem. C* **2007**, *111*, 13331.

[50] Dayeh, S. A.; Yu, E. T.; Wang, D. L. Transport coefficients of InAs nanowires as a function of diameter. *Small* **2009**, *5*, 77.

[51] Wei, W.; Bao, X. Y.; Soci, C.; Ding, Y.; Wang, Z. L.; Wang, D. L. Direct heteroepitaxy of vertical InAs nanowires on Si substrates for broad band photovoltaics and photodetection. *Nano Lett.* **2009**, *9*, 2926.

[52] Stangl, J.; Holý, V.; Bauer, G. Structural properties of self-organized semiconductor nanostructures. *Rev. Mod. Phys.* **2004**, *76*, 725.

[53] Bertness, K.A.; Roshko, A.; Mansfield, L. M.; Harvey, T. E.; Sanford, N. A. Nucleation conditions for catalyst-free GaN nanowires. *J. Cryst. Growth* 2007, *300*, 94.

[54] Dimakis, E.; Konstantinidis, G.; Tsagaraki, K.; Adikimenakis, A.; Iliopoulos, E.; Georgakilas, A. The role of nucleation temperature in In-face InN-on-GaN(0001) growth by plasma-assisted molecular beam epitaxy. *Superlattices Microstruct.* **2004**, *36*, 497.

[55] Grandal, J.; Sanchez-Garcia, M. A.; Calle, F.; Calleja, E. Morphology and optical properties of InN layers grown by molecular beam epitaxy on silicon substrates. *Phys. Status Solidi C* **2005**, *2*, 2289.

[56] Stoica, T.; Meijers, R.; Calarco, R.; Richter, T.; Luth, H. MBE growth optimization of InN nanowires. *J. Cryst. Growth* **2006**, *290*, 241.

[57] Stoica, T.; Meijers, R.; Calarco, R.; Richter, T.; Sutter, E.; Luth, H. Photoluminescence and intrinsic properties of MBE-grown InN nanowires. *Nano Lett.* **2006**, *6*, 1541.

[58] Chen, H. Y.; Shen, C. H.; Lin, H. W.; Chen, C. H.; Wu, C. Y.; Guo, S.; Davydov, V. Y.; Klochikhin, A.A. Near-infrared photoluminescence of vertically aligned InN nanorods grown on Si(111) by plasma-assisted molecular-beam epitaxy. Thin Solid Films 2006, 515, 961.

[59] Tchernycheva, M.; Cirlin, G. E.; Patriarche, G.; Travers, L.; Zwiller, V.; Perinetti, U.; Harmand, J. C. Growth and characterization of InP nanowires with InAsP insertions. *Nano Lett.* **2007**, *7*, 1500.

[60] Mohseni, P. K.; Lawson, G.; Couteau, C.; Weihs, G.; Adronov, A.; LaPierre, R.R. Growth and characterization of GaAs nanowires on carbon nanotube composite films: toward flexible nanodevices. *Nano Lett.* **2008**, *8*, 4075.

[61] Shtrikman, H.; Popovitz-Biro, R.; Kretinin, A.; Heiblum, M. Stacking-faults-free zinc blende GaAs nanowires. *Nano Lett.* **2009**, *9*, 215.

[62] Ihn, S. G.; Song, J. I. InAs nanowires on Si substrates grown by solid source molecular beam epitaxy. *Nanotechnology* **2007**, *18*, 355603.

[63] Cornet, D.M.; LaPierre, R.R. InGaAs/InP core–shell and axial heterostructure nanowires. *Nanotechnology* **2007**, *18*, 385305.

[64] Tsang, W. T. Chemical beam epitaxy of InP and GaAs. *Appl. Phys. Lett.* **1984**, *45*, 1234.

[65] http://en.wikipedia.org/wiki/Chemical_beam_epitaxy.

[66] Jensen, L. E.; Bjork, M. T.; Jeppesen, S.; Persson, A. I.; Ohlsson, B. J.; Samuelson, L. Role of surface diffusion in chemical beam epitaxy of InAs nanowires. *Nano Lett.* **2004**, *4*, 1961.

[67] Poole, P. J.; Lefebvre, J.; Fraser, J. Spatially controlled, nanoparticle-free growth of InP nanowires. *Appl. Phys. Lett.* **2003**, *83*, 2055.

[68] Chao, C. K.; Chyi, J.-I.; Hsiao, C. N.; Kei, C. C.; Kuo, S. Y.; Chang, H.-S.; Hsu, T. M. Catalyst-free growth of indium nitride nanorods by chemical-beam epitaxy. *Appl. Phys. Lett.* **2006**, *88*, 233111.

[69] Larsson, M. W.; Wagner, J. B.; Wallin, M.; Håkansson; Froberg, L. E.; Samuelson, L.; Wallenberg, L. Reine. Strain mapping in free-standing heterostructured wurtzite InAs/InP nanowires. *Nanotechnology* 2007, *18*, 015504.

[70] Eymery, J.; Rieutord, F.; Favre-Nicolin, V.; Robach, O.; Niquet, Y. M.; Fro1berg, L.; Mårtensson, T.; Samuelson, L. Effects of a shell on the electronic properties of nanowire superlattices. *Nano Lett.* **2007**, *7*, 2596.

[71] Tietjen, J. J.; Amick, J. A. Preparation and properties of vapor-deposited epitaxial $GaAs_{1-x}P_x$ using arsine and phosphine. *J. Electrochem. Soc.* **1966**, *113*, 724.

[72]   Kim, H. M.; Lee, W. C.; Kang, T. W.; Chung, K. S.; Yoon, C.S.; Kim, C. K. InGaN nanorods grown on (111) silicon substrate by hydride vapor phase epitaxy. *Chem. Phys. Lett.* **2003**, *380*, 181.

[73]   Ramdani, M. R.; Gil, E.; Leroux, Ch.; Andre, Y.; Trassoudaine, A.; Castelluci, D.; Bideux, L.; Monier, G.; Robert-Goumet, C.; Kupka, R. Fast growth synthesis of GaAs nanowires with exceptional length. *Nano Lett.* **2010**, *10*, 1836.

[74]   Fanfair, D. D.; Korgel, B. A. Bismuth Nanocrystal-Seeded III-V Semiconductor Nanowire synthesis. *Cryst. Growth Des.* **2005**, *5*, 1971.

[75]   Yang, P.D.; Yan, R.X.; Fardy, M. Semiconductor nanowires: What's next? *Nano Lett.* **2010**, *10*, 1529.

[76]   Wagner, R.S.; Ellis, W. C. Vapor-liquid-solid mechanism for single crystal growth. *Appl. Phys. Lett.* **1964**, *4*, 89.

[77]   Givargizov, E. I. Fundamental aspects of VLS growth. *J. Cryst. Growth* **1975**, *31*, 20.

[78]   Cai, X. M.; Ye, F.; Jing, S. Y.; Zhang, D.P.; Fan, P.; Xie, E.Q. A systematic study of chemical vapor deposition growth of InN. *Appl. Surf. Sci.* **2008**, *255*, 2153.

[79]   Bhunia, S.; Kawamura, T.; Fujikawa, S.; Watanabe, Y. Systematic investigation of growth of InP nanowires by metalorganic vapor-phase epitaxy. *Physica E* **2004**, *24*, 138.

[80]   Park, H. D.; Prokes, S. M.; Cammarata, R. C. Growth of epitaxial InAs nanowires in a simple closed system. *Appl. Phys. Lett.* **2005**, *87*, 063110.

[81]   Mandl, B.; Stangl, J.; Martensson, T.; Mikkelsen, A.; Eriksson, J.; Karlsson, L. S.; Bauer, G.; Samuelson, L.; Seifert. W. Au-free epitaxial growth of InAs nanowires. *Nano Lett.* **2006**, *6*, 1817.

[82]   Wang, Y. W.; Schmidt, V.; Senz, S.; Gosele, H. Epitaxial growth of silicon nanowires using an aluminium catalyst. *Nat. Nanotech.* **2006**, *1*, 186.

[83]   Kodambaka, S.; Tersoff, J.; Reuter, M. C.; Ross, F. M. Germanium nanowire growth below the eutectic temperature. *Science* **2007**, *316*, 729.

[84]   Frank, F. C. The influence of dislocations on crystal growth. *Discuss. Faraday Soc.* **1949**, *5, 48*.

[85]   Stach, E. A.; Pauzauskie, P. J.; Kuykendall, T.; Goldberger, J.; He, R.; Yang, P. D. Watching GaN nanowires grow. *Nano Lett.* **2003**, *3*, 867.

[86]   Dayeh, S. A.; Yu, E. T.; Wang, D. L. Excess indium and substrate effects on the growth of InAs nanowires. *Small* **2007**, *3*, 1683.

[87]   Wang, N.; Tang, Y. H.; Zhang, Y. F.; Lee, C. S.; Lee, S. T. Nucleation and growth of Si nanowires from silicon oxide. *Phys. Rev. B* **1998**, *58*, R16024.

[88]   Lee, S. T.; Wang, N.; Zhang, Y. F.; Tang, Y. H. Oxide-assisted semiconductor nanowire growth. *MRS Bull* **1999**, *24(8)*, 36.

[89]   Frank, F. C. Capillary equilibria of dislocated crystals. *Acta Crystallogr.* **1951**, *4*, 497.

[90]   Frank, F. C. Spontaneous breaking of chiral symmetry in the growth of crystal whiskers: An origin for giant screw dislocations. *Philos. Mag. A.* **1987**, *56*, 263.

[91]   Mayers, B.; Xia, Y. N. One-dimensional nanostructures of trigonal tellurium with various morphologies can be synthesized using a solution-phase approach. *J. Mater. Chem.* **2002**, *12*, 1875.

[92]   Xie, Y.; Yan, P.; Lu, J.; Wang, W. Z.; Qian, Y. T. A safe low temperature route to InAs nanofibers. *Chem. Mater.* **1999**, *11*, 2619.

CHAPTER 2

Growth, Properties, and Device Applications of III-Nitride Nanowire Heterostructures

This review of III-nitride nanowire heterostructures finds that their growth and properties can be well controlled and they hold great promise for applications in a range of nanoelectronic and nanophotonic devices.

Zetian Mi*

Department of Electrical and Computer Engineering, McGill University, 3480 University Street, Montreal, Quebec, Canada H3A 2A7

Abstract: This book chapter provides an overview of the recent developments of III-nitride nanowire heterostructures consisting of GaN, AlN, InN, and their alloys. The growth techniques and mechanisms for III-nitride nanowires are first briefly reviewed, followed by detailed discussions on the structural, optical and electrical transport properties of various III-nitride nanowire heterostructures. Special attention is paid to the recent achievement of high quality InN, InGaN core-shell, as well as dot-in-a-wire nanoscale heterostructures. The emerging device applications of III-nitride nanowires, including nanoscale transistors, LEDs, lasers, and solar cells are presented, and the challenges and future prospects of III-nitride nanowires are also discussed.

Keywords: Nanowire, GaN, InN, AlN, light emitting diode, transistor, solar cell.

INTRODUCTION

Group III-nitride compound semiconductors exhibit unique electrical and optical properties, including high electron mobility, large saturation velocity, large electric breakdown field, extreme chemical stability, and direct energy bandgap encompassing the entire solar spectrum [1-4]. Consequently, they have emerged as the materials of choice for ultraviolet and visible LEDs and lasers, high-power and high-temperature electronics, and future multi-junction solar cells, to name just a few [5-11]. However, due to the lack of native substrates, conventional III-nitride planar heterostructures generally exhibit very high densities of dislocations, which severely limit the device performance and reliability. Nearly defect-free III-nitride nanowire heterostructures, on the other hand, can be achieved on Si or other substrates, due to the highly effective lateral stress relaxation. Additionally, the use of nanowires provides an effective approach to scale down the dimensions of future devices and systems. In this context, tremendous progress has been made in the development of III-nitride nanowire heterostructures.

In this book chapter, we provide an overview of the recent developments of group III-nitride compound semiconductor nanowire heterostructures, as well as their emerging device applications. In Sec. 2, the growth techniques and mechanisms are discussed. In Sec. 3, the structural and optical properties of III-nitride nanowire heterostructures are presented. Special attention is paid to the recent achievement of high quality InN, InGaN core-shell, as well as dot-in-a-wire nanoscale heterostructures. The electrical transport properties of GaN and InN nanowires are described in Sec. 4. In Sec. 5, III-nitride nanowire devices, including nanoscale transistors, LEDs, lasers, and solar cells are presented. Finally, the challenges and future prospects of III-nitride nanowires are briefly discussed in Sec. 6.

GROWTH TECHNIQUES AND MECHANISMS

III-nitride semiconductor nanowires with diameters of tens of nanometers and lengths of several microns, or longer, have been realized utilizing various growth/synthesis techniques, including dry etching, direct reaction between Ga or In metals or NH_3, chemical vapor deposition (CVD), molecular beam epitaxy (MBE), chemical beam epitaxy (CBE), and hydride vapor phase epitaxy (HVPE). The underlying growth mechanisms have been extensively

*Address correspondence to Zetian Mi: Department of Electrical and Computer Engineering, McGill University, 3480 University Street, Montreal, Quebec, Canada H3A 2A7. Phone: 514 398 7114; Fax: 514 398 4470; E-mail: zetian.mi@mcgill.ca

Jianye Li, Deli Wang and Ray R. LaPierre (Eds)

investigated, which may involve the use of catalyst drops in the VLS growth mode, the spontaneous formation under nitrogen rich conditions, or the selective area epitaxy on nano-patterned substrates. Additionally, sophisticated nanowire heterostructures, including core-shell and dot-in-a-wire nanoscale heterostructures, can be achieved by using one or two of the afore-described growth processes.

Vapour-Liquid-Solid Growth

The VLS growth mode, first proposed by Wagner and Ellis in 1964 [12], has been commonly used for the growth/synthesis of III-nitride nanowires. In this growth process, metal nano-particles are first dispersed on a substrate, which induces a liquid droplet of the metal/semiconductor alloy during the growth process. Various metal elements, including Au, Ni, Fe, In, and Co have been used for the growth of III-nitride nanowires [13-16]. A steady feeding of the semiconductor reactant into the liquid droplet supersaturates the eutectic, leading to the nucleation and subsequent growth of semiconductor nanowires. The resulting nanowires are characterized by the formation of metal droplets on the wire top. Important dynamic processes during the nanowire formation include desorption, adsorption, and diffusion of atoms on the nanowire top, lateral sidewalls, and on the substrate surface, nucleation-mediated growth at the liquid-solid interface, as well as the growth on the substrate surface [17]. The position and diameter of nanowires are mainly determined by the location as well as the size of the metal catalysts, while the nanowire length is directly controlled by the growth duration. III-nitride nanowires grown using this process may exhibit either hexagonal or triangular cross-sections, illustrated in Fig. **1** [18].

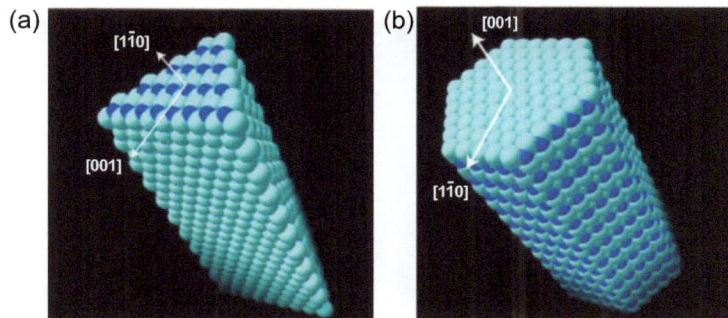

Figure 1: (a) and **(b)** Space-filling structural models for the nanowires with triangular and hexagonal cross-sections. Reprinted by permission from Macmillan Publishers Ltd: [Nature Materials] (18), copyright (2004).

Spontaneous Formation under Nitrogen Rich Conditions

III-nitride nanowires can also be spontaneously formed under nitrogen-rich conditions, wherein no external catalyst is required for the growth initiation. With the use of this growth approach, high quality III-nitride nanowires, that are free of stacking faults, have been obtained by MBE or CVD methods [19-24]. The commonly observed growth orientation for wurtzite III-nitride nanowires is along the [0001] direction (c-axis), with variations of the growth axis perpendicular to the a-planes [25] and m-planes [26,27] also being reported. Compared to the VLS growth process, however, the underlying growth mechanism for the spontaneous formation of III-nitride nanowires has remained a subject of debate.

Recent studies have suggested that the formation of III-nitride nanowires may involve a self-catalytic process, with the nanowires nucleating and growing from nanoscale group-III metal droplets on the substrate surface [28-31]. From real-time transmission electron microscopy (TEM) studies, it was observed that GaN nanowires preferentially nucleate and grow on Ga droplets, providing a direct evidence for the self-catalytic growth of GaN nanowires. However, it is important to note that this experiment was conducted in non-epitaxial growth conditions. Chang et al. have recently investigated the epitaxial growth of InN nanowires by RF-plasma assisted MBE on Si(111) substrates [32]. A direct correlation between the use of an in situ deposited thin (~ 0.5 nm) In seeding layer and the growth and structural properties of InN nanowires has been established. This observation is consistent with the VLS growth model that one of the constituting elements, i.e. In in InN, may act as a catalyst to promote the nucleation and formation of nanowires. However, it has been observed that properties of the GaN nanowires, including the size and density, were virtually unaltered when Ga droplets were deposited prior to the growth initiation [33]. In this study, a nanoscale Ga liquid droplet pattern, with droplet sizes in the range of 90 to 400 nm, were prepared in situ on Si(111) substrates by depositing Ga of thicknesses of ~ 8 to 62 ML at ~ 560 °C. This study suggests that Ga may not serve as a catalyst during the epitaxial growth of GaN nanowires.

Alternatively, a diffusion-induced mechanism has been proposed as a probable model to describe the spontaneous formation of III-nitride nanowires [34-35]. This model includes two different nanowire growth stages. During the initial stage of growth, 3-dimensional islands are formed either in the Stranski-Krastanow or the Volmer-Weber growth mode. Significant growth, with comparable growth rates in both the vertical and lateral directions, occurs on the islands, due to the reduced strain. As a consequence, nanowires preferentially nucleate on the three-dimensional islands, evidenced by the scanning electron microscopy (SEM) images shown in Fig. **2** [36]. This growth stage exhibits a nonlinear growth rate as a function of time. The second regime of nanowire evolution is characterized by a predominant growth along the vertical direction, with a commonly reported wire length and diameter growth rate ratio of ~ 32 [36-37]. The nanowire growth is largely driven by Ga (In, or Al) adatom diffusion along the nanowire sidewalls and their incorporation at the wire top, largely due to the lower free surface energies at the wire top under nitrogen rich conditions [38]. The diffusion-induced mechanism is consistent with the observation that the nanowire axial growth rate increases rapidly with the increase of substrate temperature, due to the significantly enhanced adatom surface migration, in spite of the increased Ga desorption and GaN decomposition. This growth model also suggests that the growth and properties of III-nitride nanowires can be engineered by controlling the coalescence of nucleation sites and the adatom surface migration utilizing various growth conditions.

Figure 2. 45° tilted SEM image of GaN nanowires grown on an AlN buffer layer on Si(111) substrate. Reprinted with permission from Ref. [36]. Copyright 2007 American Institute of Physics.

Selective Area Growth

An alternative approach for catalyst-free growth is the use of selective area epitaxy, wherein nanowire growth takes place on pre-patterned substrates [14, 39-47]. During this growth process, the nanowire geometry can be well-controlled without the use of any foreign metal catalysts. Various selective area growth masks, including SiN_x, Ti, Al, Au, and anodic alumina templates have been utilized. Compared to the VLS and spontaneous growth of nanowires, the selective area growth can, in principle, enable the achievement of uniform nanowire arrays with well-defined diameters and locations. Effects associated with the profile of the nanoscale masks, the anisotropic lateral overgrowth, and faceting should be considered [44, 48-50]. One of the major limitations of this growth approach is directly related to the low yield of producing nanoscale patterned template. However, with the recent development of scalable interferometric and nano-imprinting lithography techniques, this growth process may be suitable for large scale device fabrication [45, 51].

STRUCTURAL AND OPTICAL PROPERTIES OF III-NITRIDE NANOWIRES

In this section, the unique structural and optical properties of III-nitride nanowire heterostructures, including GaN, InN, AlN, and their ternary alloys, are described. Special attention is paid to the recently developed high quality InN and InGaN-based radial and axial nanowire heterostructures.

GaN Nanowires

Various substrates, including Si, sapphire, and SiO_2 layers on Si(100), have been used for the growth/synthesis of high quality GaN nanowires [21, 34, 36-38, 52-54]. GaN nanowires grown on Si(111) substrates are generally perpendicular to the substrates and exhibit a wurtzite crystal structure, with the growth direction along the c-axis, shown in Fig. **3(a)** [38]. It has been observed that the wire length (l) and diameter (d) are approximately correlated by $l=C_1(1+C_2/d)$, where C_1 and C_2 are constants related to the sample growth conditions [34, 38]. It is also noted that some of the nanowires grown on Si(111) are tilted [38, 55]. This is directly related to the presence of a thin (~ 2 nm) amorphous SiN_x layer at the GaN/Si interface, which is induced by the nitridation of the Si surface during the initial growth stage [38, 56]. Since the thin amorphous SiN_x layer formed on Si may not be perfectly flat, GaN nanowires nucleated on an inclined area generally exhibit a tilting angle. It has been demonstrated that the disorientation on Si(111) can be largely eliminated by using a thin AlN buffer layer [36]. Additionally, highly uniform GaN nanowire arrays, with controllable diameters and positions, can be realized by using the technique of selective area growth, illustrated in Fig. **3(b)** [43]. Such GaN nanowire arrays were grown on SiN$_x$ growth-mask layer, with an aperture size of ~ 221 nm.

GaN nanowires are generally free of strain and dislocations, which is evident by the often reported strong and narrow photoluminescence (PL) emission, with the presence of both donor bound excitons and free excitons [53, 57-60]. However, structural and optical properties of GaN nanowires can be significantly affected by Si and Mg dopant incorporation [61-63]. With the incorporation of Si dopant, it was observed that a thicker SiN$_x$ layer was formed at the GaN/Si interface, which led to reduced nanowire densities [62]. Additionally, the presence of Si can alter the GaN surface kinetics by possibly inducing consecutive monatomic steps along the nanowire lateral surfaces during epitaxial growth, which can lead to enhanced incorporation of Ga adatoms on the sidewalls. As a result, GaN:Si nanowires generally exhibit larger diameters and smaller lengths and, under certain conditions, show a tapered morphology as well [62]. The effect of Mg on GaN nanowire growth has also been studied. It was observed that the incorporation of Mg can greatly reduce the GaN nanowire nucleation time [62]. Enhanced growth rate on the nonpolar surfaces, due to the presence of Mg, has also been observed [62, 64]. As a result, the radial growth rate on the lateral, nonpolar planes of GaN nanowires grown on Si(111) substrates is enhanced, leading to GaN:Mg nanowires with increased diameters and reduced lengths. The incorporation of Mg can also significantly alter the crystal structure of GaN nanowires, with the formation of twins and stacking faults for relatively high Mg concentrations.

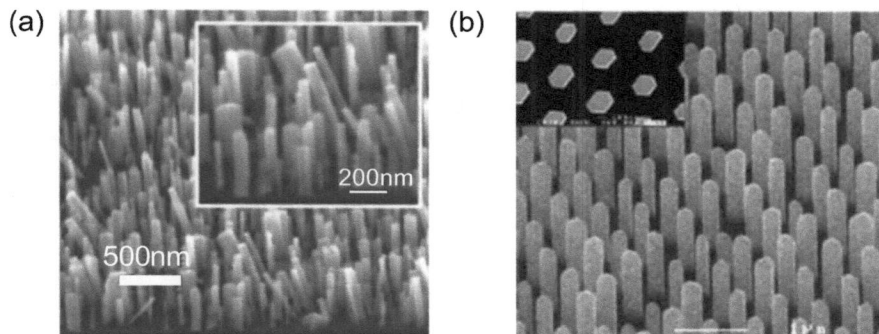

Figure 3: (a) SEM image with a zoomed-in image in the inset of GaN nanowires grown on clean Si(111) substrates. Reprinted with permission from Ref. [38]. Copyright 2008 Wiley InterScience. **(b)** SEM of a GaN nanowire array consisting of 1 μm GaN nanowires (inset shows plan view and reveals the hexagonal symmetry of the nanowires). Reprinted with permission from Ref. [43]. Copyright 2006 American Chemical Society.

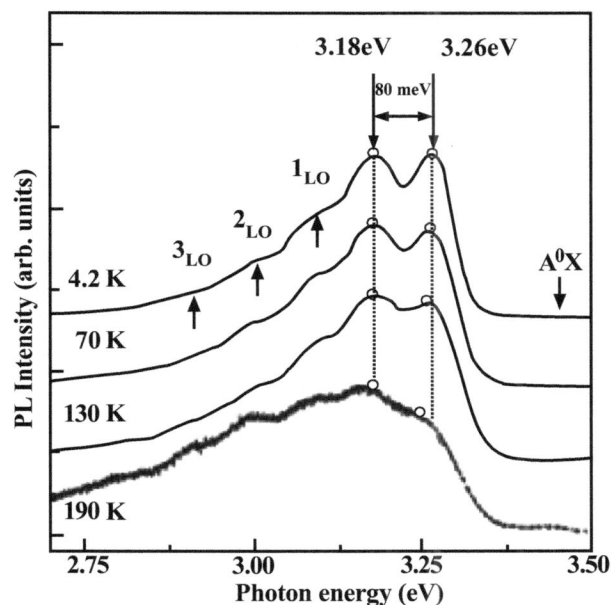

Figure 4: Temperature-dependent PL spectra from Mg-doped GaN nanowires. Reprinted with permission from Ref. [65]. Copyright 2006 Institute of Physics.

With the incorporation of Mg, GaN nanowires also exhibit significantly different optical properties, compared to those of undoped GaN nanowires. Shown in Fig. **4** is the PL spectra of Mg-doped GaN nanowires measured at various temperatures [65]. The GaN:Mg nanowires have a doping density of $\sim 1 \times 10^{17}$ cm^{-3} and diameters of ~ 100 nm. The weak emission at 3.46 eV is related to the neutral-acceptor-bound exciton emission, which becomes much stronger at relatively high excitation powers. The emission peaks at ~ 3.26 and 3.18 eV measured at 4.2 K are attributed to a conduction band to shallow acceptor transition and to transitions associated with the Mg-related structural defects, respectively. The emission peaks (1_{LO}, 2_{LO} and 3_{LO}) on the low energy side of the spectra are assigned to LO phonon assisted recombination associated with the 3.18 eV transition. The energy spacing (~ 91 meV) between these peaks agrees well with the LO phonon energy in GaN.

Al(Ga)N Nanowires

AlN exhibits unique characteristics, including a very high thermal conductivity, electrical resistivity, thermal expansion coefficient, acoustic velocity, and piezoelectric coupling coefficient. As a result, AlN is an excellent candidate for surface acoustic wave and electrical insulation applications. AlN nanowires have been grown using reactive radio-frequency magnetron sputtering and vapor-liquid-solid processes [66-69]. More recently, very promising results have been obtained by using a template- and catalyst-free chemical vapor deposition method [70], which involves the use of a controlled two-stage deposition process, including the separate growth of AlN thin film and nanowires under relatively high and low gas flow rates, respectively. To date, it has remained difficult to obtain high quality AlGaN nanowires, due to the very different growth conditions between GaN and AlN as well as the stress induced phase separation of AlGaN. Recently, single phase AlGaN nanowires have been obtained by CVD on Si substrates, wherein various alloy compositions can be achieved by changing the Al/Ga mass ratio in the mixture of Al powders and Ga droplets [71]. The resulting wires exhibit a wurtzite structure and diameters of hundreds of nanometers.

InN Nanowires

Compared to other III-nitrides, InN exhibit the highest electron mobility (~ 4400 cm^{2}V^{-1}s^{-1} at 300 K), the smallest effective mass, and the highest saturation velocity, making it an excellent candidate for a new generation of nanophotonic and nanoelectronic devices, including chip-level nanoscale lasers and high-speed field effect transistors. InN nanowires have been grown using quartz tube furnace [72], MBE [73-74], and metalorganic chemical vapour deposition (MOCVD) [75]. The resulting InN nanowires, shown in Fig. **5(a)**, generally exhibit severely tapered morphology [32, 40, 55, 73, 76-77]. It can be seen that there are large variations in the wire diameter along the wire length and amongst the wire height. The poorly defined surface morphology severely limits the structural, optical and electrical properties of InN nanowires [55, 72-73, 78-79].

Recently, a novel growth technique, with the use of *in situ* deposited In seeding layers, has been developed to achieve non-tapered and nearly homogeneous InN nanowires using plasma-assisted MBE [32]. In this growth process, a thin (~ 0.5 nm) In layer is first deposited on the substrate at elevated temperatures, prior to the introduction of nitrogen species. Shown in Fig. **5(b)**, these wires are remarkably straight, with identical top and bottom sizes. They also exhibit a homogeneous height. The wires are of wurtzite structure and well-separated, with the *c*-axis oriented vertically to the Si(111) substrate [73, 78, 80, 81]. With increasing growth temperature, InN nanowires generally exhibit reduced areal densities and larger diameters. Detailed high resolution TEM studies confirm that the entire wire exhibits a wurtzite crystal structure and is relatively free of dislocations.

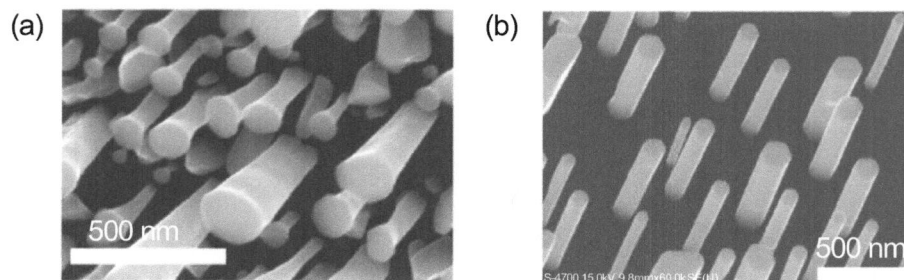

Figure 5: SEM images of InN nanowires **(a)** grown directly on Si(111) at 480 °C and **(b)** grown on Si(111) at 480 °C with the use of an *in situ* deposited In seeding layer. Reprinted with permission from Ref. [32]. Copyright 2009 Institute of Physics.

This study indicates that the formation of InN nanowires may involve a self-catalytic process, as suggested by recent experiments [29-31]. In this process, InN nanowires nucleate and grow from nanoscale In droplets created on the growing surface. During the conventional spontaneous growth of InN nanowires, there are no well-defined nucleation centers, and, consequently, the continuously random nucleation of nanowires leads to InN nanowires on Si with a large variation in height and diameter. Also due to the large diffusion rate of In adatom and their preferential incorporation near the wire top, conventional InN nanowires generally exhibit severely tapered morphology [73]. On the other hand, in the present approach, the *in situ* deposited In layer prior to growth initiation forms nanoscale liquid droplets on the Si surface at elevated temperatures, which can therefore act as seeds to promote the nucleation of InN nanowires. As a result, the nanowire density is largely pre-determined, and the size uniformity is significantly enhanced. It is therefore evident that the use of an *in situ* deposited In seeding layer can provide an additional dimension to effectively control the growth and properties of InN nanowires.

Fig. **6(a)** shows the PL spectra measured at 5 K and 300 K under excitation powers of 5 μW and 200 μW, respectively. The PL spectra are predominantly characterized by a single peak, and emission at higher energies is not observed. The measured spectral linewidths of 14 and 40 meV are significantly smaller than the commonly reported values of 60 – 120 meV for InN nanowire ensembles [73, 78, 80]. The significantly reduced inhomogeneous broadening is attributed to the minimized, or completely eliminated size variation along the wire axial direction and amongst the wires. The PL spectra of conventional InN nanowires generally exhibit high energy tails that can be described by $\sim exp(-E_{ph}/E_0)$, where E_{ph} is the photon energy and E_0 is the specific energy corresponding to both the thermal distribution of carriers as well as the nanowire inhomogeneity [78]. The values of E_0 are typically smaller for nanowires with better quality. In this experiment, any high energy tail in the PL spectra at 5 K is not measured, shown in Fig. **6(a)**, suggesting the high quality of non-tapered InN nanowires. Additionally, the high energy tail shown in the 300 K PL spectrum is characterized by a specific energy of $E_0 \approx 30$ meV, which corresponds well to the thermal energy kT, further confirming the extreme homogeneity of the non-tapered nanowires.

In order to evaluate the nonradiative recombination processes in InN nanowires, the thermal quenching behaviour of the PL emission is further measured. Fig. **6(b)** shows the temperature dependent PL peak intensity under an excitation power of ~ 1.5 mW. The intensity decreases exponentially with the increase of temperature, which can be well described by an activation energy of $E_a \approx 6.4$ meV in the temperature range of 5 K to 300 K. The integrated PL intensity also exhibits a similar trend. One of the major nonradiative recombination processes in InN is Auger recombination [82]. Recent studies suggested that Auger recombination in InN is a phonon assisted process, which is weakly dependent on temperature, with activation energy in the range of 4 to 9 meV [82]. Therefore, the thermal quenching of PL signal can be well explained by nonradiative Auger recombination in InN nanowires.

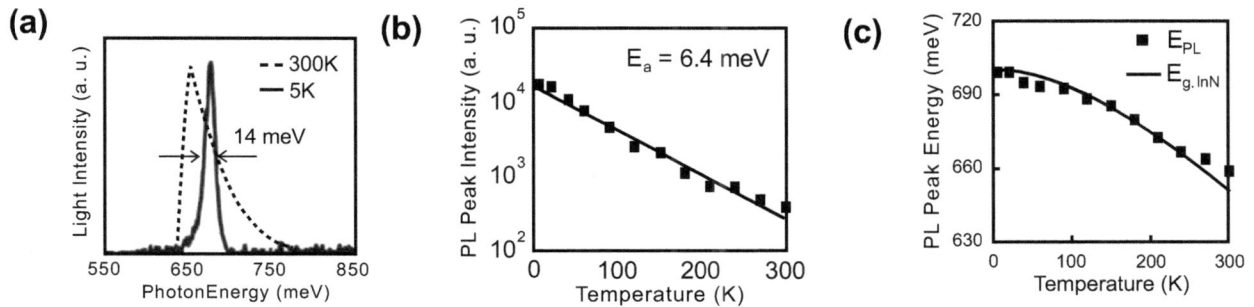

Figure 6: (a) PL emission spectra of InN nanowires on Si measured at 5 K and room temperature under excitation powers of 5 μW and 200 μW, respectively. (b) Variation of PL peak intensity with temperature. The solid line represents the curve simulated by using an activation energy of 6.4 meV. The excitation power is 1.5 mW. (c) Variation of PL peak energy of InN nanowires with temperature and the calculated InN band gap from the Varshni's equation (solid line). The excitation power is 1.5 mW. Reprinted with permission from Ref. [32]. Copyright 2009 Institute of Physics.

One of the fundamental properties of InN, the direct bandgap, is still a subject under debate. To date, the temperature dependence of the bandgap cannot be obtained from conventionally tapered InN nanowires, due to the large inhomogeneous broadening and very poor structural and optical quality [78, 80, 83]. Shown in Fig. **6(c)** is the measured PL peak position of non-tapered InN nanowires as a function of temperature. The solid line along with the data is the variation of the bandgap of InN with temperature, calculated using the Varshni's equation $E_g(T) = E_g(0) - \gamma T^2/(T+\beta)$

where $E_g(0) = 0.70\ eV$, $\gamma = 0.41\ meV/K$, and $\beta = 454\ K$ [83]. The agreement is excellent, which further confirms the extremely high quality and homogeneity of the non-tapered InN nanowires grown using a pre-deposited In seeding layer.

InGaN and InAlN Nanowires

To date, there have been very few studies on the growth and properties of $In_xGa_{1-x}N$ and $In_xAl_{1-x}N$ ternary nanowires [84-87]. The In compositions of the reported ternary nanowires are generally limited to less than 50%. In addition, a mixture of cubic and hexagonal phases have been commonly observed [85]. The most promising results, in terms of the luminescence efficiency, have been achieved from InGaN nanowires grown on GaN nanowires on Si(111) substrates by MBE. Room temperature PL emission wavelength of ~ 540 nm and internal quantum efficiency of ~ 6.3% has been obtained for such InGaN nanowire arrays with In compositions of ~ 11%. Recently, Kukendall *et al.* reported the achievement of single crystalline $In_xGa_{1-x}N$ nanowires, with *x* varying from zero to 1, using low temperature halide CVD on Si and sapphire substrates [72]. In this approach, a single-zone tube furnace with four temperature zones was utilized. The In, Ga and N sources were provided by $InCl_3$, $GaCl_3$, and NH_3, respectively. The resulting nanowires with high In compositions have diameters of ~ 100-200 nm and lengths of ~ 1-2 μm. On the other hand, nanowires with low In compositions exhibit diameters of ~ 10-50 nm and lengths of 0.5 – 1.0 μm. Such InGaN nanowires, however, are plagued with the presence of large densities of dislocations, evidenced by the measured PL peak energy of ~ 1 eV, compared to the values of ~ 0.7 eV for high quality InN. Recently, InAlN nanowires, with In compositions of ≥ 70% have been grown on Si(111) substrates by RF-plasma-assisted MBE [88]. The growth process involves an initial surface nitridation at 760 °C and the growth of dot-like GaN at 710 °C, which may serve as the nanowire nucleation seeds. Relatively low growth temperatures (376-435 °C) were utilized for InAlN nanowire growth. The resulting InAlN nanowires exhibit diameters of 40-130 nm and densities of $0.7\text{-}3.0 \times 10^{10}$ cm^{-2}, depending on the growth conditions.

III-Nitride Nanowire Heterostructures

For practical device applications, it is highly desired and essential that axial and radial nanowire heterostructures, with well-controlled layer thicknesses and compositions and well-defined interfaces, can be grown and fabricated. To achieve such heterostructures using VLS growth mechanism, it is required that the catalyst nanocluster is appropriate for the growth of the various nanowire materials under similar growth conditions. Alternatively, III-nitride nanowire heterostructures can be grown spontaneously under nitrogen rich conditions. In what follows, we describe the recent progress of III-nitride core-shell and well/dot-in-a-wire nanoscale heterostructures.

Radial Nanowire Heterostructures

The core-shell or core-multishell heterostructures arise from radial or lateral over-growth under different compositions. This lateral growth process does not involve reaction with a nanocluster catalyst and hence it competes with the VLS growth mechanism. The MOCVD growth of GaN-based core-shell-shell nanowire heterostructures has been demonstrated [16, 89]. In this approach, a GaN core was first grown in the VLS mode using nickel nanoclusters. Subsequently, an intrinsic InGaN shell was grown, which was followed by the formation of a GaN shell. The nanowires exhibit a triangular cross-section and smooth surface, and the InGaN shell has an In composition of ~ 20%. The growth and characterization of GaN-based core-multi-shell nanoscale heterostructures have been further developed [90]. Illustrated in Fig. 7(**a**), the nanowire heterostructure consists of an n-type GaN core and i-InGaN/i-GaN/p-AlGaN/p-GaN multishells. The resulting nanoscale heterostructures are relatively free of dislocations and exhibit sharp heterointerfaces. In this study, it was determined that the $In_{0.09}Ga_{0.91}N$, GaN, and $Al_{0.1}Ga_{0.9}N$ shells have thicknesses of ~ 8, 28, and 33 nm, respectively. Additionally, $(InGaN/GaN)_n$ multi-quantum-well (MQW) core/shell nanowire heterostructures, with n as large as 26, have been developed, illustrated in Figs. 7(**b**) and (**c**), which provide effective carrier confinement in the radial direction [91]. The realization of multi-color, high efficiency LEDs using such novel nanoscale heterostructures is discussed in Sec. 6.2.

Axial Nanowire Heterostructures

GaN/Al(Ga)N Disk-in-a-Wire Nanoscale Heterostructures

GaN/Al(Ga)N nanowire heterostructures have been grown on Si(111) substrates by plasma-assisted MBE under nitrogen rich conditions [92-95]. The cross-sectional TEM image of Al0.16Ga0.84N/GaN/Al0.16Ga0.84N nanowire heterostructures is shown in Fig. **8** [96]. The GaN quantum disks are ~ 4.5 nm in thickness and fully strained. GaN quantum disks (~ 2.5 nm) incorporated in AlN barriers, illustrated in Fig. 9(**a**), have also been investigated [92]. In this case, the nanowire heterostructure was grown on ~ 500 nm long GaN nanowires on Si(111) substrates. The derived strain component along the nanowire axis (ε_{zz}) is shown in Fig. 9(**b**). It is seen that ε_{zz} is nearly constant along the growth direction but exhibits a lateral gradient.

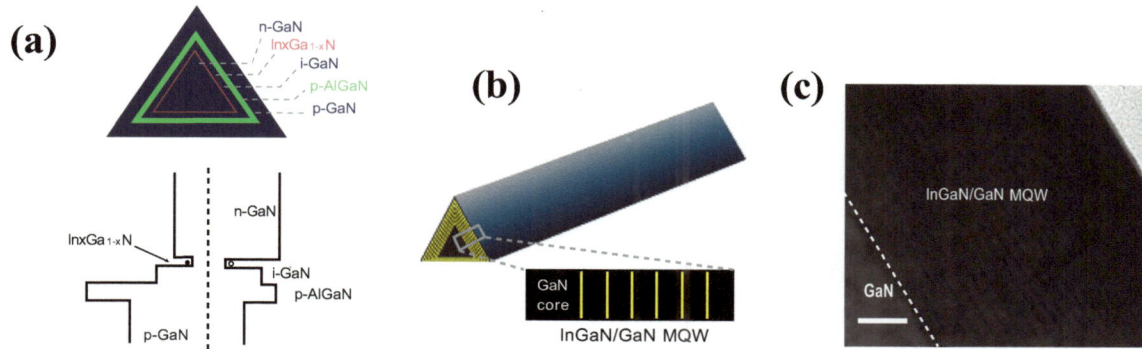

Figure 7: (a) Cross-sectional view of a core-multi-shell nanowire structure and the corresponding energy band diagram. The dashed line in the band diagram indicates the position of the Fermi level. Reprinted with permission from Ref. [90]. Copyright 2005 American Chemical Society. **(b)** Schematic diagram of an MQW nanowire and magnified cross-sectional view of a nanowire facet highlighting InGaN/GaN MQW structure. The InGaN layer is indicated in yellow color. **(c)** Bright-field HRTEM image of a typical 26 MQW nanowire cross-section. The dashed line indicates the heterointerface between the core and shell. The scale bar is 10 nm. Reprinted by permission from Macmillan Publishers Ltd: [Nature Materials] (91), copyright (2008).

GaN/Al(Ga)N quantum disks or dots exhibit unique emission characteristics. The emission spectra of a single GaN quantum disk (~ 1 nm thick) embedded in an AlN nanowire were measured under various excitation powers at 5 K [94]. Shown in Fig. **10**, both excitonic and biexcitonic recombinations can be identified. It is seen that the biexciton binding energy is ~ 20 meV. It is confirmed that the integrated PL intensities for the exciton and biexciton lines exhibit a linear and quadratic dependence on the excitation power, respectively. It is further observed such nanoscale axial heterostructures show a smaller quantum-confined Stark effect, compared to the planar heterostructures, due to the lateral stress relaxation of nanowires and the resulting reduction in the piezoelectric polarization [95].

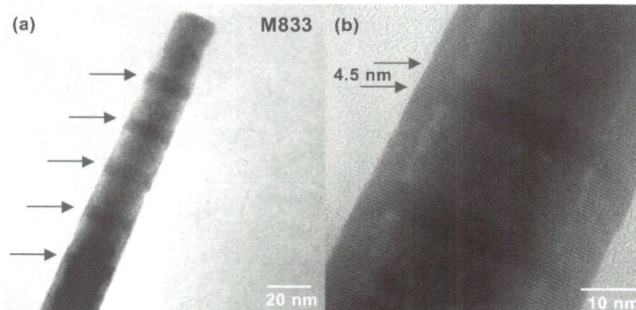

Figure 8: (a) and **(b)** Cross-sectional TEM images of a GaN/AlGaN nanowire showing GaN quantum discs embedded in an AlGaN nanowire. Reprinted with permission from Ref. [96]. Copyright 2003 American Physical Society.

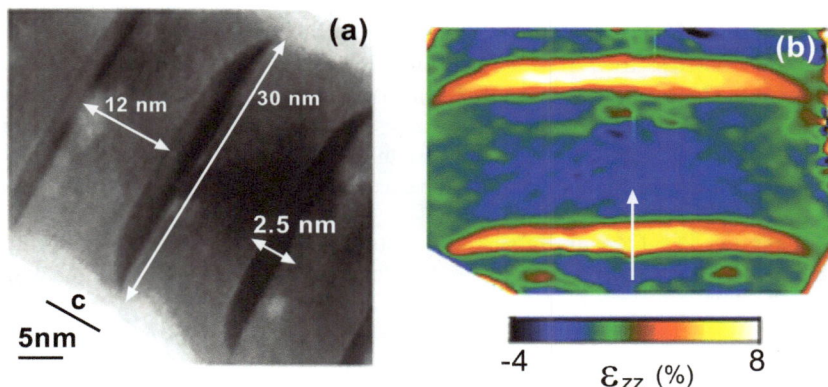

Figure 9: (a) Off-axis high resolution electron microscopy image (HREM) of GaN quantum disks (~ 2.5 nm thickness) incorporated in an AlN nanowire. **(b)** The mapping of the strain component ε_{zz} obtained from geometrical phase analysis. Reprinted with permission from Ref. [92]. Copyright 2009 Institute of Physics.

InGaN/GaN dot-in-a-Wire Nanoscale Heterostructures

InGaN/GaN dot-in-a-wire nanoscale heterostructures, with high efficiency emission in the blue, green, and amber wavelength range have been realized on Si(111) substrates [97]. Shown in Fig. **11(a)** is the TEM image of a single nanowire, which consists of a ~ 0.3 μm GaN buffer, three vertically aligned quantum dots separated by ~ 5 nm GaN barrier layers, and a ~ 0.15 μm GaN capping layer. InGaN/GaN dot-in-a-wire samples with different emission wavelengths were grown by varying the In/Ga flux ratios. The high resolution TEM image for the quantum dot active region is illustrated in Fig. **11(b)**. It is seen that the InGaN/GaN nanowire is structurally uniform, with a constant diameter of ~ 50 nm. The selected electron area diffraction pattern, shown in the inset, suggests that the wires are of a wurtzite structure and aligned along the *c*-axis. The dots exhibit a height of ~ 7 nm and base width of ~ 30 nm, which are larger than the commonly reported values for InGaN/GaN quantum dots embedded in planar heterostructures [98]. The InGaN quantum dots and surrounding GaN barrier layers are nearly free of dislocations, due to the reduced strain distribution in the nanowire heterostructures. The average In compositions in the dots are estimated to be in the range of ~ 15 – 25%, depending on the growth conditions, for the green, yellow, and amber emitting structures. It may also be noted that a small amount (~ 6%) of In atoms are incorporated in the GaN barrier layers. Detailed TEM analysis, shown in Fig. **11(c)**, confirms that each InGaN dot is not compositionally uniform, with the presence of In-rich nanoclusters. These nanoclusters are formed by phase segregation, and their sizes vary from ~ 2 to 5 nm. Illustrated in Fig. **11(d)** is the In composition distribution profile, derived from the measured lattice constants, for the selected region in Fig. **11(c)**, which shows an In-rich nanocluster with an In composition of ~ 27.9% and dimensions of ~ 3.5 nm. The reason the InGaN layers do not form perfectly flat discs, as shown in GaN/Al(Ga)N axial heterostructures, has been attributed to the strain-induced self-organization as well as the faster In–N bond decomposition rate on the nonpolar sidewalls of the nanowires than on the top *c*-plane [97, 99-102].

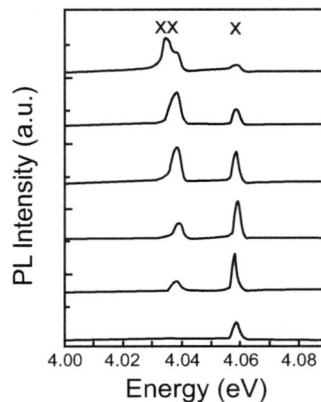

Figure 10: Power-dependent spectra of quantum dots emissions at 5 K. Each spectrum has been normalized to its maximum intensity. Excitation powers (increasing from the bottom): 10, 50, 100, 200, 300, and 500 μW. Reprinted with permission from Ref. [94]. Copyright 2008 American Chemical Society.

Fig. **12(a)** shows the normalized PL spectra measured under a pump power of ~ 100 W/cm^2 at room temperature for three InGaN/GaN nanowire heterostructures grown with different In/Ga flux ratios. The peak energy positions are at ~ 2.5, 2.2, and 2.0 eV, corresponding to the green, yellow and amber emission, respectively. The measured spectral linewidths are in the range of ~ 200 – 260 meV. For all three samples, luminescence emission from the GaN nanowires can also be observed. Their intensity, however, is significantly smaller than that of the InGaN quantum dots, in spite of the thick GaN segments surrounding the dot layers. This further confirms the excellent optical quality of InGaN quantum dots.

The variation of the integrated PL intensity versus temperature is shown in Fig. **12(b)** for the yellow emitting InGaN/GaN dot-in-a-wire heterostructures. By modeling the data using an Arrhenius plot, an activation energy (E_A) of ~ 42 meV is derived. The internal quantum efficiency at room temperature can be estimated by comparing the PL intensities measured at 300 K and 10 K. InGaN/GaN dot-in-a-wire heterostructures exhibit a large room temperature internal quantum efficiency of ~ 45%, assuming the internal quantum efficiency at 10 K is 100%, which is nearly a factor of 5 to 10 times larger than that measured in InGaN nanowires in the same wavelength range at room temperature [84]. The achievement of a large internal quantum efficiency at room temperature is attributed to the superior carrier confinement of the dots and to the use of GaN nanowires, which are nearly free of dislocations, compared to planar InGaN/GaN quantum dot or well heterostructures.

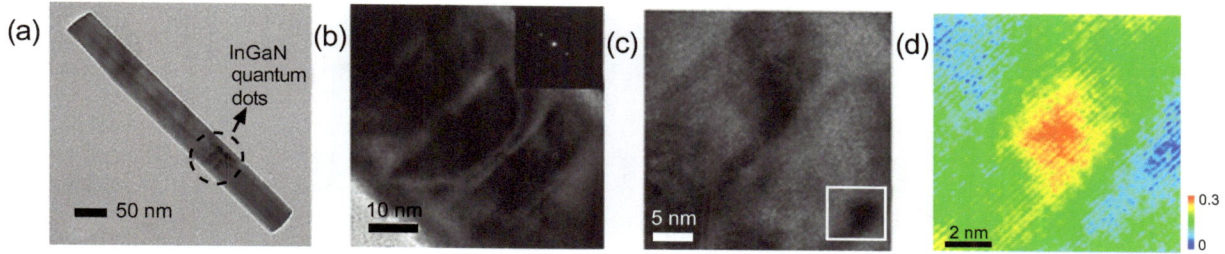

Figure 11: (a) TEM image of a GaN nanowire with the incorporation of three InGaN/GaN quantum dots. **(b)** and **(c)** HRTEM image of three vertically aligned InGaN/GaN quantum dots embedded in a GaN nanowire. The selected area electron diffraction pattern is shown in the inset of (b). The presence of In-rich nanoclusters can be seen in (c). **(d)** The In composition distribution profile for a selected area shown in (c). Reprinted with permission from Ref. [97]. Copyright 2010 American Institute of Physics.

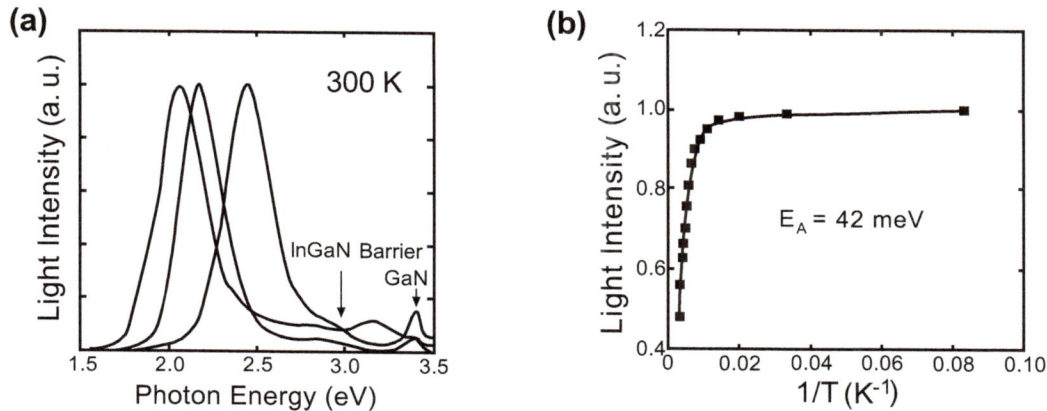

Figure 12: (a) Normalized PL spectra of green, yellow, and amber emitting InGaN/GaN dot-in-a-wire heterostructures grown on Si(111) substrates. **(b)** Variation of the integrated PL intensity with temperature for the yellow emitting InGaN/GaN nanowire heterostructures. The excitation power is ~ 100 W/cm^2. Reprinted with permission from Ref. [97]. Copyright 2010 American Institute of Physics.

ELECTRICAL TRANSPORT PROPERTIES

Compared to the conventional planar heterostructures, the surface and interface charge properties play a vital role in the electrical transport properties of III-nitride nanoscale heterostructures. It has been generally observed that there is a very high electron concentration (~ $10^{13} - 10^{14}$ cm^{-2}) on the polar and nonpolar growth surfaces of InN films and the surface Fermi-level pins deep in the conduction band [103-105]. Similar electron accumulation profile has also been measured at the lateral, nonpolar surfaces of [0001] oriented wurtzite InN nanowires, leading to a severe energy band bending in the near-surface region and a highly inhomogeneous electron distribution in the wire [106-107].

Electrical Transport Properties of GaN Nanowires

At the present time, undoped GaN nanowires generally exhibit relatively high electron concentrations, largely due to the presence of nitrogen vacancies, oxygen impurities, other defects, and/or surface states. The transport properties of unintentionally doped GaN nanowires grown by the direct reaction of gallium and ammonia vapour were investigated using four-terminal resistivity measurements [108]. The measured resistivity is in the range of $(1.0-6.2) \times 10^{-2}$ Ω cm, depending on the wire diameters. The derived electron concentration and field-effect mobility are ~ 3×10^{18} cm^{-3} and 65 cm^{2}V^{-1}s^{-1}, respectively. It is further revealed that the transport properties of these nanowires are predominantly determined by the ionized impurity scattering, with the mobility showing a ~ $T^{1.3}$ dependence near room-temperature. Transport measurements of GaN nanowires synthesized using a tube furnace in the VLS growth mode also confirm the presence of high electron densities in the range of 3-10×10^{18} cm^{-3} [109]. From the temperature dependent electrical measurements, it is suggested that, in this case, the high electron concentrations are related to unintentional defects and/or surface states. Greatly reduced carrier concentrations (~ 2×10^{17} cm^{-3}) are measured for undoped GaN wires grown by catalytic CVD, which exhibit an electron mobility of ~ 30 cm^2/V s [110].

Electrical Transport Properties of InN Nanowires

Compared to GaN nanowires, the transport properties of InN nanowires generally show a strong dependence on the wire diameters, due to the presence of surface electron accumulation. Transport properties of both undoped and Si-doped InN nanowires have been investigated using either a two- or four-terminal configuration [79, 111-112]. The room-temperature I-V characteristics of InN nanowires with various doping concentrations are shown in Fig. 13(a) [111]. A strong dependence on the wire diameters were observed for both undoped and Si-doped InN nanowires. If carrier transport through the entire volume of the wires is assumed, specific resistances of $\sim 4.2 \times 10^{-4}$ and 1.6×10^{-4} Ω cm are derived for the undoped and Si-doped wires, respectively. However, from the double logarithmic plot of the one-dimensional conductance (σ) versus the wire diameters, shown in Fig. **13(b)**, a nearly linear relationship is derived, which suggest a tubular conduction mechanism for InN nanowires. This observation can be well explained by the presence of surface electron accumulation for n-type InN that leads to carrier conduction primarily through the surface. It may also be noted from Fig. **13(b)** that the slope increases to ~ 2 for doped wires with relatively large diameters, which can be explained by the increased bulk conduction for such large wires. Additionally, it was measured that the wire resistance decreases almost linearly with decreasing temperature for both undoped and doped InN nanowires. This metal-like conductor phenomenon is consistent with the presence of degenerate electron gas at the wire surface. The mobility is estimated to be ~ 470 cm^{2}V^{-1}s^{-1} from field-effect transistor measurements, and surface electron concentrations of $\sim 7.8 \times 10^{13}$ and 2.0×10^{14} cm^{-3} are derived for the nominally undoped and Si-doped InN nanowires, respectively.

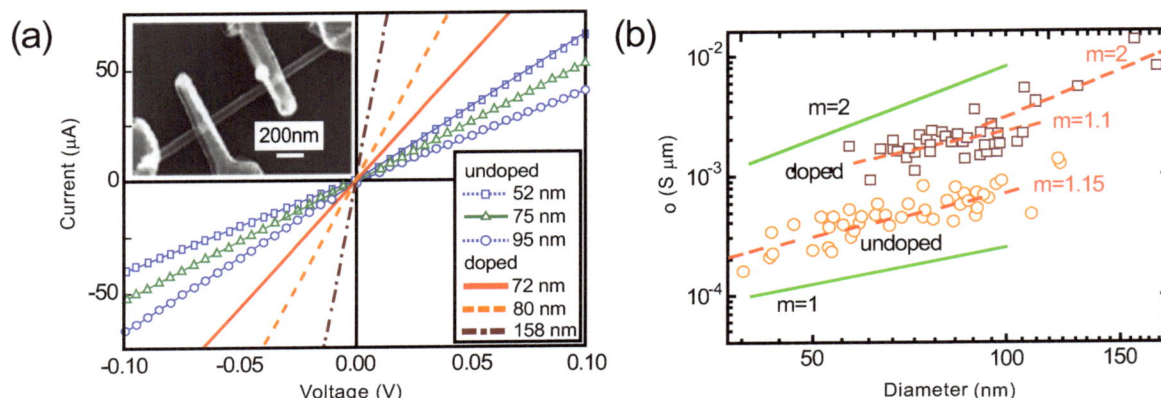

Figure 13: (a) Normalized current–voltage characteristics of three doped and three undoped wires with different diameters, but approximately the same length. The bias voltage is normalized to 1 μm wire length. The inset shows a 75 nm thick doped nanowire contacted for the four-terminal measurements. The separation between the inner two contacts is 810 nm. **(b)** Specific one-dimensional conductance σ as a function of wire diameter. The upper and lower solid lines indicate the slope of $m = 2$ and 1 for pure volume and surface conductance, respectively. The doped and undoped wires show a nearly linear current increase with diameter, showing that the transport occurs mainly in the surface accumulation layer. The doped wires with a diameter above 100 nm show a slope of 2, indicating that the bulk material becomes relevant for conduction. Reprinted with permission from Ref. [111]. Copyright 2009 Institute of Physics.

DEVICE APPLICATIONS

High quality III-nitride nanowires have emerged as promising candidates for future nanoelectronic and nanophotonic devices. With the use of such nanoscale heterostructures, nanoscale LEDs, lasers, transistors, and solar cells have been demonstrated. In what follows, the fabrication and performance characteristics of such novel devices are described.

Transistors

The use of GaN nanowires for field effect transistors (FETs) has been intensively investigated. AlGaN nanowire n-channel MOSFET fabricated on an oxidized Si substrate exhibit a current ON/OFF ratio of more than three orders of magnitude for gate bias varying from -8 to 6 V [113]. The derived field-effect mobility is in the range of 150 to 650 cm^2/V·s [113], which is comparable, or better than those measured from GaN thin film devices. With the use of an electric field assisted alignment technique, Motayed *et al.* have investigated the performance of GaN nanowire n-type depletion mode MOSFETs [114, 115]. For nanowires with relatively small (< 100 nm) diameters, the pinch-off effect and complete channel depletion were observed. Such effects, however, were absent for nanowires with relatively large diameters. A threshold voltage of ~ -15 V and a current ON/OFF ratio of five orders of magnitude were measured [115]. The low frequency noise in such nanowire transistors has also been investigated in the frequency range of 1 Hz to 50 kHz under both dark and illumination conditions [115]. It was observed that the noise spectra consisted primarily of 1/f-like noise and generation-recombination noise.

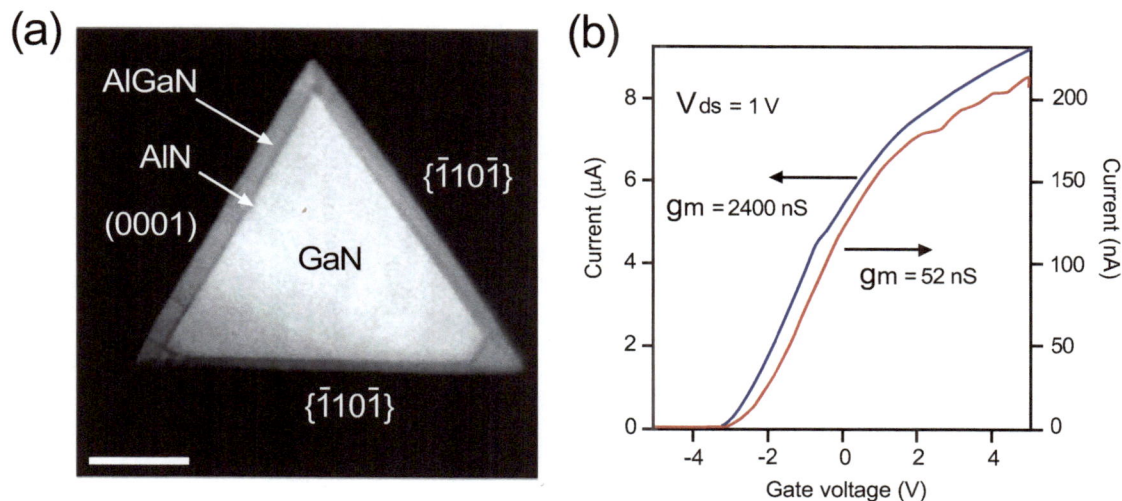

Figure 14: (a) Bright-field TEM image of a GaN/AlN/AlGaN nanowire. Scale bar is 500 nm. **(b)** I_{ds} - V_{gs} transfer characteristics of 100-nm-diameter GaN/AlN/Al$_{0.25}$Ga$_{0.75}$N (blue) and GaN (red) nanowires for V_{ds} = 1 V. Reprinted with permission from Ref. [116]. Copyright 2006 American Chemical Society.

Additionally, high electron mobility transistors have been demonstrated using dopant-free GaN/AlN/AlGaN radial nanowire heterostructures [116]. The synthesis and characterization of such radial nanoscale heterostructures is discussed in Sec. 3.5.1. The transistor device, illustrated in Fig. **14(a)**, consists of a GaN core and AlN and AlGaN shells. An electron gas is spontaneously formed at the interface of the GaN, due to the presence of large polarization fields between the GaN core and AlN/AlGaN shells. Shown in Fig. **14(b)** is the I$_{ds}$ vs. V$_{gs}$ plot (blue line). A peak transconductance of ~ 2.4 μS was measured at V$_{ds}$ = 1 V. For comparison, results for an undoped GaN nanowire device, without the incorporation of AlN/AlGaN shells, are shown as the red curve. It is seen that the undoped GaN nanowire is highly resistive and the g$_m$ values are nearly 50 times smaller. The significant difference is attributed to the presence of an electron gas in the radial nanowire heterostructures. The electron mobility is determined to be ~ 21,000 cm^2/V·s at 5 K, the highest value ever reported for semiconductor nanowires. More recently, GaN/AlN/AlGaN/GaN nanowire metal-insulator-semiconductor FETs, with gate lengths of ~ 0.5 μm, have been demonstrated, which exhibit very promising results, including an intrinsic current-gain cutoff frequency of 5 GHz and an intrinsic maximum available gain cutoff frequency of 12 GHz [117].

LEDs

To date, InGaN/GaN LEDs still exhibit very low internal quantum efficiencies in the green, yellow and red wavelength range, limited, to a large extent, by the presence of large densities of misfit-dislocations related to the large lattice mismatch (~ 11%) between InN and GaN as well as the strain-induced polarization field [118, 119]. Due to the effective lateral stress relaxation, significantly reduced defect densities and piezoelectric polarization field can be achieved in Ga(In)N nanowire heterostructures [32, 37, 56, 72]. The use of nanowires for LEDs offers additional advantages, including high light extraction efficiency and the compatibility with low cost, large area Si substrates [32, 120]. In this context, III-nitride nanowire LEDs have been intensively studied [6, 16, 90, 121-129].

Kim *et al.* reported the achievement of single GaN nanowire LEDs [129]. The GaN nanowire p-n junctions, with nominal electron and hole concentrations of ~ 10^{18} cm^{-3} and 10^{17} cm^{-3}, respectively, exhibit near-ultraviolet emission

($\lambda \sim 390$ nm) under forward bias. Emission characteristics of nanowire LEDs can be further engineered by using core-multi-shell InGaN/GaN nanowire heterostructures [16, 90]. In this regard, Qian *et al.* demonstrated multicolour, high efficiency emission from a single nanowire radial heterostructure consisting of an n-GaN core and InGaN/GaN/p-AlGaN/p-GaN shells [90]. An SEM image of the nanowire device is illustrated in the inset of Fig. **15(a)**. The current voltage characteristic is also shown. Peak emission wavelengths in the range of $367 - 577$ nm, shown in Fig. **15(b)**, were obtained by varying the indium compositions from 1% to 40%. The estimated quantum efficiencies are in the range of 3.9 to 4.8%.

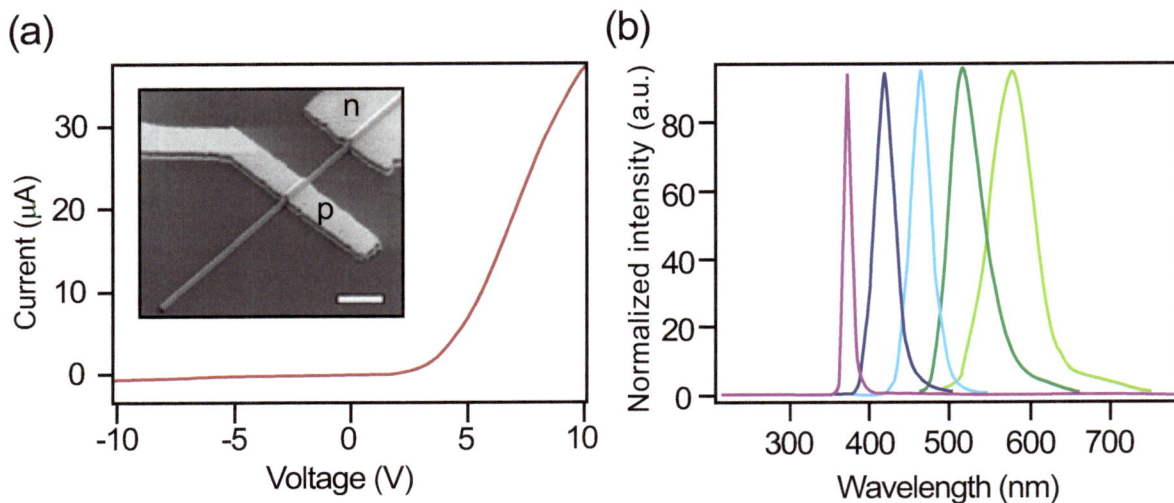

Figure 15: (a) Current versus voltage data recorded on a core-multi-shell nanowire device. Inset: field emission SEM image of a representative core-multi-shell nanowire device. Scale bar is 2 μm. **(b)** Normalized EL spectra recorded from five representative forward-biased multicolour core-multi-shell nanowire LEDs. Reprinted with permission from Ref. [90]. Copyright 2005 American Chemical Society.

Vertical nanowire LEDs monolithically grown on Si substrates have also been developed [124, 130]. Kikuchi *et al.* have fabricated GaN-nanowire-based InGaN/GaN multiple quantum disk LEDs on Sb-doped n-type Si(111) substrates [128]. The LED nanowire heterostructure consisted of Si-doped n-type GaN nanowires (750 nm), undoped GaN (10 nm), eight pairs of InGaN (2 nm)/GaN (3 nm) multiple quantum disk active layer, undoped GaN (10 nm), as well as Mg-doped p-type GaN (600 nm). The unique structure enabled p-type electrodes to be made *via* the conventional method on top of nanowire devices while, at the same time, retaining the excellent optical properties of the isolated nanowire active region. Electroluminescence (EL) was measured *via* semi-transparent electrodes with emission colors ranging from green (530 nm) to red (645 nm) [128]. High-brightness nanowire LEDs, with performance comparable to that of conventional broad area devices, have also been achieved at wavelengths of ~ 470 nm [131]. More recently, significant emission enhancement has been realized by optimizing the coalescence overgrowth of GaN nanowire heterostructures [6].

Lasers

Single III-nitride nanowire lasers have also been demonstrated [91, 132-135]. They offer potential advantages of ultralow threshold, low power consumption and high speed modulation. Jonson *et al.* first observed laser action in single GaN nanowires with diameters of $150 - 400$ nm and lengths of > 16 μm [132]. The dominant optical modes in nanowire lasers are the axial Fabry-Perot modes, with the free spectral range ($\Delta\lambda$) determined by $\lambda^2/2nl$, where n is the effective refractive index and L is the cavity length. The measured cavity Q factors are in the range of 500 to 1500, and the laser threshold power is ~ 500 nJ cm^{-2}. The near- and far-field spectra of a GaN nanowire laser with a diameter of 150 nm and length of 30 μm are shown in Figs. **16(a)** and **(b)**, respectively. A relatively narrow (~ 0.8 nm) emission peak was observed in the near-field. In the far field, multiple cavity modes, with the presence of large PL background, were measured. These modes are the longitudinal Fabry-Perot modes and have a wavelength spacing of ~ 1 nm. Choi *et al.* also reported lasing at ~ 384 nm with the use of GaN/AlGaN core-shell heterostructures formed spontaneously by phase separation [133]. A low threshold power of ~ 22 kW/cm^2 has been obtained from GaN nanowires with a triangular cross-section and a nonpolar <11-20> growth direction at room temperature [134].

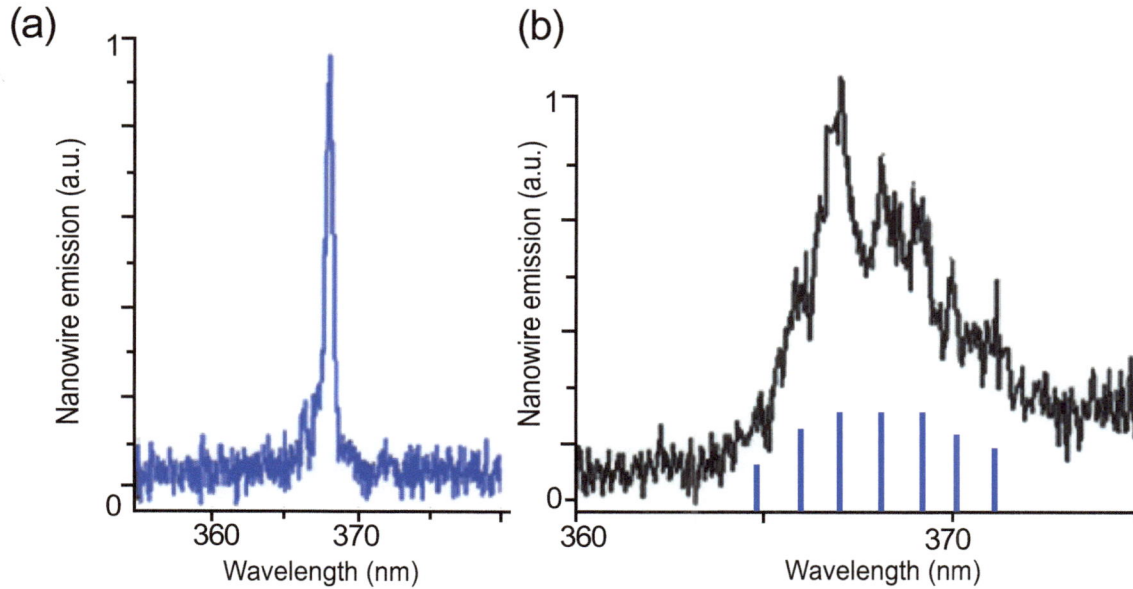

Figure 16: (a) Near-field spectrum from just beyond the GaN nanowire end. **(b)** Far-field spectrum from the wire showing longitudinal cavity modes (denoted by sticks below spectrum) appearing near the lasing threshold. Reprinted by permission from Macmillan Publishers Ltd: [Nature Materials] [132], copyright (2002).

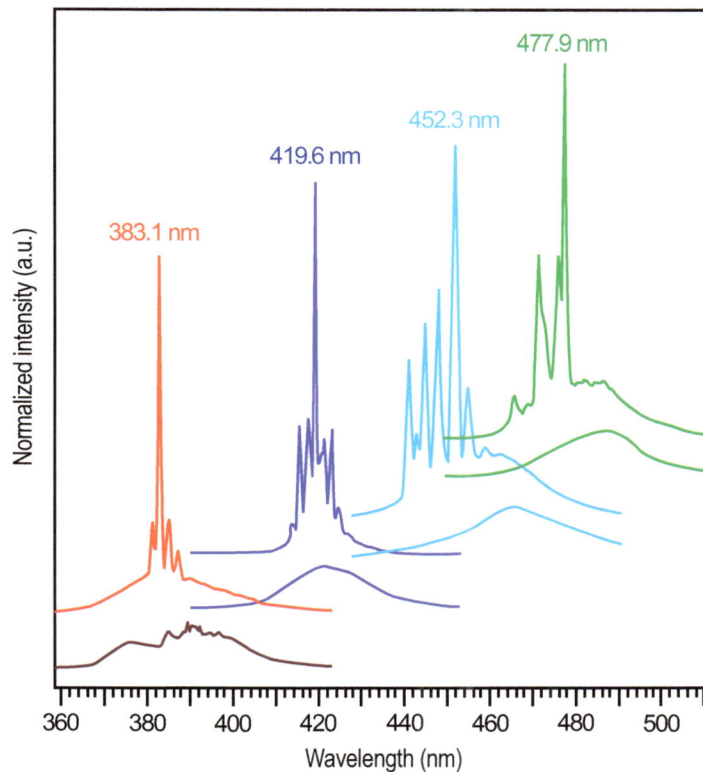

Figure 17: Normalized PL spectra collected from four representative 26 MQW nanowire structures with increasing In composition pumped at 250 kW cm^{-2} and 700 kW cm^{-2}, respectively. Spectra are offset for clarity. Reprinted by permission from Macmillan Publishers Ltd: [Nature Materials] (91), copyright (2008).

Wavelength-controlled nanowire lasers have been realized using InGaN/GaN MQW core-shell nanowire heterostructures [91]. The growth and characterization of such nanoscale heterostructures are discussed in Sec. 3.5.1. Lasing wavelengths in the range of 365 to 494 nm have been achieved using (InGaN/GaN)$_n$ quantum wells with various In compositions. PL spectra measured at ~ 250 kW cm^{-2} and 700 kW cm^{-2}, below and above threshold, respectively for MQW (n=26) nanowire lasers with various In compositions are shown in Fig. **17**. The lasing wavelengths are at ~ 383, 420, 452, and 478 nm, depending on the In compositions. The Q-factors are ~ 2,485.

Solar Cells

Vertically aligned, single-crystalline GaN/Si nanowires heterojunction solar cells have been demonstrated recently [136]. The structure is based on a p-n junction formed between p-type GaN nanowires and n-type Si(111) substrate, shown schematically in Fig. **18(a)** with the corresponding energy band diagram. A rather simple fabrication procedure is taken place by first covering the as grown nanowires with transparent, yet insulating layer of poly-methyl-methacrylate (PMMA), followed by a direct deposition of a thin layer of (30/50 nm) Ni/Au on top of the wires as well as a thin (30/50 nm) Ti/Au layer on the backside of the Si substrate. As shown in Fig. **18(b)**, a well-defined diode behaviour with a rectifying ratio of more than 10^4 at ± 0.5 V in the dark, a short-circuit photocurrent density of 7.6 mA/cm^2, and a maximum power conversion efficiency of 2.73% under simulated AM 1.5G illumination at 1 Sun was achieved.

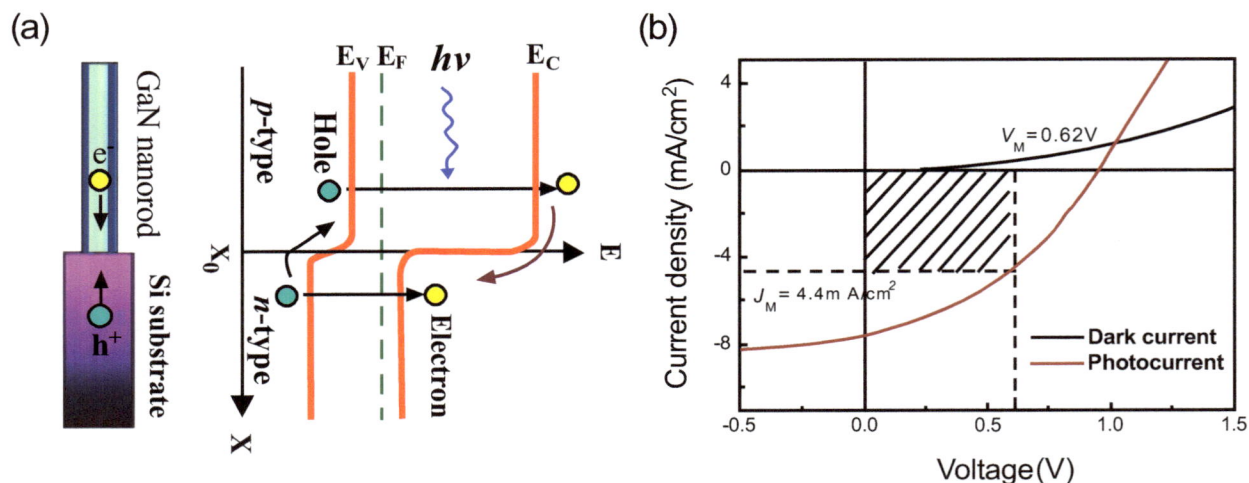

Figure 18: (a) Schematic energy band diagram of the solar cell heterojunction diode showing the photogenerated carrier transfer process. (b) Current density vs. voltage for the solar cell in the dark and under simulated AM1.5G illumination with an intensity of 100 mW/cm^2. Reprinted with permission from Ref. [136]. Copyright 2008 American Chemical Society

Single core-shell n-GaN/i-In$_x$Ga$_{1-x}$N/p-GaN nanowire solar cell devices have also been investigated lately in an attempt to explore the intrinsic limits of the III-nitride materials system and to determine the potential for the development of a new generation of solar energy conversion systems in the future [137]. The main advantage of such core-shell-shell design is the effective carrier separation in the radial direction due to the p-i-n interface extension along the entire length of the nanowire. As a result, photogenerated carriers are able to reach the p-i-n junction with enhanced efficiencies due to a relatively short diffusion length required. Fig. **19** shows the schematic illustration of the fabrication process (a), SEM image of the single wire device (b), a cross-sectional view of the core-shell-shell heterojunction structure with the corresponding energy band diagram (c), and the light-current response for a solar cell device (d). By varying the In composition in the active InGaN layer, the bandgap was tuned from 2.25 to 3.34 eV, and clear diode characteristics with ideality factors ranging from 3.9 to 5.6 were obtained. Furthermore, open-circuit voltages of 1.0-2.0 V and short-circuit current densities of 0.39-0.059 mA/cm^2 were achieved as In composition was reduced from 0.27 to 0, with a maximum efficiency of ~ 0.19%. These results confirm the ability to controllably tune the structure and composition of these core-shell III-nitride nanowires, rendering them a powerful platform for exploring nanoscale power sources in the near future.

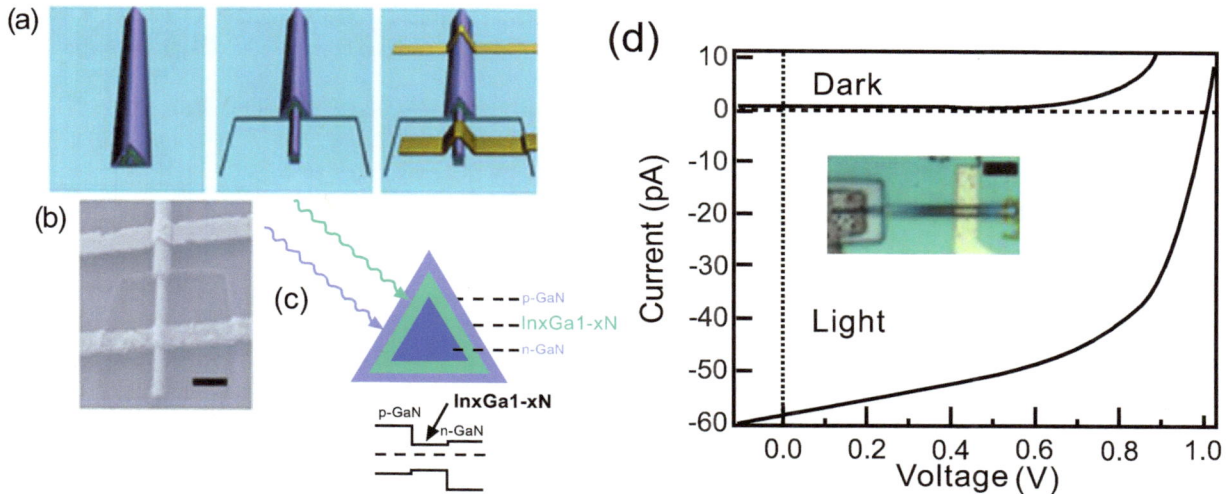

Figure 19: Nanowire solar cell device fabrication and characterizations. (a) Schematics of device fabrication: left, coaxial n-GaN/i-In$_x$Ga$_{1-x}$N/p-GaN nanowire heterostructures; middle, exposed the n-core at nanowire end following ICP-RIE etch; right, Ni/Au and Ti/Al/Ti/Au metal contacts deposited on the p-shell and n-core, respectively. (b) SEM image of a representative nanowire device. Scale bar is 2 μm. (c) Cross-sectional view of p-GaN/i-In$_x$Ga$_{1-x}$N/n-GaN heterojunction and corresponding energy band diagrams. (d) Light *I-V* curves for a green PV device. Inset, optical microscopy image of the device; scale bar is 5 μm. Reprinted with permission from Ref. [137]. Copyright 2009 American Chemical Society.

SUMMARY

In summary, with the recent advancements in various growth/synthesis techniques, significant progress has been made in the technologically important III-nitride nanowire heterostructures. To date, the growth mechanisms have been relatively well understood, and the structural, optical and electrical properties of such nanoscale semiconductors can be well controlled by varying the design and growth parameters. Some of the imminent materials challenges related to III-nitride nanowires include the demonstration of p-doping in InN and InGaN with high In compositions, the achievement of high quality InGaN and InAlN heterostructures, as well as a fundamental understanding of the surface and interface charge properties.

With the use of III-nitride nanowire heterostructures, a myriad of nanoscale electronic and photonic devices, including transistors, LEDs, lasers, and solar cells have also been intensively investigated. While these devices sometimes exhibit performance comparable to, or better than that of conventional planar devices, a fundamental understanding of the device design and operation as well as a significant improvement in the growth and fabrication process is still required in order to realize the full potential of III-nitride nanowires. Additionally, there has been a growing interest in developing high performance gas, chemical, and biological sensors using III-nitride nanowire heterostructures. The extreme chemical and temperature stability of nitrides render them suitable for sensor applications in high temperature, high pressure or corrosive ambient, compared to the conventional Si-based devices. An unprecedented sensitivity is also expected from In(Ga)N nanowire sensors, due to the surface electron accumulation.

ACKNOWLEDGEMENTS

The authors would like to thank Yi-Lu Chang for helping to collect literatures and copyright permissions during this book chapter preparation.

REFERENCES

[1]   Wu, J. Q. When group-III nitrides go infrared: New properties and perspectives. *J. Appl. Phys.,* **2009**, *106*(1), 011101.

[2]   Bhuiyan, A. G.; Hashimoto, A.; Yamamoto, A. Indium nitride (InN): A review on growth, characterization, and properties. *J. Appl. Phys.,* **2003**, *94*(5), 2779-2808.

[3]   Strite, S.; Morkoc, H. GaN, AlN, and InN - a Review. *J. Vac. Sci. Technol. B,* **1992**, *10*(4), 1237-1266.

[4] Vurgaftman, I.; Meyer, J. R. Band parameters for nitrogen-containing semiconductors. *J. Appl. Phys.*, **2003**, *94*(6), 3675-3696.

[5] Wu, J.; Walukiewicz, W.; Yu, K.; Ager, J.; Haller, E.; Lu, H.; Schaff, W. Small band gap bowing in $In_{1-x}Ga_xN$ alloys. *Appl. Phys. Lett.*, **2002**, *80*(25), 4741-4743.

[6] Tang, T. Y.; Lin, C. H.; Chen, Y. S.; Shiao, W. Y.; Chang, W. M.; Liao, C. H.; Shen, K. C.; Yang, C. C.; Hsu, M. C.; Yeh, J. H.; Hsu, T. C. Nitride Nanocolumns for the Development of Light-Emitting Diode. *IEEE Trans. Electron. Dev.,* **2010**, *57*(1), 71-78.

[7] Trybus, E.; Namkoong, G.; Henderson, W.; Burnham, S.; Doolittle, W.; Cheung, M.; Cartwright, A. InN: A material with photovoltaic promise and challenges. *J. Cryst. Growth*, **2006**, *288*(2), 218-224.

[8] Mohammad, S. N.; Salvador, A. A.; Morkoc, H. Emerging Gallium Nitride Based Devices. *Proc. IEEE,* **1995**, *83*(10), 1306-1355.

[9] Mishra, U. K.; Shen, L.; Kazior, T. E.; Wu, Y. F. GaN-Based RF power devices and amplifiers. *Proc. IEEE,* **2008**, *96*(2), 287-305.

[10] Mukai, T. Recent progress in group-III nitride light-emitting diodes. *IEEE J. Sel. Topics Quanum Electron.,* **2002**, *8*(2), 264-270.

[11] Denbaars, S. P. Gallium-nitride-based materials for blue to ultraviolet optoelectronics devices. *Proc. IEEE,* **1997**, *85*(11), 1740-1749.

[12] Wagner, R.; Ellis, W. Vapor liquid solid mechanism of single crystal growth. *Appl. Phys. Lett.,* **1964**, *4*, 89.

[13] Weng, X. J.; Burke, R. A.; Redwing, J. M. The nature of catalyst particles and growth mechanisms of GaN nanowires grown by Ni-assisted metal-organic chemical vapor deposition. *Nanotechnology*, **2009**, *20*(8), 085610

[14] Kishino, K.; Sekiguchia, H.; Kikuchi, A. Improved Ti-mask selective-area growth (SAG) by rf-plasma-assisted molecular beam epitaxy demonstrating extremely uniform GaN nanocolumn arrays. *J. Cryst. Growth*, **2009**, *311*(7), 2063-2068.

[15] Duan, X.; Lieber, C. General synthesis of compound semiconductor nanowires. *Adv. Mater.,* **2000**, *12*(4), 298-302.

[16] Qian, F.; Li, Y.; Gradecak, S.; Wang, D.; Barrelet, C.; Lieber, C. Gallium nitride-based nanowire radial heterostructures for nanophotonics. *Nano Lett.,* **2004**, *4*(10), 1975-1979.

[17] Dubrovskii, V.; Sibirev, N.; Harmand, J.; Glas, F. Growth kinetics and crystal structure of semiconductor nanowires. *Phys. Rev. B,* **2008**, *78*(23), 235301.

[18] Kuykendall, T.; Pauzauskie, P. J.; Zhang, Y. F.; Goldberger, J.; Sirbuly, D.; Denlinger, J.; Yang, P. D. Crystallographic alignment of high-density gallium nitride nanowire arrays. *Nat. Mater.,* **2004**, *3*(8), 524-528.

[19] Calleja, E.; Ristic J.; Fernandez-Garrido, S.; Cerutti, L.; Sanchez-Garcia, M.; Grandal, J.; Trampert, A.; Jahn, U.; Sanchez, G.; Griol, A. Growth, morphology, and structural properties of group-III-nitride nanocolumns and nanodisks. *Phys. Stat. Sol. (B),* **2007**, *244*(8), 2816-2837.

[20] Vajpeyi, A.; Ajagunna, A.; Tsiakatouras, G.; Adikimenakis, A.; Iliopoulos, E.; Tsagaraki, K.; Androulidaki, M.; Georgakilas, A. Spontaneous growth of III-nitride nanowires on Si by molecular beam epitaxy. *Microelectron. Eng.,* **2009**, *86*(4-6), 812-815.

[21] Bertness, K.; Roshko, A.; Mansfield, L.; Harvey, T.; Sanford, N. Mechanism for spontaneous growth of GaN nanowires with molecular beam epitaxy. *J. Cryst. Growth*, **2008**, *310*(13), 3154-3158.

[22] Furtmayr, F.; Vielemeyer, M.; Stutzmann, M.; Laufer, A.; Meyer, B.; Eickhoff, M. Optical properties of Si-and Mg-doped gallium nitride nanowires grown by plasma-assisted molecular beam epitaxy. *J. Appl. Phys.,* **2008**, *104*, 074309.

[23] Landré, O.; Fellmann, V.; Jaffrennou, P.; Bougerol, C.; Renevier, H.; Cros, A.; Daudin, B. Molecular beam epitaxy growth and optical properties of AlN nanowires. *Appl. Phys. Lett.,* **2010**, *96*, 061912.

[24] Landré, O.; Bougerol, C.; Renevier, H.; Daudin, B., Nucleation mechanism of GaN nanowires grown on (111) Si by molecular beam epitaxy. *Nanotechnology,* **2009**, *20*, 415602.

[25] Gradecak, S.; Qian, F.; Li, Y.; Park, H. G.; Lieber, C. M. GaN nanowire lasers with low lasing thresholds. *Appl. Phys. Lett.,* **2005**, *87*(17), 173111.

[26] Duan, X.; Lieber, C., Laser-assisted catalytic growth of single crystal GaN nanowires. *J. Am. Chem. Soc* **2000**, *122*(1), 188-189.

[27] Reui-San Chen, S.; Lan, Z.; Tsai, J.; Wu, C.; Chen, L.; Chen, K.; Huang, Y.; Chen, C. On-chip fabrication of well-aligned and contact-barrier-free GaN nanobridge devices with ultrahigh photocurrent responsivity. *Small,* **2008**, *4*(7), 925-929.

[28] Kang, T.; Liu, X.; Zhang, R.; Hu, W.; Cong, G.; Zhao, F.; Zhu, Q. InN nanoflowers grown by metal organic chemical vapor deposition. *Appl. Phys. Lett.,* **2006**, *89*(7), 023508.

[29] He, M.; Mohammad, S. Novelty and versatility of self-catalytic nanowire growth: A case study with InN nanowires. *J. Vac. Sci. Technol. B,* **2007**, *25*(3), 940-944.

[30] Mohammad, S. Self-catalytic solution for single-crystal nanowire and nanotube growth. *J. Chem. Phys.,* **2007**, *127*(24), 244702.

[31] Stach, E.; Pauzauskie, P.; Kuykendall, T.; Goldberger, J.; He, R.; Yang, P. Watching GaN nanowires grow. *Nano Lett.,* **2003**, *3*(6), 867-869.

[32] Chang, Y. L.; Li, F.; Fatehi, A.; Mi, Z. T. Molecular beam epitaxial growth and characterization of non-tapered InN nanowires on Si(111). *Nanotechnology,* **2009**, *20*(34),345203.

[33] Ristic, J.; Calleja, E.; Fernandez-Garrido, S.; Cerutti, L.; Trampert, A.; Jahn, U.; Ploog, K. On the mechanisms of spontaneous growth of III-nitride nanocolumns by plasma-assisted molecular beam epitaxy. *J. Cryst. Growth,* **2008**, *310*(18), 4035-4045.

[34] Debnath, R.; Meijers, R.; Richter, T.; Stoica, T.; Calarco, R.; Luth, H. Mechanism of molecular beam epitaxy growth of GaN nanowires on Si(111). *Appl. Phys. Lett.*, **2007**, *90*(12), 123117.

[35] Landre, O.; Bougerol, C.; Renevier, H.; Daudin, B. Nucleation mechanism of GaN nanowires grown on (111) Si by molecular beam epitaxy. *Nanotechnology*, **2009**, *20*(41), 415602.

[36] Songmuang, R.; Landre, O.; Daudin, B. From nucleation to growth of catalyst-free GaN nanowires on thin AlN buffer layer. *Appl. Phys. Lett.*, **2007**, *91*(25), 251902.

[37] Calarco, R.; Meijers, R.; Debnath, R.; Stoica, T.; Sutter, E.; Luth, H. Nucleation and growth of GaN nanowires on Si(111) performed by molecular beam epitaxy. *Nano Lett.*, **2007**, *7*(8), 2248-2251.

[38] Stoica, T.; Sutter, E.; Meijers, R. J.; Debnath, R. K.; Calarco, R.; Luth, H.; Grutzmacher, D. Interface and wetting layer effect on the catalyst-free nucleation and growth of GaN nanowires. *Small*, **2008**, *4*(6), 751-754.

[39] Kishino, K.; Hoshino, T.; Ishizawa, S.; Kikuchi, A. Selective-area growth of GaN nanocolumns on titanium-mask-patterned silicon (111) substrates by RF-plasma-assisted molecular-beam epitaxy. *Electron. Lett.*, **2008**, *44*(13), 819-821.

[40] Liang, C.; Chen, L.; Hwang, J.; Chen, K.; Hung, Y.; Chen, Y. Selective-area growth of indium nitride nanowires on gold-patterned Si(100) substrates. *Appl. Phys. Lett.*, **2002**, *81*(1), 22-24.

[41] Zhang, J.; Zhang, L. D.; Wang, X. F.; Liang, C. H.; Peng, X. S.; Wang, Y. W. Fabrication and photoluminescence of ordered GaN nanowire arrays. *J. Chem. Phys.*, **2001**, *115*(13), 5714-5717.

[42] Deb, P.; Kim, H.; Rawat, V.; Oliver, M.; Kim, S.; Marshall, M.; Stach, E.; Sands, T. Faceted and vertically aligned GaN nanorod arrays fabricated without catalysts or lithography. *Nano Lett.*, **2005**, *5*(9), 1847-1851.

[43] Hersee, S. D.; Sun, X. Y.; Wang, X. The controlled growth of GaN nanowires. *Nano Lett.*, **2006**, *6*(8), 1808-1811.

[44] Ishizawa, S.; Sekiguchi, H.; Kikuchi, A.; Kishino, K. Selective growth of GaN nanocolumns by Al thin layer on substrate. *Phys. Stat. Sol. (B)*, **2007**, *244*(6), 1815-1819.

[45] Hersee, S.; Sun, X.; Wang, X. The controlled growth of GaN nanowires. *Nano Lett.*, **2006**, *6*(8), 1808-1811.

[46] Ishizawa, S.; Sekiguchi, H.; Kikuchi, A.; Kishino, K. Selective growth of GaN nanocolumns by Al thin layer on substrate. *Phys. Stat. Sol. (B)*, **2007**, *244*(6), 1815-1819.

[47] Kishino, K.; Hoshino, T.; Ishizawa, S.; Kikuchi, A. Selective-area growth of GaN nanocolumns on titanium-mask-patterned silicon (111) substrates by RF-plasma-assisted molecular-beam epitaxy. *Electron. Lett.*, **2008**, *44*(13), 819-821.

[48] Lee, S.; Dawson, L.; Brueck, S.; Jiang, Y. Anisotropy of selective epitaxy in nanoscale-patterned growth: GaAs nanowires selectively grown on a SiO_2-patterned (001) substrate by molecular-beam epitaxy. *J. Appl. Phys.*, **2005**, *98*(11), 114312.

[49] Lee, S.; Dawson, L.; Brueck, S.; Stintz, A. Heteroepitaxial selective growth of $In_xGa_{1-x}As$ on SiO_2-patterned GaAs(001) by molecular beam epitaxy. *J. Appl. Phys.*, **2004**, *96*(9), 4856-4865.

[50] Tomioka, K.; Mohan, P.; Noborisaka, J.; Hara, S.; Motohisa, J.; Fukui, T. Growth of highly uniform InAs nanowire arrays by selective-area MOVPE. *J. Cryst. Growth*, **2007**, *298*, 644-647.

[51] Guo, L. J. Nanoimprint lithography: Methods and material requirements. *Adv. Mater.*, **2007**, *19*(4), 495-513.

[52] Tu, L.; Hsiao, C.; Chi, T.; Lo, I.; Hsieh, K. Self-assembled vertical GaN nanorods grown by molecular-beam epitaxy. *Appl. Phys. Lett.*, **2003**, *82*(10), 1601-1603.

[53] Park, Y.; Park, C.; Fu, D.; Kang, T.; Oh, J. Photoluminescence studies of GaN nanorods on Si(111) substrates grown by molecular-beam epitaxy. *Appl. Phys. Lett.*, **2004**, *85*(23), 5718-5720.

[54] Chen, H.; Lin, H.; Shen, C.; Gwo, S. Structure and photoluminescence properties of epitaxially oriented GaN nanorods grown on Si(111) by plasma-assisted molecular-beam epitaxy. *Appl. Phys. Lett.* **2006**, *89*(24), 243105.

[55] Grandal, J.; Sanchez-Garcia, M.; Calleja, E.; Luna, E.; Trampert, A. Accommodation mechanism of InN nanocolumns grown on Si(111) substrates by molecular beam epitaxy. *Appl. Phys. Lett.*, **2007**, *91*(2), 021902.

[56] Cerutti, L.; Ristic, J.; Fernandez-Garrido, S.; Calleja, E.; Trampert, A.; Ploog, K.; Lazic, S.; Calleja, J. Wurtzite GaN nanocolumns grown on Si(001) by molecular beam epitaxy. *Appl. Phys. Lett.*, **2006**, *88*(21), 213114.

[57] Park, Y. S.; Kang, T. W.; Taylor, R. A. Abnormal photoluminescence properties of GaN nanorods grown on Si(111) by molecular-beam epitaxy. *Nanotechnology*, **2008**, *19*(47), 475402.

[58] Robins, L.; Bertness, K.; Barker, J.; Sanford, N.; Schlager, J. Optical and structural study of GaN nanowires grown by catalyst-free molecular beam epitaxy. I. Near-band-edge luminescence and strain effects. *J. Appl. Phys.*, **2007**, *101*(11), 113505.

[59] Tchernycheva, M.; Sartel, C.; Cirlin, G.; Travers, L.; Patriarche, G.; Harmand, J.; Dang, L.; Renard, J.; Gayral, B.; Nevou, L.; Julien, F. Growth of GaN free-standing nanowires by plasma-assisted molecular beam epitaxy: structural and optical characterization. *Nanotechnology*, **2007**, *18*(38), 385306.

[60] Eliseev, P. G.; Perlin, P.; Lee, J. Y.; Osinski, M. "Blue" temperature-induced shift and band-tail emission in InGaN-based light sources. *Appl. Phys. Lett.*, **1997**, *71*(5), 569-571.

[61] Cimpoiasu, E.; Stern, E.; Klie, R.; Munden, R. A.; Cheng, G.; Reed, M. A. The effect of Mg doping on GaN nanowires. *Nanotechnology*, **2006**, *17*(23), 5735-5739.

[62]   Furtmayr, F.; Vielemeyer, M.; Stutzmann, M.; Arbiol, J.; Estrade, S.; Peiro, F.; Morante, J. R.; Eickhoff, M. Nucleation and growth of GaN nanorods on Si (111) surfaces by plasma-assisted molecular beam epitaxy - The influence of Si- and Mg-doping. *J. Appl. Phys.,* **2008,** *104*(3), 074309.

[63]   Tang, Y. B.; Bo, X. H.; Lee, C. S.; Cong, H. T.; Cheng, H. M.; Chen, Z. H.; Zhang, W. J.; Bello, I.; Lee, S. T. Controllable synthesis of vertically aligned p-type GaN nanorod Arrays on n-Type Si Substrates for Heterojunction Diodes. *Adv. Funct. Mater.,* **2008,** *18*(21), 3515-3522.

[64]   Beaumont, B.; Haffouz, S.; Gibart, P. Magnesium induced changes in the selective growth of GaN by metalorganic vapor phase epitaxy. *Appl. Phys. Lett.,* **1998,** *72*(8), 921-923.

[65]   Park, Y. S.; Na, J. H.; Taylor, R. A.; Park, C. M.; Lee, K. H.; Kang, T. W. The recombination mechanism of Mg-doped GaN nanorods grown by plasma-assisted molecular-beam epitaxy. *Nanotechnology,* **2006,** *17*(3), 913-916.

[66]   Xu, C. K.; Xue, L.; Yin, C. R.; Wang, G. H. Formation and photoluminescence properties of AlN nanowires. *Phys. Stat. Sol. (A),* **2003,** *198*(2), 329-335.

[67]   Balasubramanian, C.; Godbole, V. P.; Rohatgi, V. K.; Das, A. K.; Bhoraskar, S. V. Synthesis of nanowires and nanoparticles of cubic aluminium nitride. *Nanotechnology,* **2004,** *15*(3), 370-373.

[68]   Wu, Q.; Hu, Z.; Wang, X. Z.; Hu, Y. M.; Tian, Y. J.; Chen, Y., A simple route to aligned AlN nanowires. *Diam. Relat. Mater.,* **2004,** *13*(1), 38-41.

[69]   Lv, H. M.; Chen, G. D.; Ye, H. G.; Yan, G. J. Synthesis of monocrystal aluminum nitride nanowires at low temperature. *J. Appl. Phys.,* **2007,** *101*(5), 053526.

[70]   Yang, J.; Liu, T.; Hsu, C.; Chen, L.; Chen, K.; Chen, C. Controlled growth of aluminium nitride nanorod arrays *via* chemical vapour deposition. *Nanotechnology,* **2006,** *17*(11), S321-S326.

[71]   Hong, L.; Liu, Z.; Zhang, X.; Hark, S. Self-catalytic growth of single-phase AlGaN alloy nanowires by chemical vapor deposition. *Appl. Phys. Lett.,* **2006,** *89*(19), 193105.

[72]   Kuykendall, T.; Ulrich, P.; Aloni, S.; Yang, P. Complete composition tunability of InGaN nanowires using a combinatorial approach. *Nat. Mater.,* **2007,** *6*(12), 951-956.

[73]   Stoica, T.; Meijers, R.; Calarco, R.; Richter, T.; Luth, H. MBE growth optimization of InN nanowires. *J. Cryst. Growth,* **2006,** *290*(1), 241-247.

[74]   Calleja, E.; Sanchez-Garcia, M.; Sanchez, F.; Calle, F.; Naranjo, F.; Munoz, E.; Molina, S.; Sanchez, A.; Pacheco, F.; Garcia, R. Growth of III-nitrides on Si(111) by molecular beam epitaxy: Doping, optical, and electrical properties. *J. Cryst. Growth,* **1999,** *202,* 296-317.

[75]   Lee, S. H.; Jang, E. S.; Kim, D. W.; Lee, I. H.; Navamathavan, R.; Kannappan, S.; Lee, C. R. InN nanocolumns grown on a Si(111) substrate using Au plus In solid solution by metal organic chemical vapor deposition. *Jpn. J. Appl. Phys.,* **2009,** *48*(4), 04C141.

[76]   Cai, X.; Leung, Y.; Cheung, K.; Tam, K.; Djurisic, A.; Xie, M.; Chen, H.; Gwo, S. Straight and helical InGaN core-shell nanowires with a high In core content. *Nanotechnology*, **2006,** *17*(9), 2330-2333.

[77]   Chao, C.; Chyi, J.; Hsiao, C.; Kei, C.; Kuo, S.; Chang, H.; Hsu, T. Catalyst-free growth of indium nitride nanorods by chemical-beam epitaxy. *Appl. Phys. Lett.,* **2006,** *88*(23), 233111.

[78]   Stoica, T.; Meijers, R.; Calarco, R.; Richter, T.; Sutter, E.; Luth, H. Photoluminescence and intrinsic properties of MBE-grown InN nanowires. *Nano Lett.,* **2006,** *6*(7), 1541-1547.

[79]   Calleja, E.; Grandal, J.; Sanchez-Garcia, M.; Niebelschutz, M.; Cimalla, V.; Ambacher, O. Evidence of electron accumulation at nonpolar surfaces of InN nanocolumns. *Appl. Phys. Lett.,* **2007,** *90*(26), 262110.

[80]   Shen, C.; Chen, H.; Lin, H.; Gwo, S.; Klochikhin, A.; Davydov, V. Near-infrared photoluminescence from vertical InN nanorod arrays grown on silicon: Effects of surface electron accumulation layer. *Appl. Phys. Lett.,* **2006,** *88*(25), 253104.

[81]   Chen, C. F.; Wu, C. L.; Gwo, S. Organosilane functionalization of InN surface. *Appl. Phys. Lett.,* **2006,** *89*(25), 252109.

[82]   Jang, D.; Lin, G.; Hsiao, C.; Tu, L.; Lee, M. Auger recombination in InN thin films. *Appl. Phys. Lett.,* **2008,** *92*(4), 042101.

[83]   Wu, J.; Walukiewicz, W.; Shan, W.; Yu, K.; Ager, J.; Li, S.; Haller, E.; Lu, H.; Schaff, W. Temperature dependence of the fundamental band gap of InN. *J. Appl. Phys.,* **2003,** *94*(7), 4457-4460.

[84]   Hong, C. C.; Ahn, H.; Wu, C. Y.; Gwo, S. Strong green photoluminescence from In$_x$Ga$_{1-x}$N/GaN nanorod arrays. *Opt. Express,* **2009,** *17*(20), 17227-17233.

[85]   Ye, F.; Cai, X.; Wang, X.; Xie, E. The growth and field electron emission of InGaN nanowires. *J. Cryst. Growth,* **2007,** *304*(2), 333-337.

[86]   Kim, H.; Lee, H.; Kim, S.; Ryu, S.; Kang, T.; Chung, K. Formation of InGaN nanorods with indium mole fractions by hydride vapor phase epitaxy. *Phys. Stat. Sol. (B),* **2004,** *241*(12), 2802-2805.

[87]   Kim, H.; Lee, W.; Kang, T.; Chung, K.; Yoon, C.; Kim, C. InGaN nanorods grown on (111) silicon substrate by hydride vapor phase epitaxy. *Chem. Phys. Lett.,* **2003,** *380*(1-2), 181-184.

[88]   Kamimura, J.; Kouno, T.; Ishizawa, S.; Kikuchi, A.; Kishino, K. Growth of high-In-content InAlN nanocolumns on Si (111) by RF-plasma-assisted molecular-beam epitaxy. *J. Cryst. Growth,* **2007,** *300*(1), 160-163.

[89]   Dong, Y. J.; Tian, B. Z.; Kempa, T. J.; Lieber, C. M. Coaxial Group III-Nitride Nanowire Photovoltaics. *Nano Lett.,* **2009,** *9*(5), 2183-2187.

[90] Qian, F.; Gradecak, S.; Li, Y.; Wen, C.; Lieber, C. Core/multishell nanowire heterostructures as multicolor, high-efficiency light-emitting diodes. *Nano Lett.,* **2005,** *5*(11), 2287-2291.

[91] Qian, F.; Li, Y.; Gradecak, S.; Park, H.; Dong, Y.; Ding, Y.; Wang, Z.; Lieber, C. Multi-quantum-well nanowire heterostructures for wavelength-controlled lasers. *Nat. Mater.,* **2008,** *7*(9), 701-706.

[92] Bougerol, C.; Songmuang, R.; Camacho, D.; Niquet, Y. M.; Mata, R.; Cros, A.; Daudin, B. The structural properties of GaN insertions in GaN/AlN nanocolumn heterostructures. *Nanotechnology,* **2009,** *20*(29), 295706.

[93] Yi, S. N.; Na, J. H.; Lee, K. H.; Jarjour, A. F.; Taylor, R. A.; Park, Y. S.; Kang, T. W.; Kim, S.; Ha, D. H.; Andrew, G.; Briggs, D. Photoluminescence properties of a single GaN nanorod with GaN/AlGaN multilayer quantum disks. *Appl. Phys. Lett.,* **2007,** *90*(10), 101901.

[94] Renard, J.; Songmuang, R.; Bougerol, C.; Daudin, B.; Gayral, B. Exciton and biexciton luminescence from single GaN/AlN quantum dots in nanowires. *Nano Lett.,* **2008,** *8*(7), 2092-2096.

[95] Renard, J.; Songmuang, R.; Tourbot, G.; Bougerol, C.; Daudin, B.; Gayral, B. Evidence for quantum-confined Stark effect in GaN/AlN quantum dots in nanowires. *Phys. Rev. B,* **2009,** *80*(12), 121305(R).

[96] Ristic, J.; Calleja, E.; Sanchez-Garcia, M.; Ulloa, J.; Sanchez-Paramo, J.; Calleja, J.; Jahn, U.; Trampert, A.; Ploog, K. Characterization of GaN quantum discs embedded in $Al_xGa_{1-x}N$ nanocolumns grown by molecular beam epitaxy. *Phys. Rev. B,* **2003,** *68*(12), 085330.

[97] Chang, Y. L.; Wang, J. L.; Li, F.; Mi, Z. High efficiency green, yellow, and amber emission from InGaN/GaN dot-in-a-wire heterostructures on Si(111). *Appl. Phys. Lett.,* **2010,** *96*(1), 013106.

[98] Tachibana, K.; Someya, T.; Arakawa, Y. Growth of InGaN self-assembled quantum dots and their application to lasers. *IEEE J. Sel. Topics Quantum Electron.,* **2000,** *6*(3), 475-481.

[99] Armitage, R.; Tsubaki, K. Multicolour luminescence from InGaN quantum wells grown over GaN nanowire arrays by molecular-beam epitaxy. *Nanotechnology,* **2010,** *21*(19), 195202.

[100] Sun, Y.; Brandt, O.; Jenichen, B.; Ploog, K. In surface segregation in M-plane (In, Ga) N/GaN multiple quantum well structures. *Appl. Phys. Lett.,* **2003,** *83*(25), 5178.

[101] Koblmüller, G.; Metcalfe, G.; Wraback, M.; Wu, F.; Gallinat, C.; Speck, J. In adlayer mediated molecular beam epitaxial growth and properties of a-plane InN on freestanding GaN. *Appl. Phys. Lett.,* **2009,** *94*(9), 091905.

[102] Yamada, H.; Iso, K.; Saito, M.; Fujito, K.; DenBaars, S.; Speck, J.; Nakamura, S. Impact of substrate miscut on the characteristic of m-plane InGaN/GaN light emitting diodes. *Jpn. J. Appl. Phys.,* **2007,** *46*(46), L1117-L1119.

[103] Mahboob, I.; Veal, T.; McConville, C.; Lu, H.; Schaff, W. Intrinsic electron accumulation at clean InN surfaces. *Phys. Rev. Lett.,* **2004,** *92*(3), 036804.

[104] King, P. D. C.; Veal, T. D.; McConville, C. F.; Fuchs, F.; Furthmuller, J.; Bechstedt, F.; Schley, P.; Goldhahn, R.; Schormann, J.; As, D. J.; Lischka, K.; Muto, D.; Naoi, H.; Nanishi, Y.; Lu, H.; Schaff, W. J. Universality of electron accumulation at wurtzite c- and a-plane and zinc-blende InN surfaces. *Appl. Phys. Lett.,* **2007,** *91*(9), 202103.

[105] Darakchieva, V.; Schubert, M.; Hofmann, T.; Monemar, B.; Hsiao, C. L.; Liu, T. W.; Chen, L. C.; Schaff, W. J.; Takagi, Y.; Nanishi, Y. Electron accumulation at nonpolar and semipolar surfaces of wurtzite InN from generalized infrared ellipsometry. *Appl. Phys. Lett.,* **2009,** *95*(20), 202103.

[106] Lazic, S.; Gallardo, E.; Calleja, J. M.; Agullo-Rueda, F.; Grandal, J.; Sanchez-Garcia, M. A.; Calleja, E.; Luna, E.; Trampert, A. Phonon-plasmon coupling in electron surface accumulation layers in InN nanocolumns. *Phys. Rev. B,* **2007,** *76*(20), 205319.

[107] Liu, J.; Cai, Z. H.; Koley, G. Charge transport and trapping in InN nanowires investigated by scanning probe microscopy. *J. Appl. Phys.,* **2009,** *106*(12), 124907.

[108] Motayed, A.; Davydov, A. V.; Mohammad, S. N.; Melngailis, J. Experimental investigation of electron transport properties of gallium nitride nanowires. *J. Appl. Phys.,* **2008,** *104*(2), 024302.

[109] Simpkins, B. S.; Pehrsson, P. E.; Laracuente, A. R. Electronic conduction in GaN nanowires. *Appl. Phys. Lett.,* **2006,** *88*(7), 072111.

[110] Chang, C.; Chi, G.; Wang, W.; Chen, L.; Chen, K.; Ren, F.; Pearton, S. Electrical transport properties of single GaN and InN nanowires. *J. Electron. Mater.,* **2006,** *35*(4), 738-743.

[111] Richter, T.; Luth, H.; Schapers, T.; Meijers, R.; Jeganathan, K.; Hernandez, S. E.; Calarco, R.; Marso, M., Electrical transport properties of single undoped and n-type doped InN nanowires. *Nanotechnology,* **2009,** *20*(40), 405206.

[112] Cheng, G. S.; Stern, E.; Turner-Evans, D.; Reed, M. A. Electronic properties of InN nanowires. *Appl. Phys. Lett.,* **2005,** *87*(25), 253103.

[113] Huang, Y.; Duan, X.; Cui, Y.; Lieber, C. Gallium nitride nanowire nanodevices. *Nano Lett.,* **2002,** *2*(2), 101-104.

[114] Motayed, A.; He, M.; Davydov, A.; Melngailis, J.; Mohammad, S. Realization of reliable GaN nanowire transistors utilizing dielectrophoretic alignment technique. *J. Appl. Phys.,* **2006,** *100*(11), 114310.

[115] Rumyantsev, S.; Shur, M.; Levinshtein, M.; Motayed, A.; Davydov, A. Low-frequency noise in GaN nanowire transistors. *J. Appl. Phys.,* **2008,** *103*(6), 064501.

[116] Li, Y.; Xiang, J.; Qian, F.; Gradecak, S.; Wu, Y.; Yan, H.; Yan, H.; Blom, D.; Lieber, C. Dopant-free GaN/AlN/AlGaN radial nanowire heterostructures as high electron mobility transistors. *Nano Lett.,* **2006,** *6*(7), 1468-1473.

[117] Vandenbrouck, S.; Madjour, K.; Theron, D.; Dong, Y. J.; Li, Y.; Lieber, C. M.; Gaquiere, C. 12 GHz F_{MAX} GaN/AlN/AlGaN nanowire MISFET. *IEEE Electron. Dev. Lett.,* **2009**, *30*(4), 322-324.

[118] Barletta, P. T.; Berkman, E. A.; Moody, B. F.; El-Masry, N. A.; Emara, A. M.; Reed, M. J.; Bedair, S. M. Development of green, yellow, and amber light emitting diodes using InGaN multiple quantum well structures. *Appl. Phys. Lett.,* **2007**, *90*(15), 151109.

[119] Wetzel, C.; Detchprohm, T. Development of high power green light emitting diode chips. *Mrs Internet Journal of Nitride Semiconductor Research* **2005**, *10, 2.*

[120] Tsai, M. A.; Yu, P. C.; Chao, C. L.; Chiu, C. H.; Kuo, H. C.; Lin, S. H.; Huang, J. J.; Lu, T. C.; Wang, S. C. Efficiency enhancement and beam shaping of GaN-InGaN vertical-injection light-emitting diodes *via* high-aspect-ratio nanorod arrays. *IEEE Photon. Technol. Lett.,* **2009**, *21*(1-4), 257-259.

[121] Chen, L. Y.; Huang, Y. Y.; Chang, C. H.; Sun, Y. H.; Cheng, Y. W.; Ke, M. Y.; Chen, C. P.; Huang, J. J. High performance InGaN/GaN nanorod light emitting diode arrays fabricated by nanosphere lithography and chemical mechanical polishing processes. *Opt. Express,* **2010**, *18*(8), 7664-7669.

[122] Wu, Y. R.; Chiu, C. H.; Chang, C. Y.; Yu, P. C.; Kuo, H. C. Size-dependent strain relaxation and optical characteristics of InGaN/GaN nanorod LEDs. *IEEE J. Sel. Topics Quantum Electron.,* **2009**, *15*(4), 1226-1233.

[123] Wang, C. Y.; Chen, L. Y.; Chen, C. P.; Cheng, Y. W.; Ke, M. Y.; Hsieh, M. Y.; Wu, H. M.; Peng, L. H.; Huang, J. GaN nanorod light emitting diode arrays with a nearly constant electroluminescent peak wavelength. *Opt. Express,* **2008**, *16*(14), 10549-10556.

[124] Sekiguchi, H.; Kishino, K.; Kikuchi, A. GaN/AlGaN nanocolumn ultraviolet light-emitting diodes grown on n-(111) Si by RF-plasma-assisted molecular beam epitaxy. *Electron. Lett.,* **2008**, *44*(2), 151-152.

[125] Sekiguchi, H.; Kato, K.; Tanaka, J.; Kikuchi, A.; Kishino, K. Ultraviolet GaN-based nanocolumn light-emitting diodes grown on n-(111) Si substrates by RF-plasma-assisted molecular beam epitaxy. *Phys. Stat. Sol. (A),* **2008**, *205*(5), 1067-1069.

[126] Minot, E.; Kelkensberg, F.; van Kouwen, M.; van Dam, J.; Kouwenhoven, L.; Zwiller, V.; Borgstrom, M.; Wunnicke, O.; Verheijen, M.; Bakkers, E. Single quantum dot nanowire LEDs. *Nano Lett.,* **2007**, *7*(2), 367-371.

[127] Chiu, C. H.; Lu, T. C.; Huang, H. W.; Lai, C. F.; Kao, C. C.; Chu, J. T.; Yu, C. C.; Kuo, H. C.; Wang, S. C.; Lin, C. F.; Hsueh, T. H. Fabrication of InGaN/GaN nanorod light-emitting diodes with self-assembled Ni metal islands. *Nanotechnology,* **2007**, *18*(44), 445201.

[128] Kikuchi, A.; Kawai, M.; Tada, M.; Kishino, K. InGaN/GaN multiple quantum disk nanocolumn light-emitting diodes grown on (111)Si substrate. *Jpn. J. Appl. Phys.,* **2004**, *43*(12A), L1524-L1526.

[129] Kim, H.; Kang, T.; Chung, K. Nanoscale ultraviolet-light-emitting diodes using wide-bandgap gallium nitride nanorods. *Adv. Mater.,* **2003**, *15*(7-8), 567-569.

[130] Kawakami, Y.; Suzuki, S.; Kaneta, A.; Funato, M.; Kikuchi, A.; Kishino, K. Origin of high oscillator strength in green-emitting InGaN/GaN nanocolumns. *Appl. Phys. Lett.,* **2006**, *89*(16), 163124.

[131] Kim, H.; Cho, Y.; Lee, H.; Kim, S.; Ryu, S.; Kim, D.; Kang, T.; Chung, K. High-brightness light emitting diodes using dislocation-free indium gallium nitride/gallium nitride multiquantum-well nanorod arrays. *Nano Lett.,* **2004**, *4*(6), 1059-1062.

[132] Johnson, J.; Choi, H.; Knutsen, K.; Schaller, R.; Yang, P.; Saykally, R. Single gallium nitride nanowire lasers. *Nat. Mater.,* **2002**, *1*(2), 106-110.

[133] Choi, H.; Johnson, J.; He, R.; Lee, S.; Kim, F.; Pauzauskie, P.; Goldberger, J.; Saykally, R.; Yang, P. Self-organized GaN quantum wire UV lasers. *J. Phys. Chem. B,* **2003**, *107*(34), 8721-8725.

[134] Gradecak, S.; Qian, F.; Li, Y.; Park, H.; Lieber, C. GaN nanowire lasers with low lasing thresholds. *Appl. Phys. Lett.,* **2005**, *87*(17), 173111.

[135] Kouno, T.; Kishino, K.; Yamano, K.; Kikuchi, A. Two-dimensional light confinement in periodic InGaN/GaN nanocolumn arrays and optically pumped blue stimulated emission. *Opt. Express,* **2009**, *17*(22), 20440-20447.

[136] Tang, Y.; Chen, Z.; Song, H.; Lee, C.; Cong, H.; Cheng, H.; Zhang, W.; Bello, I.; Lee, S. Vertically aligned p-type single-crystalline GaN nanorod arrays on n-type Si for heterojunction photovoltaic cells. *Nano Lett.,* **2008**, *8*(12), 4191–4195.

[137] Dong, Y.; Tian, B.; Kempa, T.; Lieber, C. Coaxial group III-nitride nanowire photovoltaics. *Nano Lett.,* **2009**, *9*(5), 2183-2187.

III-Phospide Nanowires

Erik P.A.M. Bakkers[1,2,*], Moira Hocevar[2], Rienk E. Algra[3,4], Lou-Fé Feiner[1] and Marcel A. Verheijen[5]

[1]*Eindhoven University of Technology, 5612 AZ Eindhoven, The Netherlands;* [2]*Delft University of Technology, 2628CC Delft, The Netherlands;* [3]*Materials Innovation Institute (M2i), 2628CD Delft, The Netherlands;* [4]*IMM, Solid State Chemistry, Radboud University Nijmegen, Heijendaalseweg 135, 6525AJ Nijmegen, The Netherlands and* [5]*Philips Research Laboratories Eindhoven, High Tech Campus 11, 5656AE Eindhoven, The Netherlands*

Abstract: In this chapter we review the latest developments in the field of InP and GaP nanowires. A major part of the chapter is devoted to the vapor-liquid-solid (VLS) growth (mechanism) of nanowires, but also 'catalyst-free' growth mechanisms will be discussed. The chapter starts with methods to obtain control of nanowire diameter and position. Such control is essential to understand the details of the growth mechanism. It will be shown that the liquid particle can truly catalyze the growth of nanowires. The other main theme considers the structural and optical nanowire properties. Nanowires can have a different (wurtzite) crystal structure than the bulk (zinc blende) materials. Parameters, which can affect the formation of specific crystal structures, will be discussed. Closely related to the crystal structure is the formation of stacking faults, such as twin planes. Single twins, paired twin and twin superlattices are characterized and their occurrence is explained by a kinetic growth model. Photoluminescence has been used to assess the optical properties of single nanowires. It is shown that wurtzite InP has higher bandgap energy than zinc blende InP. Control of the crystal structure allows for a new type of band-engineering. Finally, the photonic properties of wire ensembles are discussed.

Keywords: Nanowire, InP, GaP, VLS growth mechanism, stacking fault, superlattice, optical properties.

INTRODUCTION

The success of the semiconductor industry is based on bandgap engineering. By varying the chemical composition spatially, the electronic properties of a layer stack are tweaked. However, lattice matching requirements limit the possibilities of combining bulk materials with different crystal parameters. Nanowires show an unprecedented ability to combine different semiconductor materials due to their small dimensions. Different materials can be stacked and even different classes of materials can be combined. Already new and unexpected piezoelectric [1], thermoelectric [2,3], optical [4] and electrical [5,6] nanowire properties have been reported and promising results on applications such as sensors [7], LEDs [8], solar cells [9-11] and transistors [12,13] have been published.

Nanowires are grown from a metal nanoparticle by the vapor-liquid-solid (VLS) mechanism (Fig. 1). Originally discovered by Wagner and Ellis at Bell Labs [14] for the growth of Si wires, the field was further developed by Hiruma at Hitachi labs [15] for III-V nanowires, Prof. Lieber at Harvard [16], Prof. Yang at Berkeley [17] and Prof. Samuelson at Lund [18]. As the name 'vapor-liquid-solid' suggests three phases are involved in nanowire growth. By heating, the metal particles on the substrate form an alloy with the substrate material and specific (group III) precursors from the gas phase. The formation of a eutectic turns the seed particle from solid into liquid, following the binary phase diagrams of the different materials systems. Subsequently, by continuous supply of precursor material from the gas phase, the particle saturates and eventually reaches supersaturation. If the supersaturation is sufficient, crystal nucleation starts at the droplet-substrate interface. The (solid) wire grows layer by layer lifting the metal particle. During growth there is a dynamic equilibrium between materials *supply* from the gas phase, *transport* through the droplet, and crystal *growth* at the droplet- nanowire crystal interface. Group IV, III-V and II-VI semiconductors can be synthesized with the VLS mechanism. During growth the composition of the gas phase can be changed, inducing a change in material or doping along the axis of the nanowire. It has been shown by several groups that junctions in the length direction as well as in the radial direction can be formed [16-19]. The great freedom in the material composition allows envisioning new applications in chemistry, physics, engineering science and bioscience.

*Address correspondence to Erik P.A.M. Bakkers: Eindhoven University of Technology, 5612 AZ Eindhoven, The Netherlands; E-mail: e.p.a.m.bakkers@tue.nl

Jianye Li, Deli Wang and Ray R. LaPierre (Eds)

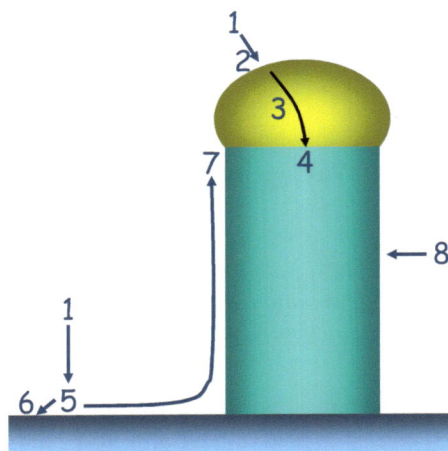

Figure 1: Schematic of the processes that are relevant for the vapor-liquid-solid (VLS) nanowire growth mechanism. 1: precursor transport, 2: precursor dissociation, 3: diffusion through Au, 4: formation of GaP/GaAs, 5: adsorption on substrate, 6: film deposition on substrate, 7: surface diffusion, 8: film deposition on nanowire surface.

The growth of nanowires is a complex mechanism that involves many parameters as illustrated in Fig. **1**. Amongst these parameters are the nature of the substrate (surface), substrate temperature, total pressure, precursor partial pressure, III-V ratios, and impurity concentration, such as dopants. As will be shown throughout this chapter these parameters can affect the crystal structure, morphology and defect density of the nanowires.

In this chapter we will focus on the III-phosphide nanowires, such as InP and GaP. InP based materials are interesting for many optical applications, such as LEDs, lasers and solar cells, but also for electronic devices, because of the relatively high mobility of electrons [20]. GaP, in general, is less interesting for applications, since it has a rather low carrier mobility, and more importantly an indirect optical bandgap [21]. However, GaP seems to be a 'model' system for investigating the VLS growth mechanism, since GaP nanowires can be grown in a wide parameter range. In addition, GaP has a small lattice mismatch with Si (0.4%) making it a natural epitaxial interface between III-V's and silicon. We will discuss different growth techniques to synthesize InP and GaP wires. We will present in detail several fundamental processes affecting the nanowire growth, the resulting crystal structure and morphology. We will discuss recent progress in fabricating more complex structures such as axial and radial heterostructures and dopant profiles (doped segments). Finally, the optical properties of the different structures studied with photoluminescence will be presented.

GROWTH TECHNIQUES

Different growth techniques are used by research groups to fabricate nanowires, which all make use of the VLS growth mechanism. The growth techniques mainly differ in the way the source material is brought into the gas phase. Some examples are thermal evaporation, Laser Ablation (LA) [22], Molecular Beam Epitaxy (MBE) [23], Chemical Beam Epixtay (CBE) [24] and Metal-Organic Vapor Phase Epitaxy (MOVPE)[15]. Thermal evaporation and LA offer a very limited control of the partial pressures with poor uniformity across the sample. MBE, MOVPE, and CBE are established techniques to make high quality layered structures. Typically, the growth rates are very low for MBE making it a suitable technique mainly for academics. MOVPE offers high deposition rates and is used in the commercial production of LEDs and lasers. CBE combines some (dis)advantages of both techniques. For the growth of phosphide compounds, which have a high vapor pressure, MOVPE is convenient.

In MOVPE, metalorganic precursor molecules are used for the group III and group V atoms. Typical partial pressures used for phosphine are between 10^{-2} mbar and several mbar and for the group III (tri(m)ethylgallium or tri(m)ethylindium) between 10^{-5} mbar and 10^{-3} mbar. The total reactor pressure is typically between 10-500 mbar. In a horizontal reactor lay-out, a laminar flow of the precursors over the substrate feeds the growth. Hydrogen or nitrogen is used as a carrier gas for the precursors. The substrates are placed on a graphite susceptor which is heated by, for instance, radio frequency (RF) heating. At high temperatures, the precursors decompose in an irreversible way close to the substrate surface, where it will be deposited. It is important to prevent decomposition reactions in the gas phase since they can pollute the reactor walls and can cause material memory effects in future experiments.

III-phosphide nanowires are also grown by solid source molecular beam epitaxy (SS-MBE). The MBE systems are equipped with solid sources providing indium and gallium atoms, and cracker phosphorous and arsenic sources producing dimers. In such systems Au diffusion cell can be used to deposit in-situ gold thin films. The main advantage of MBE comes from the ultra high vacuum (10^{-10} Torr) in which the materials are grown. Consequently, this fabrication method allows for highly pure, carbon-free materials. Moreover, reflection high-energy electron diffraction (RHEED) is a powerful in-situ characterization tool generally used in MBE systems, which follows the crystalline evolution of nanowires in real time during growth. The main disadvantage of using this system for nanowires concerns first the low growth rates. Growth rates are typically 50 to 100 nm/min for InP [25], whereas nanowires can grow in MOVPE reactors up to micrometers/min[19]. Another issue of MBE growth concerns the beams that are not ideally normal to the surface of the samples but at 75°. This geometry provides additional species for lateral growth, making it difficult to obtain a zero axial growth rate [26]. Finally, the growth processes are slightly different using MBE technique, as it is assumed that the atoms only diffuse along the nanowire walls to feed the Au nanoparticle, whereas for MOVPE growth, the precursors also decompose at the Au particle to form the eutectic. InP nanowires growth by MBE is generally realized on InP substrates [23] at growth temperatures ranging between 350 and 420°C, or Si substrates [27] at 380°C. So far, GaP NWs grown by MBE have not been reported.

ALTERNATIVE GROWTH MECHANISMS

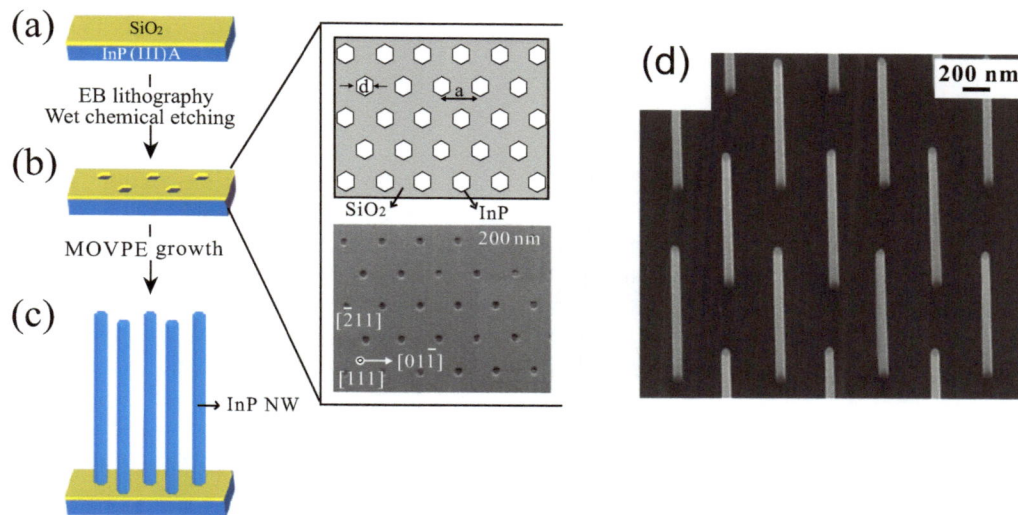

Figure 2: Fabrication sequence of InP nanowire arrays. **(a)** 30 nm SiO_2 thin film deposited on InP(111)A substrate by plasma sputtering. **(b)** Hexagonal openings formed on the masked substrate by electron beam lithography and wet chemical etching. The inset shows the schematic illustration and SEM image of the periodically patterned masked substrate. **(c)** SA-MOVPE growth on the masked substrate giving rise to **(d)** vertically aligned InP nanowire arrays. Reprinted with permission from Ref. [28]. Copyright 2005 IOP Publishing.

The fabrication of nanowires is not restricted to VLS growth with Au catalysts. The main reason why research is active in growing nanowires without gold is related to its highly contaminating effect in semiconductor technology, such as unwanted deep levels and gold migration on the substrates.

Fukui *et al.* developed a growth method called selected area metalorganic vapor phase epitaxy (SA-MOVPE) to grow InP [28,29] and other material nanowires [30-32]. With this technique the nanowire growth is spatially controlled through an opening in a masking material as shown in Fig. 2. The first reported work dealt with GaAs nanowires [33] and one year later the same group published a work on InP nanowires grown using SA-MOVPE [28]. The advantages of SA-MOVPE are that the nanowires are grown without the use of a seed nanoparticle, with a uniform size and a controllable positioning.

The nanowires are grown on (111)A InP or Si substrates and the mask patterning realized by electron beam lithography on SiO_2 masks. The growth temperatures are ranging between 550°C and 650°C, these temperatures being much higher than when using Au catalysts. The typical work pressure and partial pressure of trimethylindium (TMI) are 0.1 atm and 4.4×10^{-6} atm, respectively. The tertiary butyl phosphine (TBP) pressure can be varied from 5.5×10^{-3} to 1.6×10^{-4} atm: at low phosphorous partial pressure, In atoms are not attached to the side wall of the nanowires and can migrate to the top of the nanowires, giving rise to thin and long nanowires. On the contrary, at high P pressure, In atoms are strongly absorbed on the (110) facets of the nanowires, and thus thicker wires can be grown. It has been shown that a high temperature decreases the growth rate. P sites on an In-terminated (111) substrate surface are weaker adsorption sites, thus In adatom desorption rate increases causing a decrease in the growth rate. Finally, InP nanowires grown by SA-MOVPE exhibit a hexagonal cross-section and distinct {110} vertical facets. It has to be noted that these nanowires exhibit a pure wurtzite crystal structure along the whole length.

Nanowires can also be grown 'catalyst-free' using In droplets as seeds. In 2003, Yang *et al.* [34] showed the possibility of growing GaN nanowires using Ga droplets. Using In droplets, catalyst-free InP nanowires are grown on InP [35-37], (111) Si [38] and (001) Si [39]. Generally, growth is performed in a MOVPE reactor and the In droplets are deposited *in situ* on the substrates. Mattila *et al.* investigated the influence of the substrate orientation on InP growth direction: the nanowires always grow in the [111]B direction, leading either to different growth orientation with respect to the substrate surface or to kinks during growth in case another substrate orientation than (111)B InP is used. InP nanowires grown on Si substrates from In seeds point off the plane of the substrate, but they are never completely vertical.

Catalyst-free growth of nanowires is a very promising field of research, however we will mainly focus now on VLS mechanism in which gold particles are involved.

ROLE OF THE METAL PARTICLE

In principle, during VLS growth, metal atoms from the particle are not incorporated in the resulting nanowire. As an example, the solubility of Au in InP is very low (6×10^{14} cm^{-3}) [40], suggesting that only a few Au atoms can be incorporated in a typical wire volume. In addition, the wire surface is always nearby enabling the out-diffusion of incorporated impurity atoms. Analytical tools, such as atom-probe microscopy [41] or cross sectional STM [42], to measure the impurity concentration in nanoscale structures are just becoming available. It has been shown for InAs wires that Au atoms are present [41] in the wire crystal lattice. The VLS growth mechanism is characterized by the metal particle, which has multiple functions. First of all, the particle determines the diameter and the position of the nanowire as will be discussed below. The particle can catalyze the decomposition of the precursor molecules, which makes it possible to grow the wires at significantly lower temperatures than the corresponding bulk materials. Nanowires have even been grown below the eutectic temperature. It is debated whether the particle is in the liquid [43] or in the solid phase [44]. Furthermore, several groups have shown that the shape of the particle [45], and more specifically the contact angle of the particle with the nanowire side facets [46-48] determines the nanowire morphology and crystal structure.

Diameter Control

We will start with discussing how to control the size, or diameter of the wires. For many applications it is essential to be able to predefine the wire diameter, since it affects the electronic band structure [49] and electron mobility [50] especially for the smaller diameters. It has been found by the Korgel [51] and Lieber [52] groups that the size of the metallic nanocluster determines the size of the wire during growth as illustrated in Fig. **3a**, and thus one can envision creating wires with a narrow size distribution by exploiting monodisperse metal nanoclusters. Well-defined gold colloids were used in these initial studies by Gudiksen [52] as seeds for the growth of nearly monodisperse samples of single crystalline GaP nanowires with diameters of 10, 20, and 30 nm.

Transmission electron microscopy (TEM) analysis of nanowire diameters demonstrates the good correlation with the colloid diameters and dispersion (Fig. **3b-d**); that is, for wires grown from 28.2±2.6, 18.5±0.9, and 8.4±0.9 nm colloids, mean diameters of 30.2±2.3, 20.0±1.0, and 11.4±1.9 nm, were observed respectively. The mean nanowire diameter is generally 1-2 nm larger than that of the colloids. This increase has been explained by alloying of the Ga and P reactants with the colloids before nucleation of the nanowire occurs. it is clear that the width of the nanowire distributions mirrors those of the colloid, suggesting that the uniformity of the wire diameter is limited only by the disparity of the colloids. For the 10 nm diameter wires (Fig. **3d**), a small broadening (1 nm) of the wire distribution can be attributed to aggregation of the colloids. This work demonstrates clearly for the first time an ability to exert systematic control over the diameter of semiconductor nanowires for a variety of colloid sizes. Note that in these studies the colloids are randomly distributed on a SiO_2 substrate and neither the position nor the orientation of the wires is defined.

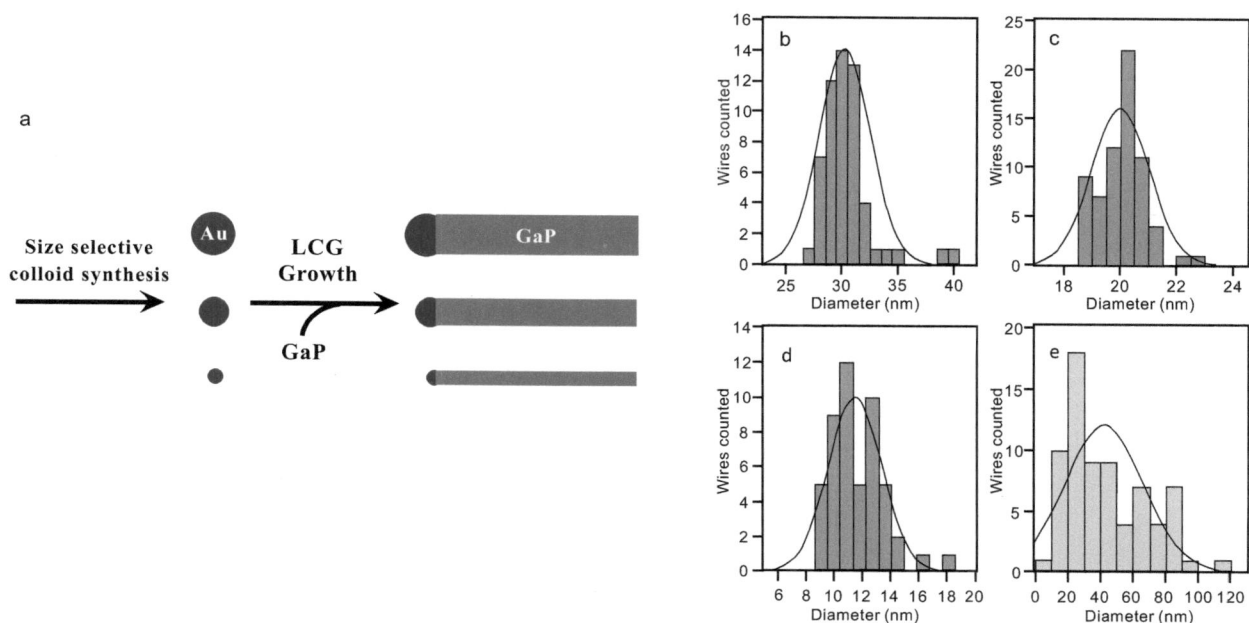

Figure 3: **(a)** Schematic depicting the use of monodisperse gold colloids as catalysts for the growth of well-defined GaP semiconductor nanowires. Histograms of measured diameters for wires grown from **(b)** 28.2 nm, **(c)** 18.5 nm, and **(d)** 8.4 nm colloids. The solid lines show the wire distributions. **(e)** Histogram of diameters for wires grown by the laser-assisted-growth, in which the laser is used to both generate the Au nanoclusters and the GaP reactants. The distribution is very broad (SD=23.9 nm) and the mean diameter (42.7 nm) greater than those synthesized using the predefined colloid catalyst. Reprinted with permission from Ref. [52]. Copyright 2000 American Chemical Society.

Position Control

The next step is to control both the diameter and the position of the wires. Position control is important for applications in which as-grown (vertical) wires will directly be used and contacted to form larger circuits. In addition, as we will discuss in the next paragraph, the nanowire growth rate sensitively depends on the density and the dimensions of the metal particles. Such a variation of the growth rate can result in non uniform optical or electrical properties. It is likely that, for instance, the dispersion of the electron mobility in InAs/InP core-shell nanowires [50], is related to the large spread in shell thickness of these nanowires grown from randomly deposited colloidal gold particles. Besides the axial growth rate, most probably also the radial (shell) growth is affected by the catalyst density. Hence, in order to improve the nanowire uniformity, it is important to control the position and density of the catalyst particles on the substrate. Different approaches have already been reported, such as e-beam lithography [53], gold deposition through an anodic aluminium oxide (AAO) template [54] and nanosphere lithography [55]. In general, these techniques do not allow for large-area structuring because of either high cost (e-beam) or lack of long-range order (AAO and nanosphere).

Alternatively, nanoimprint lithography has been reported by Mårtensson [56] to define InP nanowire positions. This technique enables patterning of large surface areas at relatively low-cost. We discuss here [57] an optimized process to control the position of InP and GaP nanowires on wafer scale, called Substrate Conformal Imprint Lithography (SCIL). In order to fabricate uniform, defect free nanowire arrays the effect on nanowire growth of chemical and thermal treatment of the substrate which contained gold catalyst particles has been studied systematically.

The substrate conformal imprint lithography method used by Pierret *et al.* [57] makes use of a flexible patterned stamp made from Poly-Di-Methyl-Siloxane (PDMS), which is molded from a silicon master pattern which contained arrays of holes and was fabricated using e-beam lithography. The pattern is transferred into the photoresist by pressing the mold into the resist layer. Au has been deposited by evaporation. Finally, a lift-off process is used, which dissolves the photoresist layer and releases the gold metal layer on top. This leaves precisely placed gold dots of controlled diameter and thickness on the semiconductor material, but also (organic) residues on the substrate.

(a)   (b)

(c)   (d)

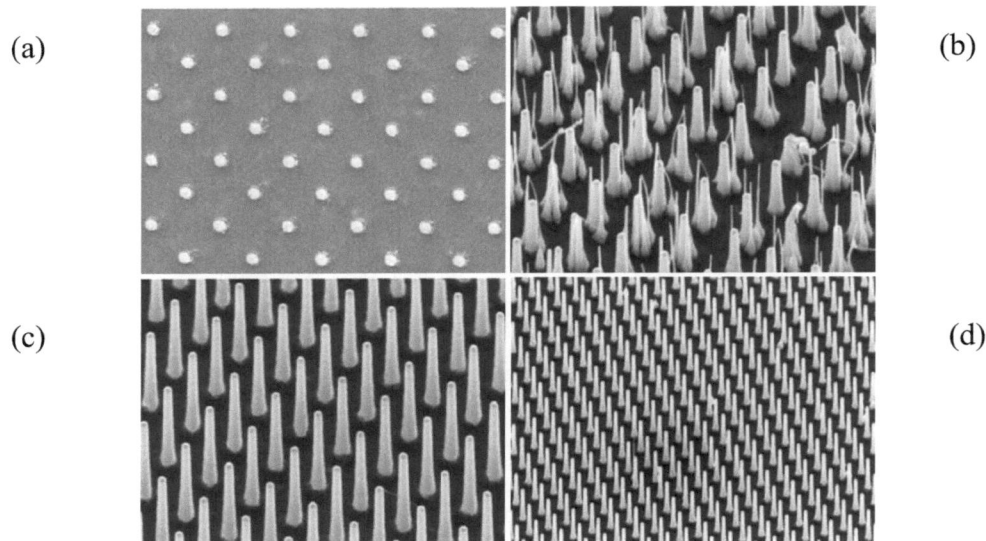

Figure 4: InP nanowire arrays by soft nano-imprint lithography. **(a)** Pattern of Au dots after lift-off process, prior to the growth of nanowires. **(b)** Nanowire array without suitable cleaning step prior to growth resulting in extra 'grass-like' wire growth between the patterned arrays. **(c)** and **(d)** Nanowire array with a piranha cleaning step and 550 °C anneal before growth. Reprinted with permission from Ref. [57]. Copyright 2010 IOP Publishing.

Fig. **4a** shows a Scanning Electron microscope (SEM) image of the surface after the lift-off process without further treatment. We can clearly see the high fidelity of the pattern transfer, as the Au particles are arranged in an ordered pattern. However, extra material is present around the Au islands. These contaminants have been identified with Energy Dispersive X-ray (EDX) measurements, carried out within the SEM, and are carbon-based residues from the PMMA layer. The SEM image in Fig. **4b** is taken after InP nanowire growth by MOVPE on a (111)B oriented InP substrate. Besides nanowires grown from the predefined catalyst particles in a regular pattern, a lot of undesired thin nanowires are obtained. These extra nanowires arise if no or improper pre-treatment was used after the lift-off. This shows that the organic material observed in Fig. **4a** initiates growth of these extra nanowires. An extra cleaning step is necessary to remove the organics prior to growth.

Samples exposed to a piranha solution at 20°C prior to nanowire growth almost have no undesired nanowires and all the intended nanowires are at the predefined positions, as shown in Fig. **4c** and **4d**. The overview SEM image in Fig. **4d** shows that the InP nanowire dimensions are uniform across the sample. The obtained InP nanowires have an interspacing of 500 nm, as defined by the mold, and a length of 1761 ± 19 nm which is only a spread of 1.1%. Furthermore the nanowires are tapered and have a top diameter of 95 ± 4 nm and base diameter of 171 ± 4 nm. A similar treatment with a piranha solution, albeit with a longer exposure time to this solution, is best also for GaP wire growth. This chemical treatment, hoewever, is not sufficient to obtain defect-free nanowire arrays. A thermal anneal step is essential to obtain the best result.

The effect was studied of a thermal annealing after piranha treatment, but prior to the growth. It is well known that a thermal anneal at high enough temperatures can remove In_2O_3 or Ga_2O_3 from the substrate surface [58]. Clearly, too low annealing temperatures, or no annealing at all, results in undesired extra wire growth and irregular growth of the patterned nanowires. An annealing temperature of at least 550 °C proves to be sufficient for fabrication of defect-free arrays of InP wires. At higher temperatures nanowires with a thicker base are obtained. Probably the surface chemistry changes at these temperatures and promote the lateral growth at the base of the nanowires. Similar results were obtained for the growth of GaP nanowires. Whereas at an annealing temperature of 550 °C still undesired GaP nanowires are obtained, at a temperature of 700 °C the defect density of the array is very low. The GaP nanowires show slight tapering, but do not have a thick base as observed for the InP nanowires when annealed at this temperature. Growth without chemical cleaning but with an annealing step prior to growth resulted in many undesired nanowires. This shows that the combination of piranha treatment with a sufficiently high annealing temperature is essential for both InP and GaP wire growth. The different optimum annealing temperatures for InP and GaP shows that the removal process is substrate-dependent and suggests that the metal oxides play an important role in removing the organic residues.

Catalytic effects of the metal particle

By using the diameter and position control, we can now systematically study the effect of particle size and wire-to-wire spacing on the nanowire growth rate. The nanowire growth mechanism is a complex process with many parameters, such as temperature, partial pressures, and III/V ratio, which often have a cross relation. In addition, several distinguished processes, such as materials competition [59], and the Gibbs Thomson effect [60] have a, sometimes opposite, effect on the nanowire growth rate. In addition, the metal particle is often named a 'catalyst', but catalytic action by the particle, *i.e.* a decrease of the activation energy for growth has just been shown unambiguously in a few cases [61]. In this chapter we will discuss a systematic study unravelling several processes and indicating their effect on the nanowire growth [62].

GaP nanowires were grown epitaxially on GaP(111)B substrates by MOVPE to induce vertical growth from predefined gold catalyst arrays. The SEM images in ref 62 show that thin wires next to a thick wire are taller than those in the middle of the field surrounded by thin wires. The growth rate of one wire is enhanced by the presence of another one, demonstrating that the wires interact during growth.

In Fig. **5a**, the GaP nanowire growth rates (R_{GaP}) for various catalyst sizes have been plotted as a function of the interdistance. It quantitatively presents the three different growth regimes as a function of interdistance, which have been marked *1*, *2*, and *3*. In the following, we distinguish those and discuss the different contributions to the nanowire growth rate:

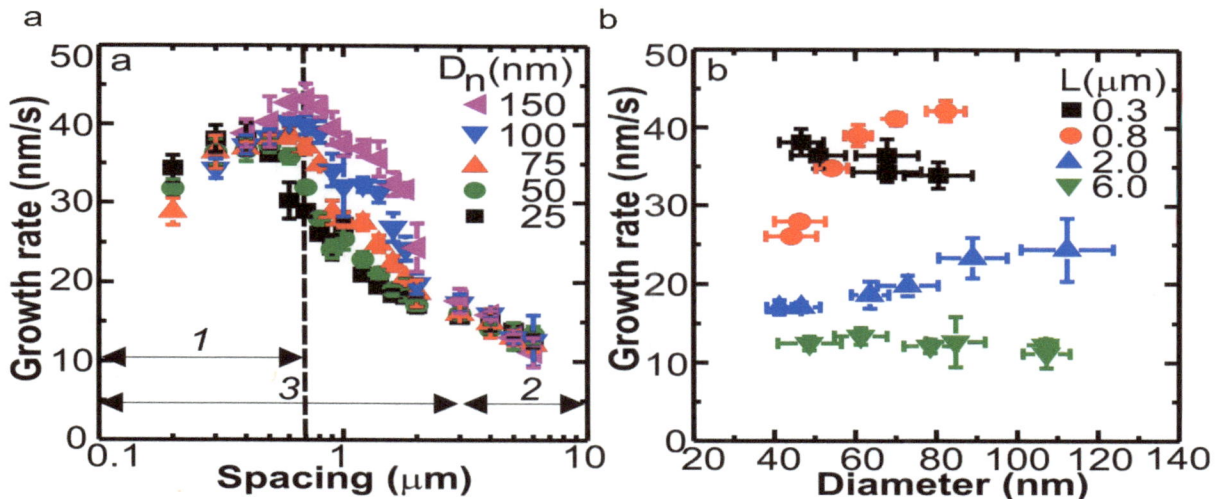

Figure 5: (a) Growth rate as a function of nanowire spacing and nominal diameter on a semi-logarithmic scale. The different growth regimes are indicated: 1: competitive, 2: independent and 3: synergetic, as discussed in the text. Regime 3 and 1 overlap owing to the nature of the synergy. **(b)** Growth rate as a function of diameter for varied spacing. From ref. 62. Copyright 2007 Nature Publishing Group.

1) *Materials competition regime.* For small nanowire interdistance $(P < 0.7 \, \mu m$, Fig. **5a**), the precursor material is shared between the wires. At the smallest available interdistance, small diameter wires

grow faster than large diameter wires (Fig. **5a** and black squares in Fig. **5b**) due to mass conservation and materials distribution over a constant number of growing wires. Part of the precursor material does not directly impinge on the catalyst particle, but lands on the substrate surface. From there it can diffuse, over the surface and along the wire side facets, towards the metal particle and induce wire growth. Thus, a wire collects material from the surroundings and the area is determined by the diffusion length of the specific precursor species. If the collection areas of neighbouring wires overlap, the material is shared between the wires, *i.e.* materials competition. Since the growth rate under these conditions is determined by the materials flux, the total growth volume is constant. The growth rate increases linearly with the surface collection area. This effect has been observed before by Jensen *et al.* [59] for InAs nanowires grown by CBE.

2) *Independent regime.* At relatively large interdistance ($P > 3$ µm) wires cannot interact *via* diffusion and can be treated as independent isolated islands. Importantly, for the largest interdistance, $P = 6$ µm, the nanowire growth rate is independent of the particle diameter (Fig. **5b**). This demonstrates that neither materials competition (which would result in higher growth rates for thinner wires as discussed above) nor the Gibbs Thomson effect is rate limiting under these growth parameters. The Gibbs-Thomson effect is often used to explain a decrease of the nanowire growth rate with decreasing diameter. A decreasing particle size leads to an increase of the pressure inside the metal particle due to an increase of the surface curvature. Therefore, species from the gas phase will be absorbed in the particle at a lower rate, resulting in a lower nanowire growth rate. Strikingly, R_{GaP} becomes diameter dependent for smaller interdistance (Fig. **5b**) and increases with particle size until materials competition becomes relevant. This peculiar phenomenon will be discussed in the following paragraph.

3) *Catalyst surface fraction enhanced (synergetic) growth regime.* At intermediate and small interdistance, $P < 3$ µm at a temperature of 500 °C, R_{GaP} increases with decreasing nanowire interdistance, strongly contradicting surface diffusion models that predict a decrease in growth rate with decreasing nanowire interdistance. [59] These results show that another effect plays a crucial role in determining the nanowire growth rate, that is, the nanowire growth rate is enhanced by nearest neighbours. R_{GaP} increases from 11 to 43 nm/s with interdistance decreasing from 6 to 0.7 µm for the largest particles. In this range, the surface collection areas for nanowire material do not overlap (Fig. **5a**), but the wires are still in close proximity such that interaction *via* diffusion can occur. R_{GaP} increases linearly from 10 to 40 nm/s by a catalyst surface fraction increase from 0 to 0.01, corresponding to a 300% increase of material available for nanowire growth. This indicates the catalytic action of the Au alloy.

In the model (Fig. **6**) Borgström *et al.* [62] propose that the gallium precursor, trimethylgallium (TMG), decomposes on the surface of a gold alloy catalyst particle into a secondary species believed to be monomethylgallium, MMG. TMG decomposes stepwise *via* dimethylgallium (DMG) and MMG, which is known to be relatively stable [63-65]. This catalysed decomposition of TMG on the Au alloy surface leads to a locally enhanced vapour pressure of MMG around the catalytic particles such that MMG can either fully pyrolyse and incorporate in the connected wire, or can desorb (Fig. **6a**). The irreversible decomposition process in an MOVPE system leads to an almost unavoidable re-adsorption of desorbed species, which incorporate by competing thin film growth. Alternatively, they can diffuse to a neighbouring wire where they are re-adsorbed by another catalyst particle and synergetically contribute to wire growth (Fig. **6b**). An increased surface fraction results in a higher local MMG partial pressure, and stronger diffusion. Obviously, in this regime (Fig. **6c**), the probability that TMG and MMG will be adsorbed outside the nanowire collection areas increases with smaller surface fractions, and the nanowire growth rate is limited by competing thin film growth on the GaP substrate.

The Arrhenius activation energies were determined for different surface fractions. For the largest surface fractions (in the competitive regime) E_a is 66 kJ/mole, which is significantly below 95 kJ/mole as observed for random, medium density, GaP nanowire growth on GaP substrates [extracted from Arrhenius plot in reference 66]. These results show the true catalytic action of the Au alloy, and that although the growth rate is limited by material competition, the synergy between the gold particles plays an important role for the nanowire growth rate. We argue that in this regime, almost all of the Ga species pyrolyse completely into Ga atoms *via* one or more Au catalyst particles, and are incorporated in a nanowire.

In this paragraph we have shown that the metal particle catalyses the growth of GaP wires and the indispensable need for surface patterning in order to correctly interpret catalyst-size related nanowire growth phenomena has been demonstrated.

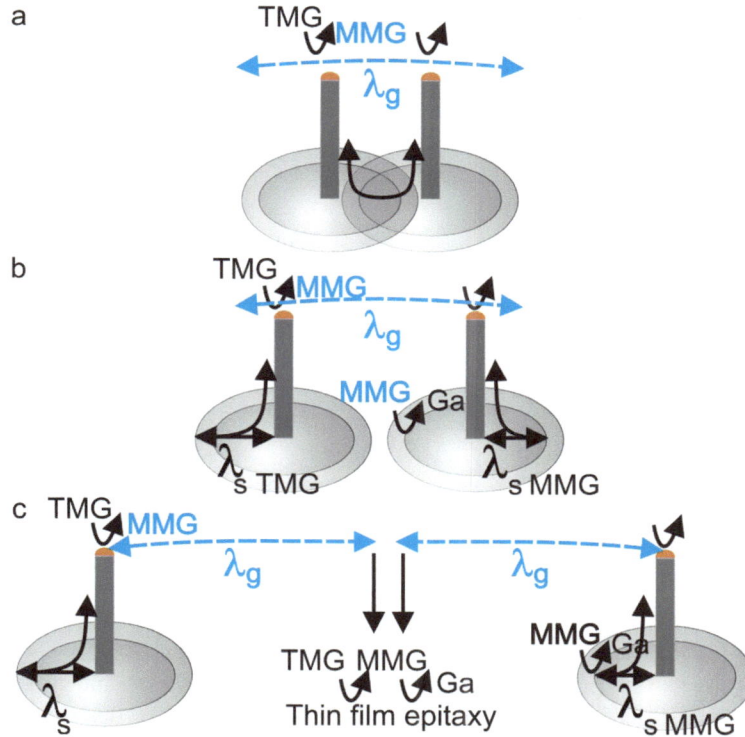

Figure 6:. Schematic nanowire growth model. **(a)** Competitive regime, where surface collection areas are defined by the surface diffusion length ls of TMG and partly decomposed species overlap. **(b)** Synergetic regime, where wire-to-wire gas-phase interaction occurs and surface collection areas become fully separate. TMG is catalytically decomposed to MMG at a Au particle. Part of the MMG is desorbed and diffuses through the gas phase to another Au particle, enhancing wire growth. **(c)** Independent regime, where the growth rate is determined by material surface diffusion to the wire. TMG and MMG landing outside the respective surface collection areas of the wires contribute to competitive substrate epitaxy. From ref. 62. Copyright 2007 Nature Publishing Group.

CRYSTAL STRUCTURE

InP and GaP have a cubic (zinc blende) crystal structure in the bulk. However, with the VLS mechanism zinc blende, but also wurtzite InP and GaP nanowires have been grown. Different crystal structures have different optoelectronic properties, such as their bandgap [67, 68]. Crystal structure and planar defects are closely related, since a twinned layer (stacking fault) in a zinc blende (wurtzite) crystal can be considered as a monolayer of the wurtzite (zinc blende) crystal structure. There has been significant progress in the understanding of the nanowire growth mechanism in recent years [45-48, 69]. It has been shown that the crystal structure and the defect density can be tuned by the wire diameter, impurity dopants, growth temperature, the III-V ratio or by a combination of these parameters. It is not yet clear, however, *how* these parameters affect the structural properties.

Control of the crystal structure can be exploited to fabricate homojunctions, *i.e.* junctions of the same material, but with a different crystal structure. This is a new type of band engineering, which has not been possible in bulk materials, and, for example, ZB segments could be fabricated in a WZ wire or vice versa. A WZ/ZB interface in InP has been categorized as a type II junction [68, 70, 71], and superlattice and QD structures could be fabricated with interesting and novel optical properties. In the following we will discuss three experimental routes to control the crystal structure of InP and GaP nanowires: by controlling the wire diameter, the chemical composition of the catalyst particle and by the temperature.

Effects of the Wire Diameter on the Crystal Structure

Bulk InP has the ZB crystal structure, because the free energy is slightly lower ($\Delta E = 6.8$ meV/III-V atom pair) [72] for ZB than for WZ InP. However, nominally undoped InP nanowires grown from Au particles commonly exhibit the wurtzite crystal structure. A possible explanation for the formation of WZ nanowires is the lower surface energy of the parallel side facets of WZ wires compared to that of ZB wires [72]. This effect would make crystallisation in the WZ phase especially favourable for thin wires that have a large surface to bulk ratio.

Algra *et al.* have grown InP nanowires with different diameters from colloidal gold particles (10, 20, 50, 100, and 200 nm) [47]. Undoped InP wires (with diameters from 10 up to 250 nm) consistently have the WZ structure, though in general contain many stacking faults. With increasing diameter the number of stacking faults decreases, leading to wires with a larger fraction of WZ. Other studies [73] suggest that thicker (50 nm) wires have a slightly higher ZB crystal structure fraction compared to thinner (20 nm) wires. We note, however, that for both studies randomly positioned colloidal Au particles have been used with possibly a variation in catalyst density. As shown in paragraph 3.4.3 the supersaturation strongly depends on the catalyst density and may thus vary across the substrate for these studies, inducing spread in the data.

These observations at least indicate that the surface energies of the wire side facets do not ultimately determine the crystal structure of the wires. The proposed model by Akiyama [72] is based on the surface energies of fully developed nanowires, *i.e.* after wire growth. Dubrovskii *et. al.* [74] claimed that the transformation of the wire crystal structure is controlled by the nucleation kinetics instead of the thermodynamics. They find that the crystal structure strongly depends on the supersaturation in the liquid particle.

Effects of Chemical Composition of the Catalyst Particle on the Crystal Structure

The main difference between bulk and VLS growth is the presence of the catalyst particle from which the crystal is precipitated, and therefore the atomic interactions at the liquid–solid interface should be considered. It was found that a parameter that critically determines the nanowire crystal structure and the stacking fault density is the chemical composition of the catalyst particle near the liquid-solid interface. The high number of planar stacking faults in the undoped InP WZ nanowires can be reduced by adding sulphur (S) to the gas phase. At the highest S partial pressure of $4.2*10^{-3}$ mbar ($8.3*10^{-7}$ mbar corresponds to a free electron concentration of $3 \cdot 10^{18}$ cm^{-3} in our nanowires) [75] a perfect WZ crystal without stacking faults was obtained as shown in Fig. **7**.

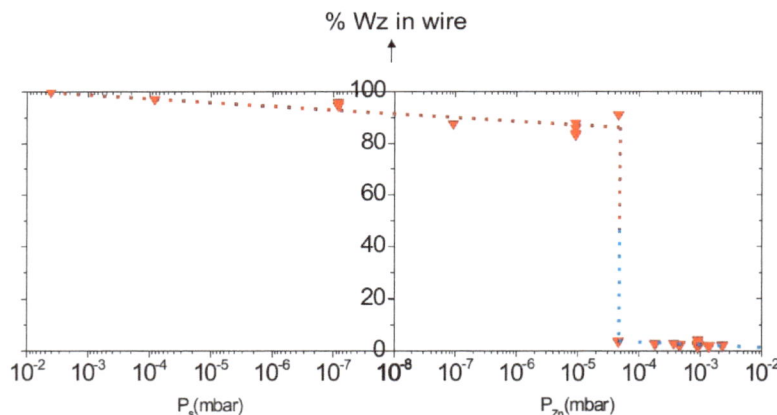

Figure 7: Graph showing the fraction of the wurtzite crystal structure in InP wires as a function of the Zn (right side) and S (left side) concentration. Intrinsic InP wires exhibit a WZ crystal structure with stacking faults. At a Zn partial pressure of $4.6*10^{-5}$ mbar the crystal structure of the nanowire abruptly changes from Wz to ZB. High sulfur concentrations lead to a perfect Wz structure without any stacking faults. From Ref. 47. Copyright 2008 Nature Publishing Group.

Diethylzinc (DEZn) is introduced in the growth system to establish p-type doping in the nanowires. Importantly, when sufficient Zn is added to the system, the nanowires precipitate in the ZB crystal structure (Fig. **7**) [47]. We find a sharp transition from WZ to ZB at a DEZn partial pressure of $4.6*10^{-5}$ mbar ($4.6*10^{-4}$ mbar corresponds to a free hole concentration of 10^{18} cm^{-3} in the InP nanowires) [75]. Very similar effects have been obtained by using another Zn-precursor, dimethylzinc (DMZn) [76], indicating that the Zn, and not carbon from the ligands, induces the change of crystal structure. Upon addition of ZnSe also zinc blende InP wires have been obtained [77].

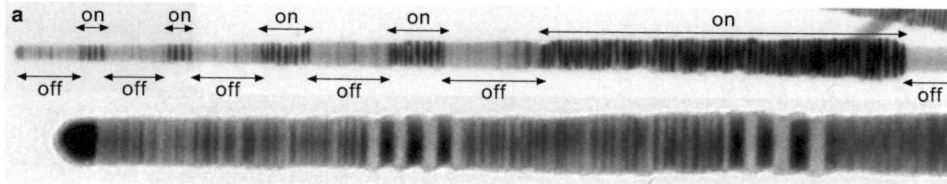

Figure 8: (a) An overview TEM image of a wire containing segments of intrinsic InP (containing randomly distributed stacking faults in a wurtzite structure) and Zn doped segments (with different lengths) with periodic twin planes in a zinc-blende structure. The 'on' represents the p-doped segments where diethylzinc is added, whereas the 'off' represents the intrinsic, undoped, segments. **(b)** Higher magnification TEM image of the segments closest to the gold particle. From Ref. 47. Copyright 2008 Nature Publishing Group.

Control of the crystal structure by impurities allows for the formation of more complex structures by varying the Zn concentration during growth. Without Zn, random stacking faults in a WZ crystal structure should occur, and with Zn present the zinc blende structure should be obtained. This is indeed the case as demonstrated by the TEM images in Figs. **8a** and **8b**, showing a wire for which intermittently a Zn partial pressure of $9.2*10^{-4}$ mbar has been used during growth. The tapering of the nanowire is due to sidewall growth, which preferentially occurs on the ZB sections.

The main effect of adding zinc during growth is a decrease in $\Delta\gamma/\gamma_{sl,ZB}$, which means a lowering of the liquid-solid step energy for ZB as compared to WZ. This suggests a strong interaction of the zinc atoms with the InP growth interface as was also observed during electrical resistance measurements of Au(Zn)-InP contacts [78,79]. In general, for nanowires grown by the VLS mechanism the crystal structure may be intrinsically different from that of the bulk material and will depend on the combination of semiconductor with catalyst material.

Importantly, Mattila *et al.* [36] reported that InP wires grown from Au/In particles have the ZB or WZ crystal structure (depending on growth temperature), whereas wires from In-droplets exhibit the pure (stacking-fault free) wurtzite crystal structure. Another indication that impurities affect the solid-liquid interface comes from the fact that hollow zinc blende InP tubes can be grown by the VLS mechanism at high temperatures [80-82]. The transition temperature from solid to hollow depends on the type and concentration of the added dopants, such as Zn, S, Se, and Te. This substantiates that the chemical composition of the catalyst particle near the liquid-solid interface is important for the wire crystal structure.

Effects of Temperature on the Morphology and Crystal Structure

To our best knowledge there is not (yet) a systematic study on how the InP wire crystal structure depends on the temperature. For GaP a detailed study on the structure and morphology, depending on V/III ratio and temperature, using TEM tomography has been reported [83]. GaP wires grown at different temperatures and precursor flows were characterised. All wires were studied in TEM rotating them around their long axis. Upon viewing along the <11-2> direction, straight sidewalls running parallel to the long axis of the wire were observed for all growth conditions, in agreement with previous studies [84]. However, upon viewing along the <-110> direction a growth parameter dependent morphology was observed. The obtained morphologies are summarised in the structure diagram in Fig. **9**.

At temperatures below 500°C, the GaP segments show the zinc blende crystal structure, whereas at higher temperatures, a mixture of domains with zinc blende and wurtzite are observed. In total, five different types of nanowire morphologies could be discriminated, which are all represented in Fig. **9** and will be discussed separately below:

Type 1

For wires grown at low growth temperatures (400°C<T≤460°C) and low V/III precursor flow ratios (≤9), a zig-zag pattern of non-parallel facets bounded at the twin boundaries is visible upon viewing along the <-110> direction, The wire is terminated by {111} facets, in accordance with literature [84].

Type 2

At higher growth temperatures (460°C<T<530°C) wires with straight edges were observed for both the <-110> and <11-2> viewing directions. In this case, additional HRTEM imaging of cross-sectional samples showed that the GaP nanowire core has six {112} side facets [61]. This sixfold symmetry cross-section is predominant only at 460 and 480°C, and intermediate V/III ratios.

Types 2' and 2"

Interestingly, for higher V/III ratios at 460°C and at higher growth temperatures (460°C<T<530°C) one of the two types of {112} facets becomes more stable, leading to a threefold symmetry with three larger and three smaller {112} facets. This morphology was confirmed by HAADF imaging upon viewing along the <-110> direction close to the top. The intensity profiles across the wire were consistent with a threefold symmetry shape [83].

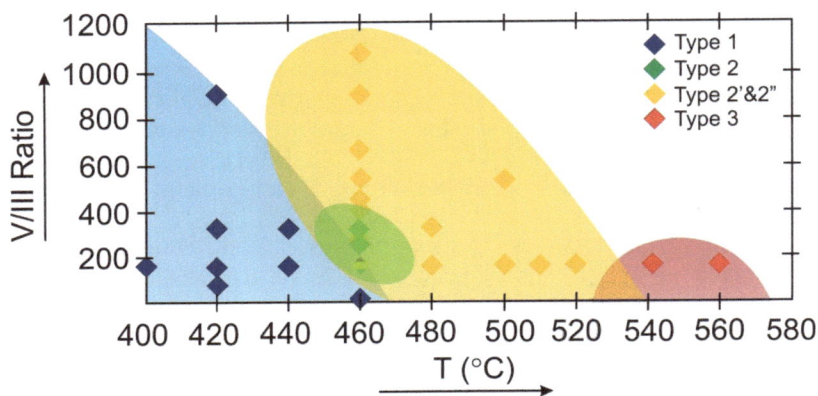

Figure 9: GaP wire morphologies as a function of temperature and V/III precursor flow ratio. All type 1 and 2 wires have the zinc blende crystal structure, and type 3 has the wurtzite structure. Reprinted with permission from Ref. [83]. Copyright 2007 American Chemical Society.

The variation in stability of the two types of {112} side-facets as a function of group V overpressure can be explained by the polar nature of these facets. An increased concentration of active P, induced by the temperature and/or PH_3 flow, will stabilise the P-terminated $\{112\}_B$ facets, resulting in a cross-sectional shape with larger $\{112\}_B$ than $\{112\}_A$ planes. This is consistent with the sixfold-to-threefold morphology change as a function of temperature. In the temperature regime used for our experiments, the PH_3 pyrolysis in presence of III-V material (in this case the nanowire) increases superlinearly with temperature [85], which indicates that the V/III ratio is the only parameter that shows a systematic correlation to the morphology changes. This is indicated in Fig. **9** by the two alternatives 2' and 2".

Type 3

In the temperatures range 530°C<T<560°C, the GaP segments contain a mixture of zinc blende and wurtzite domains. Both have sixfold symmetry and they are bounded by six {112} facets (Fig. **9**) and six {100} facets respectively. The facet orientation of the wurtzite segments was deduced from the equivalent orientation in the zinc blende domains.

The morphology determining parameter for the formation of type 1, 2 or 3 is temperature. This can be understood from the temperature dependence of the dominant mechanism for thin film growth: In the temperature range 400-460°C, sidewall growth is limited by a thermally activated process [15], most likely being the incorporation of growth units at step edges, leading to the formation of flat facets. In this case the most stable facets, being the {111} facets, will dominate the growth form. In the temperature range 460-520°C sidewall growth becomes a diffusion-limited process [61]. Facets parallel to the growth direction are now more likely to develop.

One should keep in mind that the morphologies described above refer to the 'pure-VLS' morphology, *i.e.* without sidewall growth. Generally, sidewall growth on ZB nanowires favours the development of flat {111} side-facets, independent on the type of morphology (1, 2, 2', 2"), yielding the 'zig-zag' pattern of non-parallel facets described above. Sidewall growth occurs in an epitaxial manner, *i.e.* stacking faults present in the core are continued in the shell, also for hetero-epitaxy.

TWINNING IN NANOWIRES

Crystal imperfections or defects may affect the opto-electronic properties of nanowires [67-71]. Twin planes are the dominating stacking fault for III-V nanowires, which preferentially grow in the [111]B direction. Randomly positioned twin planes form naturally as a consequence of the relatively low twin-plane energy (about 10 mJ m^{-2} for InP and 20 mJ m^{-2} for GaP) [86]. This energy barrier is easily overcome at the typical nanowire growth temperatures and the relatively high supersaturations used in general. Twins planes have been reported in nanowires made of various materials, independent of the fabrication method, and are also abundant in bulk III-V semiconductors. The presence and distribution of such defect planes can potentially be employed to tailor the band structure of a nanowire consisting of a single material with a superstructure [87, 88]. This implies that it is of substantial value to explore and control the mechanism behind the formation of twin planes.

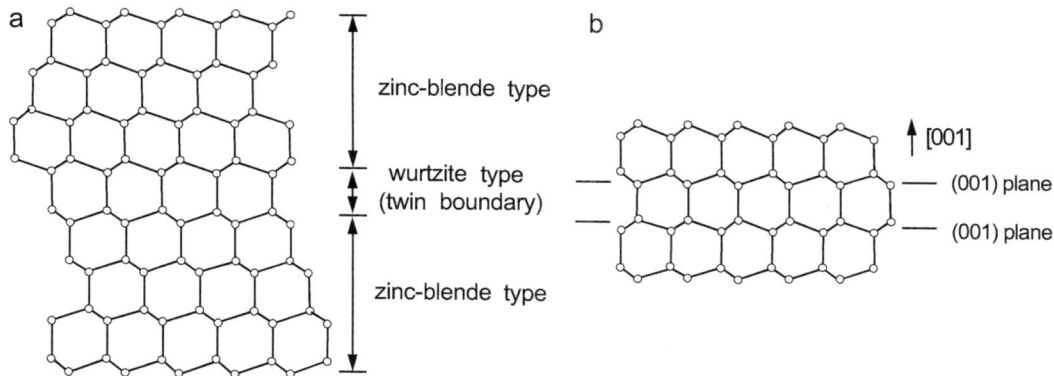

Figure 10: Crystal structure model of **(a)** a twin plane in a zinc blende crystal, showing that a monolayer of the wurtzite structure is formed at the twin plane, **(b)** wurtzite crystal structure formed by a twin at every (111) zinc blende plane. Reprinted with permission from Ref. [15], Journal of Applied Physics, Vol. 77, 447-462. Copyright 1995 American Institute of Physics.

A twin is a rotational-symmetric defect with respect to the [111] axis. The crystal structure is rotated by 180° around this axis at a twin boundary as shown in Fig. **10a**. Hiruma *et al.* [15] suggested that a [111] rotational twin contributes to the mechanism of crystal structure change. Note that a monolayer of the wurtzite type crystal is generated at the twin boundary, as indicated in Fig.**10a**. Thus, if such a rotational twin occurs at every (111) plane of zinc-blende, the crystal type changes completely from zinc-blende to wurtzite, as shown in Fig. **10b**. As a result, (111) zinc blende planes are transformed into (001) wurtzite planes. Because zinc-blende and wurtzite are based on face-centered cubic and hexagonal closed packing, respectively, both types are maximally packed and the potential energy in the former's (111) planes and in the latter's (001) planes are very close to each other.

In the zinc-blende lattice, four symmetrically equivalent <111> directions are available for rotational twin formation. These directions can be subdivided in i) the <111> direction parallel to the <111>$_B$ growth direction and ii) the three other <111> directions making an angle of 109° with the growth direction. Although both types of twin geometries have been observed, the former one appears far more frequent than the latter one. This is consistent with a layer-by-layer growth mechanism. For wurtzite, the <001> growth direction has no symmetry equivalent; stacking faults of the close-packed layers can only be present perpendicular to this direction.

Using TEM, twin boundaries in the zincblende lattice can be visualized by imaging the nanowires slightly off the <-110> zone axis; an alternating pattern of bright and dark regions is visible, as shown in Fig.**11b**.

Single and Paired Twins

A twin plane can be formed because the formation energy of a twinned layer in InP and GaP is only slightly higher than that of a normal, untwinned, layer. The formation of a twin plane is a statistical process. The statistics of random twinning have been investigated by Johansson *et al.* [84]. They found that the shortest segment lengths (number of monolayers between neighbouring twins) are the most abundant. Usually a twin plane occurs in an isolated position, *i.e.* without another one present in its immediate vicinity. These *single* twin planes (indicated by the yellow arrows in Fig. **11a**) separate nanowire sections of opposite crystal orientation, as the high resolution TEM micrograph of Fig. **11f** shows, which are clearly visible as the alternating bright and dark sections in the bright-field image (Fig. **11b**). (note: Fig. **c-d-e** not cited in the text).

Besides random, single twins also *'paired'* twin planes can occur. In this case, a first twin plane is formed at a random position (referred to as 'random twin plane'), which is followed by the formation of a second twin plane of which the position is directly related to that of the first one. Twin pairs are present in reported data for GaP [89,90] and also in other materials systems like GaAs [91-93]. At many positions within a long wire section, a change of contrast occurs, visible as a thin stripe in Fig. **11b**. These stripes represent short segments between two twin planes in close proximity, which we define as a twin *pair*. The distance between the twins planes of a 'pair' is not constant but scales with the nanowire diameter D, indicating that the occurrence of the second twin plane is not caused by some physical interaction, like a strain field, with the first twin plane, but is induced by the evolution of the wire cross sectional shape initiated by the occurrence of the first twin [48].

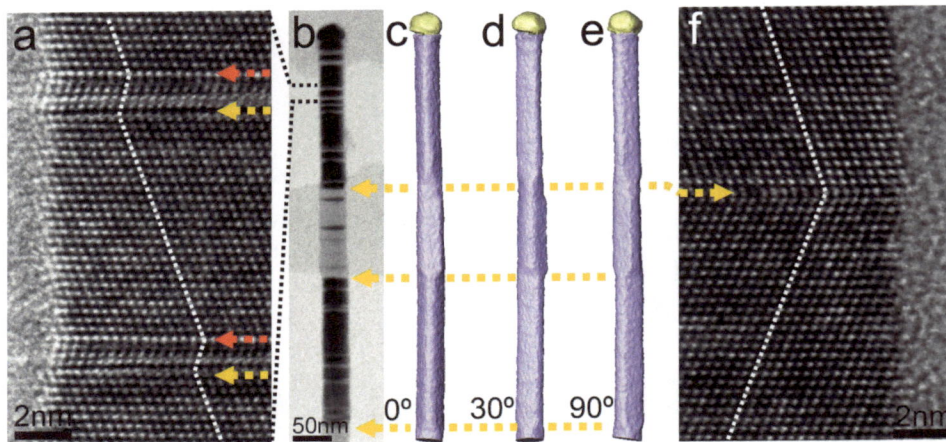

Figure 11: TEM images of a multiply twinned GaP nanowire. a,f. HRTEM micrographs of two paired twin planes **(a)** and one single **(f)** twin plane imaged along the <110> zone axis, showing the zinc blende crystal structure. The first (random) and the second (paired) twin planes are indicated by yellow and red arrows, respectively. **(b)** Bright field TEM image of the nanowire revealing single and paired twins. The transitions in contrast indicate the positions of the twin planes. The bright and dark stripes are short segments between two twins spaced closely together, forming a 'twin pair'. **(c),(d),(e):** A surface rendering representation of a tomographic reconstruction of a STEM/HAADF tilt series. Three viewing directions are displayed, ranging from parallel to perpendicular to {11-2} side facets. For 0°, 30° and 90° this results in the [11-2], [01-1], and [-110] viewing directions, respectively. Clearly, the single twins induce 180° rotations of the truncated triangular cross-section, whereas in the regime of twin pairs the overall morphology remains unchanged. Reprinted with permission from Ref. [48]. Copyright 2010 American Chemical Society.

Modeling of Nanowire Morphology and Twins

VLS nanowire growth is characterized by the presence of a liquid droplet which is in contact with the nanowire top facet. Twin planes in nanowires are atomically flat, suggesting that each crystal plane originates from a single nucleation event. It has been argued [84] that, since the group V species have a very low solubility in the Au particle, the nucleation starts at the triple (vapor-liquid-solid) phase interface. Following Glas *et al.* [46], formation of the nucleus at the triple phase junction is favorable for another reason: it eliminates a portion of the *preexisting* droplet surface. This largely outweighs the replacement of part of the lateral nucleus-liquid interface by a possibly slightly costlier nucleus-vapor interface.

Thus, growth proceeds *via* a layer-by-layer process, starting with the formation of a two-dimensional nucleus at one of the edges of the nanowire-droplet interface, followed by rapid lateral growth of the layer [46,47,74]. Addition of a layer with a specific crystal orientation and specific external facets thus requires that an edge nucleus is formed with corresponding crystal orientation and corresponding outer (solid-vapour) facet. The free energy of formation, ΔG^*_{XY}, of an edge nucleus formed at nanowire side facet X, depends on the contact angle δ_{XY} of the liquid catalyst droplet with respect to the outer facet Y, with tilt angle θ_Y, of the nucleus, according to [46,47,74,94],

$$\Delta G^*_{XY} = b\,h\,\frac{[(1-\alpha)\gamma_{SL} + (\alpha/\cos\theta_Y)(\gamma_{SV} - \gamma_{LV}\cos\delta_{XY})]^2}{[\Delta\mu - \gamma_{nN}/h]}, \qquad (1)$$

where γ_{SL}, γ_{SV}, and γ_{LV} are the solid-liquid, solid-vapour, and liquid-vapour surface energies, γ_{nN} is the interface energy between nucleus and nanowire which is equal to the twinning energy γ_T for a nucleus inducing a twin plane and is zero otherwise, $\Delta\mu$ is the supersaturation (difference in chemical potential per unit volume between liquid and solid), α is the fraction of the perimeter of the nucleus which is in contact with the vapour and b is a geometrical constant (for a hexagonal nucleus $\alpha=1/6$ and $b = 2\sqrt{3} \approx 3.46$). The crucial point is that the droplet contact angles at the various nanowire side facets depend on the shape of the (droplet-nanowire) growth interface. As the shape of the growth interface changes, the droplet distorts in order to minimize its surface energy. It has been found [47] that the contact angles, which characterize the droplet distortion, depend linearly on the interface deformation. Specifically, the contact angle is smaller at the narrow side facet edges than at the broad side facet edges, *i.e.* the droplet leans over at the broad edges.

Nucleation probabilities. The probability of formation for each type of nucleus, p_{XY}, in general obeys

$$p_{XY}(\phi) \propto \exp\left[\frac{-\Delta G^*_{XY}(\phi)}{k_B T}\right]. \qquad (2)$$

So the probability depends on the shape of the growth interface and therefore changes slowly during growth when the interface shape evolves [47,48]. The interface shape changes when the side facets are not parallel to the growth direction. Quite generally, this leads to a competition between several different nuclei. The gradual evolution of the various probabilities explains why the nanowire may switch between different types of side facets during growth. To appreciate this competitive layer-by-layer growth process one should keep in mind that each probability $p_{XY}(\phi)$ reaches its maximum value when $\delta_{XY}(\phi) = 0$ [see equation (1)], *i.e.* when the droplet surface is tangent to the outer facet of the nucleus. In addition the formation of a nucleus which induces a twin plane costs an extra twin energy, γ_T, which increases ΔG^*_{XY} [see equation (1)], leading to an overall reduction of the associated probability. It is possible, combining equations (1) and (2), to calculate the formation probabilities of the potentially relevant nuclei as a function of the deformation ϕ of the nanowire-droplet interface. With this model the nanowire morphology, the occurrence of single and paired twins and the formation of a twinning superlattice has been explained.

Twinning Superlattice

A significant feature would be to have a constant spacing between rotational twins in the wires such that a twinning superlattice is formed, as this is predicted to induce a direct bandgap in normally indirect bandgap semiconductors [87,88], such as silicon and gallium phosphide. Optically active versions of these technologically relevant semiconductors could have a significant impact on the electronics and optics industry. Twinning superlattices in nanowires have been reported for several systems, such as ZnSe [95], ZnS [96], HgSe [97], InP [47,77], GaP [47], and InAs98. As will be discussed below the zinc blende crystal structure and the non-parallel {111} wire side facets are essential for the formation of the periodic structures [47].

A route to induce this specific morphology in InP and GaP nanowires is by addition of Zn to the gas phase, as discussed above. Having adequate control over the nanowire crystal structure, we can now address the formation of twinning superlattices. In Fig. **12**, transmission electron micrographs (TEM) are shown of Zn-doped ($9.2*10^{-4}$ mbar) InP nanowires with nominal diameters of 10, 20, 50 and 100 nm. It is clear from the overview images in Fig. **12a** that the periodically twinned structure is general, although not all of the wires have the optimal orientation with respect to the electron beam. The segment length is uniform (Fig. **12b**) throughout the wires. From the high-resolution images in Fig. **12c** we observe that the periodic nanowires have the ZB crystal structure with {111} side facets, which are not parallel with respect to the long nanowire axis. From the high-resolution images the number of monolayers between successive twin planes was counted. Segment lengths of 7 ± 2, 13 ± 2, 25 ± 3 and 33 ± 6 monolayers were found for wires with a diameter of 10, 20, 50 and 100 nm, respectively. Importantly, the periodicity in twinning is demonstrated by the relatively narrow distributions in segment lengths.

For the formation of the twinning superlattice it is crucial that the {111}A and {111}B side facets are tilted in opposite directions (by $\theta = \theta_B = -\theta_A \approx 19.5°$) with respect to the nanowire axis (see Fig. **13a**). As shown schematically in Fig. **13**, at a certain moment during growth (situations 1 and 3) the top surface of the nanowire is a hexagon, and the shape of the catalyst droplet, connected to this surface, is close to spherical. When the wire grows, the {111}A edges move inward and their length increases, while the {111}B edges move outward and their length decreases. Thus the shape of the nanowire-droplet interface becomes increasingly triangle-like, as shown for situations 2 and 4 in Fig. **13b**. This induces the catalyst droplet to distort so as to minimize its surface area, leaning over towards the long {111}A edges. At a certain point it becomes more favourable to form a twin plane and to start reducing the distortion of the catalyst particle by re-growth towards a hexagonal shape, rather than to continue growth towards a completely triangular shape. This mechanism of inverting triangularly shaped interfaces repeats itself continuously and produces the periodically structured wire.

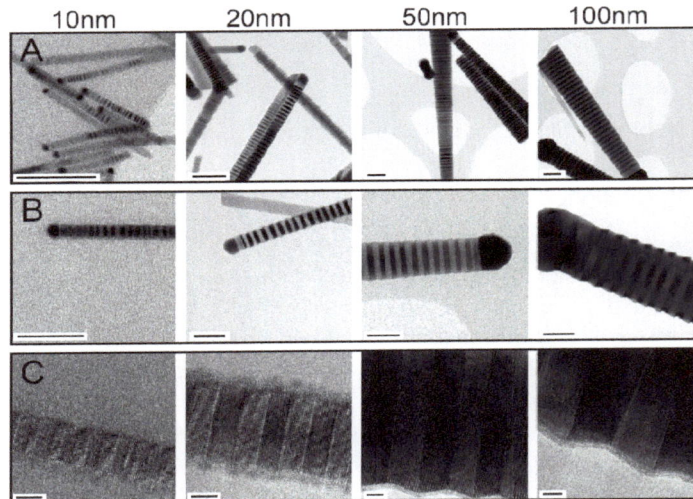

Figure 12: (a), (b) Overview and **(c)**, high resolution TEM images of Zn-doped InP nanowires with a diameter of nominally 10, 20, 50, and 100nm. The scalebars correspond to (a), 100 nm, (b), 50 nm, and (c), 5 nm. From Ref. 47. Copyright 2008 Nature Publishing Group.

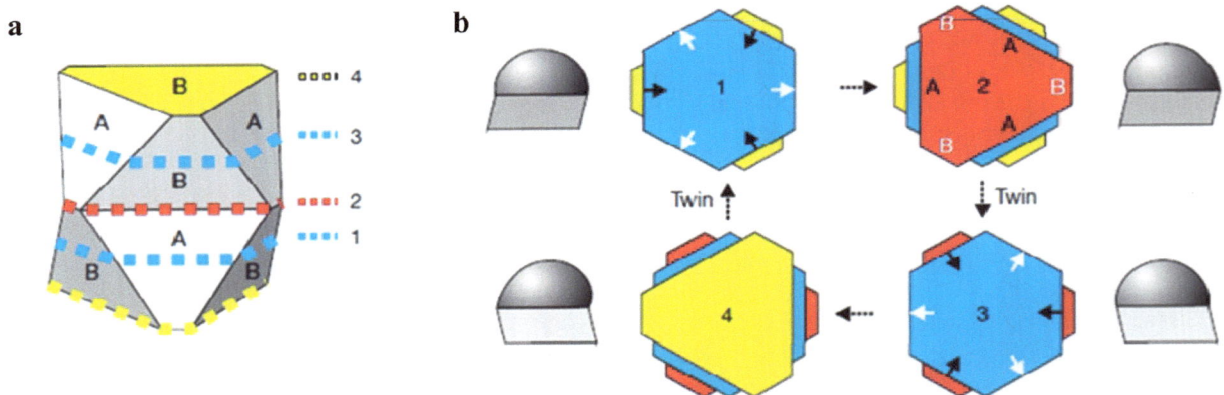

Figure 13: (a) Schematic representation of the morphology of a twinned nanowire with the zinc blende crystal structure with non-parallel {111} side facets. **(b)** The cross-sectional shapes of the top facet of the nanowire crystal at the solid–liquid interface during growth. The numbers correspond to the positions indicated in a). The corresponding calculated shape of the catalyst particle on a hexagonal (1 and 3) and a triangularly deformed (2 and 4) interface is depicted. From Ref. 47. Copyright 2008 Nature Publishing Group.

In order to understand this process in more detail, Algra *et al.* [47] developed a quantitative model based upon a specific mechanism relating the distortion of the droplet to the growth process. Nanowire growth proceeds layer by layer with a single nucleation event per layer, initiated at the nanowire edge [46,47,74]. The free energy of formation of the nucleus depends upon the difference between the liquid-solid contact angle and the tilt angle of the external facet of the nucleus. This dependence has been analysed quantitatively, making use of the simulator program "Surface Evolver" [99,100]. It follows that the formation of nuclei with an external (solid-vapour) A-facet is strongly suppressed. As a result, the dominant processes are nucleation of external-B-facet nuclei at either B edges, adding another facet-conserving ZB layer, or at A edges, introducing a layer that involves a twin plane and initiating re-growth. During facet-conserving growth the liquid-solid contact angles increase (decrease) at the A edges (B edges) due to the progressing droplet distortion illustrated in Fig. **13**. We find that the contact angles depend linearly on the ratio of the wire height $H = Nh$ (with N the number of monolayers measured from the last hexagonal cross section, and h the layer thickness) and the wire diameter D. From this, it follows that the difference in free energy of the two competing nucleation processes, $\delta\Delta G^*$, decreases linearly with wire height as $\delta\Delta G^* = \delta\Delta G_0^* - A\,Nh/D$. As a result, we find that the critical number of monolayers, N_c, at which twin formation becomes the more favourable process, is proportional to the wire diameter D.

DEFECT-FREE NANOWIRES

Twin planes are typical for wires grown along the [111] direction. Defect-free nanowires are defined as nanowires presenting a monotypic crystal structure without any rotational twins or stacking faults along the length of the nanowire. GaP and InP nanowires grown by MOVPE under standard growth conditions usually present a mixed crystal structure along the [111] growth direction composed of wurtzite and zinc blende segments. Nanowires may exhibit a defect-free crystal structure under certain growth conditions. As an example, Mattila *et al.* reported the growth of pure wurtzite InP nanowires grown above 430°C using Au catalyst [36]. Naji *et al.* [27] for InP nanowires grown on (001) Si by MBE with SrTiO$_3$ buffer layer reported pure wurtzite nanowires.

Stacking-fault free InP nanowires can be grown on (001)InP [101]. These nanowires can grow in the [001] direction if the catalyst is only saturated with In and P coming from the vapor phase and not from the substrate. The [001] InP nanowires have a high crystalline perfection in comparison with [111] InP nanowires. The reason for a high number of stacking faults in usual [111] InP nanowires comes from the small energetic differences for hexagonal or cubic stacking sequences in the <111> direction. Consequently, the stacking faults can freely end at the nanowires' side facets as planar defects perpendicular to the growth direction. The formation of similar defects during growth in [001], however, needs to overcome an activation barrier for the creation of Frank partial dislocation [101].

IMPURITY DOPING

For most semiconductor devices, quantitative control of impurity doping is essential. Impurity doping has not been studied extensively for nanowires in general. There are a few papers reporting intentional impurity n- and p-type doping of InP nanowires [75, 102-104]. Naturally, intrinsic InP tends to be n-type and it is challenging to make p-type InP wires. For p-type doping the group II element Zn [75,76,102-104] has been used and for n-type doping group IV elements Si [25], Sn [76], and group VI elements S [75], Se [80], and Te [102, 103] have been used. Precursors are used for Zn (DEZn [75] and DMZn [76]), S (H$_2$S) [75], and Sn (TESn)[76]. Si, Se and Te are produced by evaporation from elemental sources.

The dopant type has been analysed by contacting nanowires as a field-effect transistor. Successful n- and p-type doping has been reported. In addition, p-n junctions have been shown for crossed [102] InP wires and within a single wire [75,76,103] (see Fig. **14**). These devices show the typical diode behavior expected from a p-n junction and electroluminescence when put in the forward direction. The electroluminescence spectrum shows an emission band corresponding to the bandgap energy of InP, confirming that the light arises from the p-n junction. In addition, electric field microscopy, performed with an AFM tip, shows that the potential drop and the electroluminescence are at the same position [75].

It is not yet clear *via* which mechanism the different impurities are incorporated in nanowires. It has been shown that during vertical (VLS) growth also radial (VS) growth can take place that leads to tapering. Dopants can be incorporated *via* these two different growth mechanisms. By employing infrared near-field optical microscopy, the free-carrier distribution in InP nanowires with doping modulation (intrinsic/sulfur) along the axial and radial directions has been mapped with nanoscale resolution [105]. It was found that the incorporated free electron concentration is higher in the shell than in the core, similar to what has been observed for Ge NW [106]. This shows that S is incorporated more effectively *via* the VS than the VLS growth mechanism.

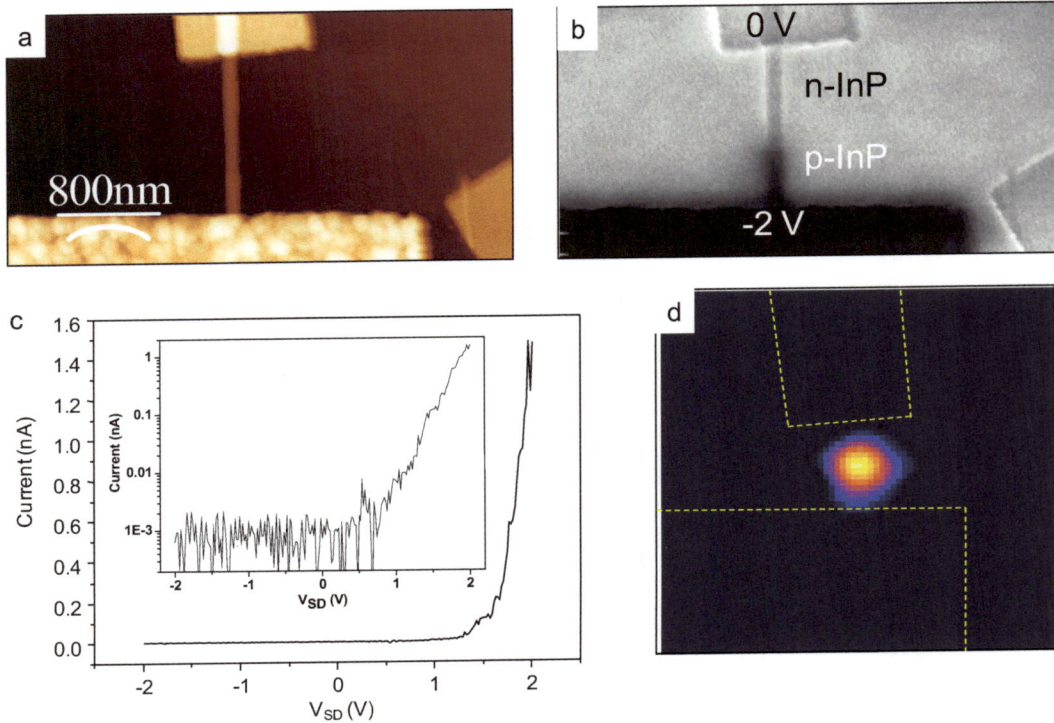

Figure 14: (a) AFM height image of a LED device. The p-InP/n-InP NW is electrically contacted on the left and right side by Ti/Zn/Au and Ti/Al, respectively. The surface roughness at the p-type electrode is due to Zn grain formation. A step-like change in NW diameter, from 55 to 45 nm, corresponds to the location of the p-n junction. **(b)** Surface potential map of a device in reverse bias (top) and line trace along NW axis (bottom). The step in surface potential corresponds to the depletion region of the p-n junction. **(c)** Room temperature *I-V* characteristic of a NW LED device. The inset shows data on a log plot. **(d)** Electroluminesence from the nanowire when put in the forward bias. Reprinted with permission from Ref. [75]. Copyright 2007 American Chemical Society.

The incorporation mechanism of Zn atoms in InP wires has been investigated by using p-n (Zn-S) doped wires [107]. The wires were radially etched to remove the shell, which was grown by the VS mechanism during VLS growth. After etching the wires were contacted and a clear diode characteristic was observed. This shows that, at least, Zn atoms are incorporated *via* the VLS mechanism.

The electron and hole concentrations are estimated from the same type of field-effect devices yielding n_e >3×10^{18}cm^{-3} for sulphur doped (H$_2$S molar fraction=1.7×10^{-6})[75] wires and $n_h\approx10^{18}$cm^{-3} for Zn-doped (DEZn molar fraction=3.6×10^{-4})[75] InP wires. For similar wires n_e was quantified with SNOM measurements at 7.6×10^{18} cm^{-3} [105]. The electron concentration has been quantitatively modulated in the range 10^{17}-10^{19} cm^{-3} with Sn dopants (TESn molar fraction in the range 1.0×10^{-6} - 3×10^{-5}) [76]. We are not aware of reports on doping of GaP wires.

OPTICAL CHARACTERIZATION

Nanowire optics is a particularly exciting topic thanks to the ability to control the light emission by changing the composition along the nanowire axis. In particular, nanowires synthesized from direct bandgap semiconductors are promising materials for subwavelength nanophotonic devices for the generation, waveguiding, and detection of light at the nanoscale. There are more details on optics and optoelectronics in semiconducting nanowires in ref. [108]. Here, we will focus on the optical properties of InP nanowires.

Single Wires

Photoluminescence spectroscopy is a classical instrument to assess the quality and impurity concentration of semiconductors. In 2001, Wang *et al.* [109] investigated the polarization dependence of the photoluminescence from single InP nanowires. InP nanowires exhibit a giant polarization anisotropy for diameters ranging from 10 to 50 nm as shown in Fig. **15**. When the excitation polarization is parallel to the nanowire axis, the PL emission of the nanowire is maximum, whereas the PL emission is minimum when the excitation is perpendicular to the nanowire [109].

The polarized light emission from single semiconductor nanowires has been modified by coupling this emission to surface plasmon polaritons on a metal grating [110] The polarization anisotropy of the emitted photoluminescence from single nanowires is compared for wires deposited on silica, a flat gold film, and a shallow gold grating. By varying the orientation of the nanowire with respect to the grating grooves, the large intrinsic polarization anisotropy can be either suppressed or enhanced. This modification is interpreted by the appearance of an additional emission channel induced by surface plasmon polaritons and their conversion to *p*-polarized radiation at the grating [110].

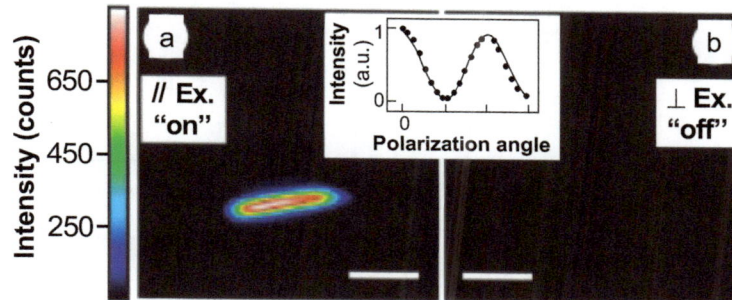

Figure 15: (a) PL image of a single 20-nm InP nanowire with the exciting laser polarized along the wire axis. Scale bar, 3μm. **(b)** PL image of the same nanowire as in (a) under perpendicular excitation. Intensity scale is identical to (a). Inset: variation of overall photoluminescence intensity as a function of excitation polarization angle with respect to the nanowire axis. The PL images were recorded at room temperature with integration times of 2 s. Reprinted with permission from *Science* (www.sciencemag.org), ref 109. Copyright 2001 AAAS.

Photoluminescence measurements have been performed by van Weert *et al.* [104] on individual nanowires at low temperatures and at low excitation intensities. *P*-type InP wires show a considerable redshift ($\Delta\lambda$=200 meV) and a large linewidth (FWHM≈70meV) compared to bulk InP emission. The observed redshift and large linewidth are a result of a built-in electric field in the nanowires. This bandbending is induced by Fermi-level pinning at the nanowire surface. Upon increasing the excitation intensity, the typical emission from these *p*-InP wires blueshifts with 70 meV/ decade, due to a reduction of the bandbending induced by an increase in the carrier concentration. For intrinsic and *n*-type nanowires, several impurity- related emission lines were found.

Several groups have used photoluminescence spectroscopy to investigate intrinsic InP nanowires in more detail at both room temperature and low temperature. In general, undoped InP wires show a large spectral line width (80-120 meV), even at low temperatures, and emission at energies above the bulk zinc blende InP bandgap has been observed. In the first instance this blue shift of InP wires, with diameters ranging from 10-50 nm, has been explained by size quantization in the radial direction of the wires [111]. Importantly, quantum confinement is ruled out by Mohan *et al.* [28] and Mattila *et al.* [38] for InP nanowires grown on an InP substrate with a mean diameter in the range 50–150 nm, which show a PL blue-shift of about 60-80 meV compared to the substrate. Based on x-ray diffraction and transmission electron microscopy measurements, it is found that the as-grown vertically oriented nanowires have crystallized in the wurtzite lattice instead of in the zinc blende structure, which results in a blue-shifted PL.

A clear example of the emission of pure phase nanowires is shown in Fig. **16** which presents the emission spectra of (stacking-fault free) InP nanowires in the zinc blende structure and wurtzite structure taken at 4 K. The nanowire, showing only strong band-gap-related luminescence at 1.41 eV (thick line, Fig.**16a**) is clearly identified as a [001] wire in the zinc blende crystal structure [101]. On the other hand, a pure wurtzite InP nanowire arrays exhibited a single intense peak at 1.49 eV (Fig.**16b**) [28] Note that both pure crystal phases exhibit strong luminescence compared to WZ/ZB structures, suggesting stronger band-gap-related luminescence due to the higher crystalline quality. InP nanowires generally exhibit stacking fault and present a zinc blende/wurtzite mixture. Theorically, the band alignment is type II, *i.e.* the band extrema are found in different layers of the heterostructure. Several authors carefully studied the properties of these nanowires. Bao *et al.* [70] have developed a technique to perform both transmission electron microscopy and microphotoluminescence on the same nanowire, thus allowing to directly correlate structural and optical properties. The excitation power-dependent blue-shift of the photoluminescence of a InP wire was explained in terms of the predicted type-II staggered band alignment of the zinc blende/wurzite InP heterostructure and of the indirect transitions between localized electron and hole states responsible for radiative recombination.

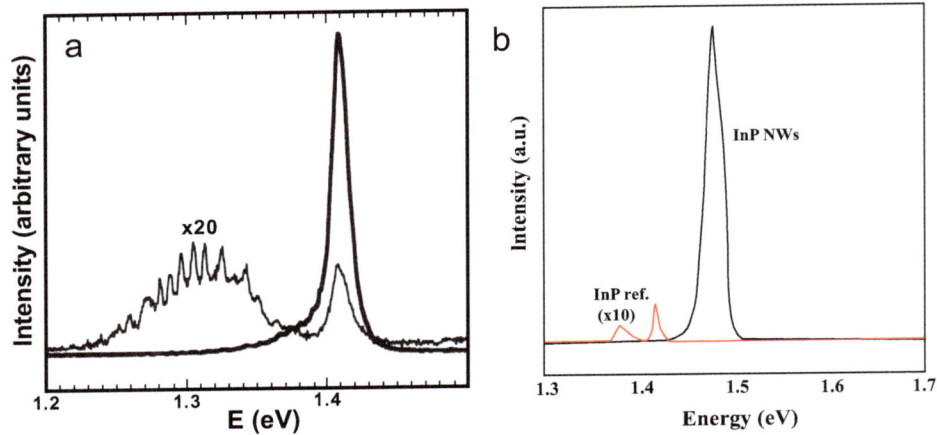

Figure 16: Low temperature (4 K) phololuminescence spectra of **(a)** a pure zinc-blende InP nanowire and **(b)** a pure wurtzite InP nanowire. Reprinted with permission from Ref. [101], Applied Physics Letters, Vol. 85, 2077-2079. Copyright 2004 American Institute of Physics.Reprinted with permission from Ref. [28]. Copyright 2005 IOP Publishing.

The carrier dynamics of quantum confined excitons in InP nanowires containing both wurtzite (WZ) and zincblende (ZB) crystalline phases have been investigated in detail by Pemasiri *et al.* [71] using time-resolved photoluminescence. It was found that the recombination lifetime is nearly 2 orders of magnitude longer for electrons and holes that are strongly confined in quantum wells defined by zinc blende monolayers in an overall wurtzite nanowire (indirect transition) than for excitons from pure wurtzite segments (direct transition).

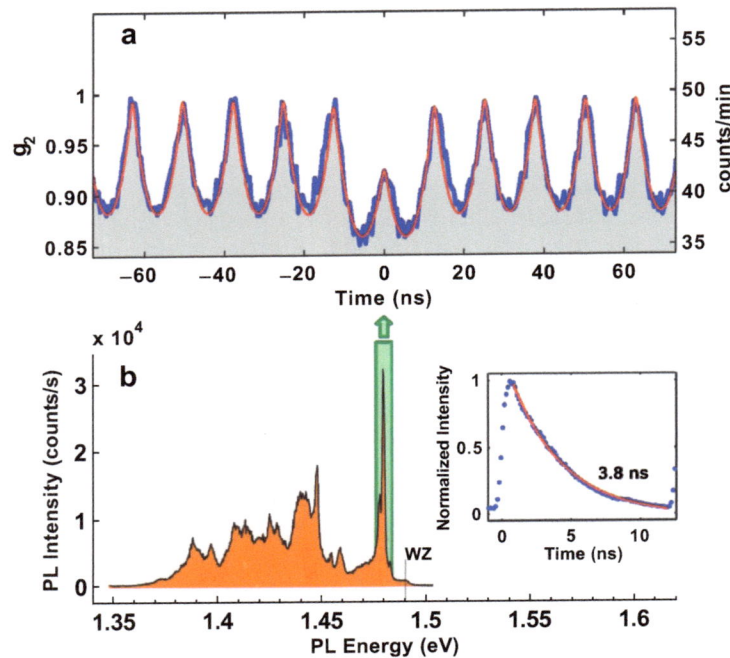

Figure 17 : Crystal phase quantum dot. **(a)** Second-order autocorrelation measurement of photons emitted from a few crystal phase QDs under pulsed excitation. The red line is a fit to the model discussed in the text. **(b)** PL of a single NW. The green area schematically represents a band-pass transmission filter with spectral width of 6 meV. This part was spectrally selected from the rest of the PL for autocorrelation measurements in (a). The inset shows the decay curve measured from a QD emitting at 1.48 eV.

The red solid line is a monoexponential fit to the experimental data. Reprinted with permission from Ref. [68]. Copyright 2010 American Chemical Society.

The ability to control rotational twinning and to introduce a heterostructure in a chemically homogeneous (all InP) nanowire material and alter its optical properties opens new possibilities for band-structure engineering. Akopian *et al.* [68] have fabricated and studied single quantum dot devices defined solely by crystal phase in a chemically homogeneous nanowire, so-called homodots, and observed single photon generation as shown in Fig. **17**. They demonstrated the quantum nature of crystal phase QDs, by performing photon correlation measurements under pulsed excitation (Fig. **17(a)**). Correlation events missing at time 0 suggest that the emission originates from a few single photon emitters. In this experiment, the number of single photons emitted in a single spot is 4 with a lifetime of 5 ns. Fig. **17(b)** shows photoluminescence of a single NW, where the green area schematically represents the emission peak spectrally selected for autocorrelation measurements in Fig. **17(a)**. The inset shows the lifetime measurements from a QD emitting at 1.48 eV with a lifetime of 3.8 ns. This new type of single photon emitters allows for applications like quantum memories and solar cells, benefiting from this clean, defect-free system with sharp monolayer interfaces.

There are several reports [112,113] on surface effects in the optical properties of InP nanowires. It is shown that the room-temperature photoluminescence intensity is increased by 2-3 orders of magnitude after the hydrofluoric acid surface treatment, and that there is also a significant increase in the double-exponential photoluminescence decay time.

Wire Ensembles

Semiconductor nanowires are nanostructures with a diameter much smaller and a length much larger than the wavelength of visible light. This large anisotropy, combined with the high refractive index of some semiconductors, leads to fascinating optical phenomena. Giant polarization anisotropy for the absorption and emission of light was first reported by Wang *et al.* [109] by measuring the photoluminescence of individual InP nanowires.

Optical birefringence describes the difference in refractive index of light with different polarizations traveling inside an anisotropic material. Next to the natural birefringence of crystals, artificial birefringence occurs in materials containing aligned macroscopic scatterers. The phenomenon has been observed in various systems such as aligned carbon nanotube films [114] and porous silicon [115]. Muskens *et al.* [116] demonstrate giant birefringence in ensembles of randomly grown, vertically aligned semiconductor nanowires. The birefringence in the material is shown to depend on the nanowire length. Short nanowires with a volume filling fraction around 50% exhibit the strongest birefringence, in agreement with effective medium theories. A large difference is found between the in-plane and out-of-plane refractive indices of $n=0.8$, exceeding by a factor of 75 the natural birefringence of quartz and by more than a factor of 2 that of inverted artificial materials.

An array of nanowires is strongly photonic, resulting in a significant coupling mismatch with incident light due to multiple scattering. A design principle has been found by Muskens *et al.* [117] for the effective suppression of reflective losses, based on the ratio of the nondiffusive absorption and diffusive scattering lengths. Using this principle, suppression of the hemispherical diffuse reflectance of InP nanowires to below that of the corresponding transparent effective medium has been demonstrated. The design of light scattering in nanowire materials is of large importance for optimization of the external efficiency of nanowire-based photovoltaic devices. In addition, Diedenhofen *et al.* [118] demonstrated by measuring the transmission and reflection of GaP nanorods on top of a GaP substrate that the reduction of the reflection in these layers is mainly caused by a graded-refractive-index coating and interference in the nanorod film. The broadband reduction of the reflection is demonstrated by direct transmission and specular reflection measurements at wavelengths in the visible and near-infrared regions, and by angle-integrated total reflection and transmission measurements.

Muskens *et al.* [119] demonstrate that GaP nanowire arrays are one of the strongest optical scattering materials to date Fig. **18**. Strong Mie-type internal resonances of the nanowires have been found, which can be tuned over the entire visible spectrum. The tunability of nanowire materials opens up exciting new prospects for fundamental and applied research ranging from random lasers to solar cells, exploiting the extreme scattering strength, internal resonances, and preferential alignment of the nanowires. Although the sudy has been focused on gallium phosphide nanowires, the results can be universally applied to other types of group III-V, II-VI, or IV nanowires.

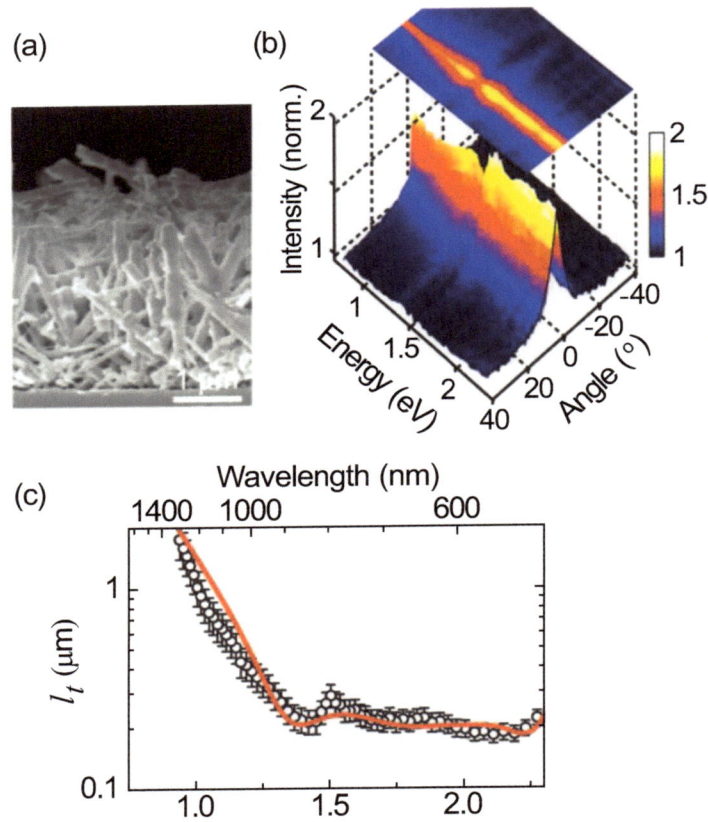

Figure 18: Investigation of strongly scattering, compressed nanowire material. **(a)** SEM image of the compressed nanowire layer. Scale bar 1 μm. **(b)** Enhanced backscattering (EBS) spectrum from 0.9 to 2.2 eV of a 4.5 μm thick nanowire layer. Internal nanowire resonances are observed as a modulation of the width and amplitude around 1.4 eV. A projection of the EBS spectrum is shown in the color-density map above **(b)**. Reprinted with permission from Ref. [119]. Copyright 2009 American Chemical Society.

HETEROSTRUCTURES

Heterostructured nanowires based on GaP and InP are mainly fabricated with their arsenic counterparts. The phosphides have higher bandgaps making them ideal as a surface passivation layer in a radial configuration or as an electronic barrier in the longitudinal direction.

GaP/GaAs heterostructures have been used to study the fundamentals of the VLS growth mechanism and to distinguish radial from axial growth [16,61,120]. The transition sharpness between the different materials has been investigated as a function of the growth parameters. [121] Borgstrom *et al.* [122] demonstrated for the first time that a short segment in a wire can be an optical quantum dot. This was shown by exciton/biexciton emission and anti-bunching behaviour from short GaAs segments in GaP wires.

InP/InAs heterostructured wires have been grown with great precision [18] and used for resonant tunneling and memory devices. Quantum dot emission and access to spin polarization has been demonstrated from InAsP dots in InP wires.

REFERENCES

[1]   Wang Z.L. and Song. J.H. *Science* **2006**, 14, 242.

[2]   Hochbaum A. *et al. Nature* **2008**, 451, 163.

[3]   Boukai Y., Bunimovich J., Tahir-Kheli J.K, Goddard W.A. III, Heath J.R., *Nature* **2008**, 451, 168.

[4]   Huang M.H., Mao S., Feick H., Yan H.Q.,. Wu Y.Y, Kind H., Weber E., Russo R., Yang P., *Science* **2001**, 292, 1897.

[5]   Doh Y.J., van Dam J. A., Roest A. L., Roest E. L., Bakkers E. P. A. M., Kouwenhoven L.P., De Franceschi S., *Science* **2005**, 309, 272.

[6]   J. A. Van Dam, Y. V. Nazarov, E. P.A.M. Bakkers, S. De Franceschi and L.P. Kouwenhoven, *Nature* **2006**, 442, 667.

[7]   Cui Y., Wei Q., Park H., Lieber C.M., *Science* **2001**, 293, 1289.

[8]   Bao J., Zimmler M.A., Capasso F., Wang X.and Ren Z. F., *Nano Lett.* **2006**, 6, 1719.

[9]   Tian B.Z., Zheng X.L., Kempa T.J., Fang Y., Yu N.F., Yu G.H., Huang J.L., Lieber C.M., *Nature* **2007**, 449, 885.

[10]   Colombo C., Hei M., Grätzel M., and Fontcuberta i Morral A., *Appl. Phys. Lett.* **2009**, 94, 173108.

[11]   Kelzenberg M.D., Boettcher S.W., Petykiewicz J.A., Turner-Evans D.B., Putnam M.C., Warren E.L., Spurgeon J.M., Briggs R.M., Lewis N.S., Atwater H.A. *Nature Materials* **2010**, 9, 239.

[12]   Bryllert T., Wernersson L.-E, Fröberg L.E., and Samuelson L., *IEEE Electron Dev. Lett.* **2006**, 27, 323.

[13]   Huang Y., Duan X., Cui Y., Lauhon L. J., Kim K.H., Lieber C.M., *Science* **2001** 294, 1313.

[14]   Wagner R. S., Ellis W. C, *Appl. Phys. Lett.* **1964**, 4, 89.

[15]   Hiruma, K. *et al. J. Appl. Phys.* **1995**, 77, 447-462.

[16]   Gudiksen M.S., Lauhon L.J, Wang J., Smith D.C., Lieber C.M., Nature *2002*, 415, 617.

[17]   Kuykendall T, Ulrich P., Aloni S. and Yang P., *Nature Materials* **2007**, 6, 951.

[18]   Bjork M. T., Thelander C., Hansen A. E., Jensen L. E., Larsson M. W., Wallenberg L. R., Samuelson L., *Nanoletters* **2004**, 4, 1621.

[19]   Verheijen M.A., Immink G., de Smet T., Borgstrom M.T., and Bakkers E.P.A.M., *J. Am. Chem. Soc.* **2006**, 128, 1353.

[20]   http://www.ioffe.rssi.ru/SVA/NSM/Semicond/InP/index.html

[21]   http://www.ioffe.rssi.ru/SVA/NSM/Semicond/GaP/index.html

[22]   Duan, X. & Lieber, C. M. *Adv. Mat.* **2000**, 12, 298-302.

[23]   Tchernycheva, M.; Cirlin, G. E.; Patriarche, G.; Travers, L.; Zwiller,V.; Perinetti, U.; Harmand, J.-C. *Nano Lett.* **2007**, 7, 1500–1504.

[24]   Björk, M.T.; Ohlsson, B.J.; Sass, T.; Persson, A.I.; Thelander, C.; Magnusson, M.H.; Deppert, K.; Wallenberg, L.R.; Samuelson, L. *Nano Lett.* **2002**, 2, 87

[25]   Rigutti,L,Bugallo,AD,Tchernycheva,M,Jacopin,G, Julien, FH, Cirlin, G, Patriarche, G, Lucot, D, Travers, L, Harmand, JC, *J. of Nanomaterials* **2009**, 435451.

[26]   Tchernycheva M, Travers L, Patriarche G, Glas F, Harmand JC, Cirlin GE, Dubrovskii VG, *J. Appl. Phys.* **2007**, 9, 094313.

[27]   Cheng J, Regreny P, Largeau L, Patriarche G, Mauguin O, Naji K, Hollinger G, Saint-Girons G., *J. Cryst. Growth 2009*, 311, 1042-1045.

[28]   Mohan P., Motohisa J. and Fukui T., *Nanotechnology* **2005,** 16, 2903–2907.

[29]   Kitauchi Y., Kobayashi Y., Tomioka K., Hara S., Hiruma K., Fukui T., and Motohisa J., *Nano Letters* **2010**,10, 1699-1703.

[30]   Motohisa, J.; Noborisaka, J.; Takeda, J.; Inari, M.; Fukui, T. *J. Cryst.Growth* **2004**, 272, 180–185.

[31]   Noborisaka, J.; Motohisa, J.; Fukui, T. *Appl. Phys. Lett.* **2005**, 86, 213102.

[32]   Tomioka, K.; Motohisa, J.; Hara, S.; Fukui, *Jpn. J. Appl. Phys.* **2007**,46, L1102–L1104.

[33]   Ooike N, Motohisa J, Fukui T, *J. Cryst. Growth.* **2004**, 272, 175

[34]   Stach E.A., Pauzauskie P.J., Kuykendall T., Goldberger J., He R.R., Yang P.D., *Nano Letters* **2003**,3, 867-869.

[35]   Novotny C.J., Yu P.K..L, *Appl. Phys. Lett* **2005**, 20, 203111

[36]   Mattila M., Hakkarainen T., Lipsanen H., Jiang H., Kauppinen E.I., *Appl. Phys. Lett.* **2006**, 89, 063119

[37]   Mattila M., Hakkarainen T., Lipsanen H., *J. Cryst. Growth* **2007**, 298, 640-643

[38]   Mattila M., Hakkarainen T., Jiang H., Kauppinen E.I., Lipsanen H., *Nanotechnology* **2007**, 18, 155301

[39]   Yu S.Z., Miao G.Q., Jin Y.X., Zhang L.G., Song H., Jiang H., Li Z.M., Li D.B., Sun X.J., *Physica E low dimensional systems & nanostructures* **2010**, 42, 1540-1543.

[40]   Parguel, V.; Favennec, P. N.; Gauneau, M.; Rihet, Y.; Chaplain, R. L'Haridon, H.; Vaudry, C. *J. Appl. Phys.* **1987**, *62*, 824-827

[41]   D. E. Perea, J. E. Allen, S. J. May, B. W. Wessels,D. N. Seidman, L. J. Lauhon, *Nano Letters* **2006**, 6, 181-185

[42]   Yakunin A.M., Silov A.Y., Koenraad P.M., Wolter J.H., Van Roy W., De Boeck J., Tang J.M., Flatte M.E., *Phys. Rev. Lett.* **2004**, 21, 216806

[43]   Tchernycheva M., Harmand J.C., Patriarche G., Travers L. and Cirlin G.E., *Nanotechnology* **2006**, 17, 4025–4030

[44]   Persson A.I., Larsson M.W., Stenström S., Ohlsson B.J., Samuelson L., Wallenberg L.R., *Nat. Materials* **2004**, 3, 677

[45]   Ross F.M., Tersoff J., Reuter M. C., *Phys. Rev. Lett.* **2005**, 95, 146104.

[46]   Glas, F., Harmand J.C. and Patriarche G., *Phys. Rev. Lett.* **2007**, 99, 146101-1-4

[47]   Algra R.E., Verheijen M.A., Borgström M.T., Feiner L.F., Immink G., van Enckevort W.J.P, Vlieg E., and Bakkers E.P.A.M., *Nature* **2008**, 456, 369-372

[48]   Algra R.E., Verheijen M.A., Feiner L.-F., Immink G.W, Theissmann R., van Enckevort W.J.P., Vlieg E., Bakkers E.P.A.M., *Nano Letters* **2010**, ASAP

[49]   Lind E., Persson M.P., Niquet Y.M., Wernersson L.E., *IEEE Transactions on electron devices* **2009**, 56, 201-205

[50]   van Tilburg J.W.W., Algra R.E., Immink W.G.G.,. Verheijen M.A, Bakkers E.P.A.M., Kouwenhoven L.P., *Semicond. Sci. Techno.* **2010**, 25, 024011

[51]   Holmes J. D., Johnston K.P., Doty R. C., Korgel B.A., *Science* **2000**, 287, 1471

[52]   Gudiksen M. S., Lieber C.M., *J. Am. Chem. Soc.* **2000**, *122,* 8801-8802

[53]   Mårtensson T., Borgstrom M., Seifert W., Ohlsson B.J., Samuelson L.,. *Nanotechnology.* **2003**, 14, 1255-1258.

[54]   Fan H.J., Lee W., Scholz R., Dadgar A., Krost A., *et al.*, *Nanotechnology.* **2005**, 16, 913.

[55]   Fan H.J., Fuhrmann B., Scholz R., Syrowatka F., Dadgar A., *et al.*, *Journal of Crystal Growth.* **2006**, 287, 34-38.

[56]   Mårtensson T., Carlberg P., Borgström M.T, Montelius L., Seifert W., *et al.*, *Nano Letters.* **2004**, 4, 699-702.

[57]   Pierret A., Hocevar M., Diedenhofen S. L., Algra R.E., Vlieg E., Timmering E.C., Verschuuren M.A., Immink G.W.G., Verheijen M.A., Bakkers E.P.A.M., *Nanotechnology* 2010, 21, 065305.

[58]   Stringfellow G. B., Organometallic Vapor-Phase Epitaxy:Theory and Practice Academic Press(New York), **1989**.

[59]   Jensen L.E., Bjork M.T., Jeppesen S., Persson A.I., Ohlsson B.J, *et al.*, *Nano Letters.* **2004**, 4, 1961-1964.

[60]   Givargizov E.I., *J. Cryst. Growth* **1975**, 31, 20.

[61]   Verheijen M. A., Immink G., deSmet T., Borgstrom M. T, Bakkers E. P. A. M., *J. Am. Chem. Soc.* **2006**, 128, 1353–1359.

[62]   Borgström M.T., Immink G., Ketelaars B., Algra R., Bakkers E.P.A.M., *Nature Nanotechnology* **2007**, 2, 541-544.

[63]   Denbaars, S. P., Maa, B. Y., Dapkus, P. D., Danner, A. D. & Lee, H. C.. *J. Cryst. Growth* **1986** 77, 188–193.

[64]   Nishizawa, J., Sakuraba, H. & Kurabayashi, T. *J. Vac. Sci. Technol. B* **1996**, 14, 136–146.

[65]   Shogen, S., Matsumi, Y., Kawasaki, M., Toyoshima, I. & Okabe, H. *J. Appl. Phys.* **1991**, 70, 462–468.

[66]   Seifert, W. *et al. J. Cryst. Growth* **2004**, 272, 211–220.

[67]   Spirkoska D., Arbiol J., Gustafsson A., Conesa-Boj S., Glas F., Zardo I., Heigoldt M., Gass M. H., Bleloch A. L., Estrade S., Kaniber M., Rossler J., Peiro F., Morante J. R., Abstreiter G., Samuelson L.and Fontcuberta i Morral A. *Physical Review B* **2009**, 80, 245325.

[68]   Akopian N., Patriarche G., Liu L., Harmand J.C., and Zwiller V. *Nano Letters* **2010**, 10, 1198-1201.

[69]   Dubrovskii V. G., Sibirev N. V. *Phys. Rev. B* **2008**, 77, 035414.

[70]   Bao J., Bell D.C., Capasso F., Wagner J.B. Mårtensson T., Trägårdh J., and Samuelson L., *Nano Letters* **2008**, 8, 836-841.

[71]   Pemasiri K., Montazeri M., Gass R., Smith L.M., Jackson H.E., Yarrison-Rice J., Paiman S., Gao Q., Tan H.H.,. Jagadish C, Zhang X.and Zou J., *Nano Lett.* **2009**, 9, 648-654.

[72]   Akiyama, T. *et al.* J. J. Appl. Phys. **2006**, 45, L275-L278.

[73]   Paiman S., Gao Q., Tan H.H., Jagadish C., Pemasiri K., Montazeri M.,. Jackson H.E, Smith L.M., Yarrison-Rice J.M., Zhang X., Zou J., *Nanotechnology* **2010**, 20, 225606.

[74]   Dubrovskii V.G., Sibirev N.V., Harmand J.C., Glas F., *Phys. Rev. B* **2008**, 78, 235301.

[75]   Minot E., Kelkensberg F., van Kouwen M., Borgström M.T., Zwiller V., Kouwenhoven L.P., Bakkers E.P.A.M., *Nano Letters* **2007**, 7, 367.

[76]   Borgström M.T., Norberg E., Wickert P., Nilsson H.A., Trägårdh J.,. Dick K.A, Statkute G., Ramvall P., Deppert K., Samuelson L., *Nanotechnology* **2008**, 19, 445602.

[77]   Shen G.Z., Bando Y., Liu B.D., Tang C.C., Golberg D., *J. Phys. Chem.* **2006**, 110, 20129-20132.

[78]   Malina, V. *et al.. Semicond. Sci. Technol.* **1994**, 9, 1523-1528.

[79]   Weizer, G.W. *et al.* NASA Technical Memorandum **1994**, 106590.

[80]   Bakkers E.P.A.M., Verheijen M.A., *J. Am. Chem. Soc.* **2003**, 125, 3440-3441.

[81]   Tang C.C., Bando Y., Liu Z.W., Golberg D., *Chem. Phys. Lett.* **2003**, 376, 676-682.

[82]   Yin L.W., Bando Y., Golberg D., Li M.S., *Appl. Phys. Lett.* **2004**, 85, 3869-3871.

[83]   M.A. Verheijen, R.E. Algra, M.T. Borgström, G. Immink, E. Sourty, W.J.P.van Enckevort, E.Vlieg, and E.P.A.M. Bakkers, *Nano Letters* **2007**,7, 3051-3055.

[84]   Johansson, J. *et al.* Nat. Mater. **2006**, 5, 574-580.

[85]   Larsen, C. A.; Buchan, N. I.; Stringfellow, G. B. *J. Cryst. Growth* **1987**, *85*, 148-153.

[86]   H. Gottschalk, G. Patzer, and H. Alexander, *Phys. Stat. Sol (a)* **1978**, *45*, 207-217.

[87]   Ikonic, Z. *et al.. Phys. Rev. B* **1993**, 48, 17181–17193.

[88]   Ikonic, Z. *et al. Phys. Rev. B* **1995**, 52,14078–14085.

[89]   Dick K.A., Kodambaka S., Reuter M.C., Deppert K., Samuelson L., Seifert W., Wallenberg L.R. and Ross F.M., *Nano letters* **2007**, 7, 1817-1822.

[90]   Xiong Q., Wang J., and Eklund P. C., *Nano Lett.* **2006**,*6*, 2736–2742.

[91]   Zou J., Paladugu M., Wang H., Auchterlonie G.J., Guo Y.N., Kim Y., Gao Q., Joyce H.J., Tan H.H., Jagadish C., *Small* **2007**,*3*, 389-393.

[92]   Mikkelsen A., Skold N., Ouattara L., Borgstrom M., Andersen J.N., Samuelson L., Seifert W. and Lundgren E., *Nature Materials* **2004**,*3*, 519-523.

[93]   Joyce H.J., Gao Q., Tan H.H., Jagadish C., Kim Y., Fickenscher M.A., Perera S., Hoang T.B., Smith L.M., Jackson H.E., Yarrison-Rice J.M., Zhang X.and Zou J., *Nano Lett.* **2009**, *9*, 695–701.

[94]   Joyce H.J., Wong-Leung J., Gao Q., Tan H.H. and Jagadish C., *Nano Lett.* **2010**, *10*, 908–915.

[95]   Li Q. *et al. Adv. Mater.* **2004**, 16, 1436–1440.

[96]   Hao, Y., Meng, G., Wang, Z. L., Ye, C. & Zhang, L. *Nano Lett.* **2006**, 6, 1650–1655.

[97]   Qin, A. *et al. Adv. Mater.* **2008**, 20, 768–773.

[98]   Caroff P. *et al.*, Nature Nanotechnology **2009**, 4, 50.

[99]   Brakke, K.E. The surface evolver. *Experimental Mathematics* **1992**, 1, 141-165.

[100]   http://www.susqu.edu/brakke/evolver/evolver.html.

[101]   Krishnamachari U. *et al. Appl. Phys. Lett.* **2004**, 85, 2077.

[102]   Duan X., Huang Y., Cui Y., Wang J., Lieber C.M., *Nature* **2001**, 409, 66.

[103]   Gudiksen M.S., Lauhon L.J., Wang J., Smith D.C., Lieber C. M., *Nature* **2002**, 415, 617.

[104]   van Weert M. H. M., Wunnicke O., Roest A. L., Eijkemans T. J., Silov A. Y, Haverkort J. E. M., t' Hooft G. W, and. Bakkers E. P. A. M, *Appl. Phys. Lett.* **2006**, 88, 043109.

[105]   Stiegler J.M. *et al. Nano Letters* **2010**, *10*, 1387-1392.

[106]   Perea, D. E.; Hemesath, E. R.; Schwalbach, E. J.; Lensch-Falk, J. L.; Voorhees, P. W.; Lauhon, L. J. *Nat. Nanotechnol.* **2009**, *4*(5), 315.

[107]   van Weert M.H.M. *et al. J. Am. Chem. Soc.* **2009**, *131*, 4578.

[108]   Agarwal R.and Lieber C.M. Applied Physics A: Materials Science & Processing 2006, 85, 3.

[109]   Wang J., Gudiksen M.S., Duan X., Cui Y., Lieber C.M., *Science* **2001**, *293*, 1455.

[110]   Muskens O. L., Treffers J., Forcales M., Borgström M. T., Bakkers E. P. A. M., and Gómez Rivas J., *Optics Letters* **2007**, *32(15)*, 2097.

[111]   Gudiksen M. S., Wang J. W., and Lieber C. M., *J. Phys. Chem. B* **2002**, 106, 4036.

[112]   van Vugt L. K., Veen S. J., Bakkers E. P. A. M., Roest A. L., and Vanmaekelbergh D., *J. Am. Chem. Soc.* **2005**, 127, 12357.

[113]   Mattila M. *et al. Appl. Phys. Lett.* **2007**, 90, 033101.

[114]   de Heer W. A., Bacsa W. S., Chatelain A., Gerfin T., Humphrey-Baker R., Forro L., and Ugarte D., *Science* **1995**, 268, 845.

[115]   Genereux F., Leonard S. W., Van Driel H. M., Birner A., and Gösele U., *Phys. Rev. B* **2001**, 63, 161101.

[116]   Muskens O. L., Borgström M. T., Bakkers E. P.A.M., Gómez Rivas J., *Appl. Phys. Lett.* **2006**, *89*, 233117.

[117]   Muskens O.L., Rivas J.G., Algra R.E., Bakkers E.P.A.M., and Lagendijk A., *Nano Letters*, **2008**, 8 (9), 2638–2642.

[118]   Diedenhofen S.L., Vecchi G., Algra R.E., Hartsuiker A., Muskens O.L., Immink G., Bakkers E.P.A.M., Vos W.L., Rivas J.G., *Advanced Materials* **2009**, 21, 1-6.

[119]   Muskens O. L., Diedenhofen S. L., Algra R., Bakkers E. P. A. M., Kaas B., Lagendijk A. *Nano Letters* **2009**, 9, 930-934.

[120]   Mohseni P.K. *et al. J. Appl. Phys.* **2009**, 106, 124306

[121]   Borgström M.T., Verheijen M.A., Immink G., de Smet T., Bakkers E.P.A.M. *Nanotechnology* **2006**, *17*, 4010.

[122]   Borgström M. T. *et al. Nano Letters* **2005**, 5, 1439.

CHAPTER 4

Growth of III-Arsenide/Phosphide Nanowires by Molecular Beam Epitaxy

Jean-Christophe Harmand*, Frank Glas, Gilles Patriarche, Ludovic Largeau, Maria Tchernycheva, Corinne Sartel, Linsheng Liu and Fauzia Jabeen

Laboratoire de Photonique et de Nanostructures, CNRS Route de Nozay, 91460 Marcoussis, France

Abstract: We describe and analyze the growth of III-V semiconductor nanowires by molecular beam epitaxy activated with gold particles. We focus on (Al)GaAs(P) and InP(As) compounds. Optimal conditions of substrate surface preparation and adequate growth parameters are reported. The catalyst particles are shown to be rich in group-III atoms and liquid when nanowires form. The favorable growth temperatures thus depend on which group-III element is used. We explain why the nanowires often adopt the unusual wurtzite structure, pointing out that the crystalline phase is determined at the nucleation stage and that high liquid supersaturation is necessary. The kinetics of nanowire elongation is investigated with an original method based on modulations of the incident fluxes. The group-III flux intercepted by the nanowire sidewall facets is the main contribution to axial growth. Conditions leading to a short diffusion length of the group-III adatoms on these facets, produce lateral growth that we use to form abrupt core-shell heterostructures. We also present a method to bury vertical freestanding nanowires by a planar epitaxial growth. This process induces the layer-by-layer transformation of the nanowire phase from wurtzite to zinc-blende. Finally, the statistics of nucleation at the liquid-solid interface is revealed by using the flux modulation method. We show that the diluted concentration of group-V atoms in the nano-sized catalyst drop is at the origin of the self-regulation of the nucleation events.

Keywords: Semiconductor nanowire, molecular beam epitaxy, vapor liquid solid growth, nucleation, catalyst-assisted growth, crystal phase, growth kinetics, core-shell heterostructures.

INTRODUCTION

To a large extent, making one dimensional (1D) semiconductor materials is much more difficult than fabricating quantum wells or quantum dots. For all these objects, one needs to control at least one of their dimensions at the nanometer scale. For a 1D material, the difficulty is to achieve uniform shaping and uniform material composition along the third and unconfined dimension. Indeed, small fluctuations along the long axis of a 1D crystal can result in the formation of quantum dots and thus move us away from the ideal 1D object from which new and largely unexplored behaviors are expected [1]. This challenging semiconductor quantum wire fabrication has been explored for several decades with different approaches [2-5]. In the late 1990s, growth assisted by nano-sized catalyst particles emerged as a very powerful means to obtain 1D-like structures. Since that time, the term "nanowires" (NWs) has superseded those employed by the early pioneers [5-9] of catalyst-assisted growth: filamentary crystals, whiskers, rods. Exciting developments from leading groups [10,11] have stimulated extended investigations in this field worldwide.

The present chapter concentrates on the epitaxial growth of III-V NWs elaborated by molecular beam epitaxy (MBE) in presence of nano-sized Au catalyst particles deposited on the substrate prior to growth. Most of our investigations rely on detailed structural analyses of the samples by transmission electron microscopy (TEM) and related techniques: transmission electron diffraction (TED), high resolution TEM (HRTEM), energy dispersive X-ray spectroscopy (EDXS), high angle annular dark field (HAADF) scanning TEM (STEM). We derive basic information on the NW formation which allows us to discuss thermodynamic, kinetic and statistical aspects of this peculiar growth. Several mechanisms are highlighted. Among them, some are specific to the MBE technique and some are valid independently of the growth method. In terms of materials, the scope of this study is restricted to arsenide and phosphide compounds. Nevertheless, some aspects can be transposed to other material systems. As will be seen, the presence of at least two semiconductor constituents, the group-III and group-V elements, leads to subtle

---

***Address correspondence to Jean-Christophe Harmand:** Laboratoire de Photonique et de Nanostructures, CNRS Route de Nozay, 91460 Marcoussis, France. E-mail: jean-christophe.harmand@lpn.cnrs.fr

Jianye Li, Deli Wang and Ray R. LaPierre (Eds)

mechanisms, more complex than those involved in Si or Ge NW formation, but sometimes very beneficial. All our samples are fabricated on ($\bar{1}\bar{1}\bar{1}$) oriented GaAs or InP substrates. This orientation is by far the most commonly used because it corresponds to the most favorable growth direction of catalyst-assisted NWs. Efforts to obtain different NW orientations did not encounter much success so far and will not be addressed here.

HOW TO ACTIVATE NANOWIRE FORMATION

Catalyst Droplet Formation

The III-V substrates are deoxidized in the growth chamber under a beam pressure of their group-V constituents. Subsequently, a two-dimensional (2D) buffer layer is systematically grown. Note that obtaining smooth buffer layers on the ($\bar{1}\bar{1}\bar{1}$) orientation is much more critical than with the standard (001) orientation. Narrow ranges of growth temperature and V/III ratio produce flat surfaces and pyramids quickly develop outside these optimal ranges. An example of smooth GaAs buffer obtained at 630 °C and observed by atomic force microscopy (AFM) is shown in Fig. **1a**. One can clearly see wide and atomically flat terraces. Our growth chamber is equipped with a gold effusion cell which is used to deposit a small amount of Au directly onto this very clean and flat surface under ultra-high vacuum. After Au deposition at low temperature (200 °C), terraces are still visible at the sample surface (Fig. **1b**). After annealing the sample at 500 °C, many round-shaped features with diameter of a few tens of nm appear (Fig. **1c**), strongly suggesting that liquid droplets were formed. By annealing at higher temperature, larger and fewer droplets are obtained. The process can be speeded up if the Au deposition is performed at the temperature to be further used for NW growth. We used *in situ* reflection high energy electron diffraction (RHEED) to monitor a GaAs surface coated with Au, during annealing and subsequent cooling [12]. After 5 min under arsenic flux at 550 °C the RHEED pattern shows a (1x1) streaky pattern, which switches to a (2x2) reconstruction upon cooling at 520 °C (Fig. **2a**). These reconstructions are those of a clean ($\bar{1}\bar{1}\bar{1}$) GaAs surface, and nothing in the diffraction pattern indicates the presence of the Au deposit. However, when the temperature is lowered down to 340 °C, additional diffraction features appear (Fig. **2b**), namely spots superimposed on the GaAs (2x2) pattern and situated outside the surface streaks. Upon further heating, the spots disappear around 400 °C and the (2x2) streaky pattern of Fig. **2a** is restored. The same observations can be repeated by cycling the sample temperature up and down. We attribute the additional spots to the diffraction through the volume of the circular particles revealed by AFM. Our observations therefore indicate phase transitions of these particles: they become solid below 340 °C, presenting a specific crystalline structure, and they return to the liquid phase with a hysteretic behavior, at 400 °C, when they do not diffract electrons anymore. TEM analysis, performed at room temperature, confirms the presence of truncated-sphere-like particles on the GaAs surface (Fig. **3a**). The TED diffraction spots were indexed precisely and most of the particles were found to correspond to the β' phase of the definite compound Au_7Ga_2 (Fig. **3b**). The particle composition was confirmed by EDXS chemical analysis. Arsenic atoms are not detected for two reasons. First their solubility in Au is very low (a few atomic percent at most). Second, since their equilibrium pressure is rather high, they probably escape from the droplet during cooling. Thus, a significant amount of Ga is present in these particles. At higher temperature, the Ga composition is probably more than the 22% contained in the β' Au_7Ga_2 phase since some Ga might condensate as GaAs during cooling. Fig. **3a** suggests that the Au-Ga particles form by dissolving a few GaAs substrate monolayers in their immediate vicinity. Our RHEED observations indicate that these particles are liquid from 340-400 °C. Consistently, the lowest eutectic points of the Au-Ga phase diagram (339.4 °C and 348.9 °C), for which liquid alloys can form, place in this temperature range and correspond to Ga concentrations close to 0.34 [13].

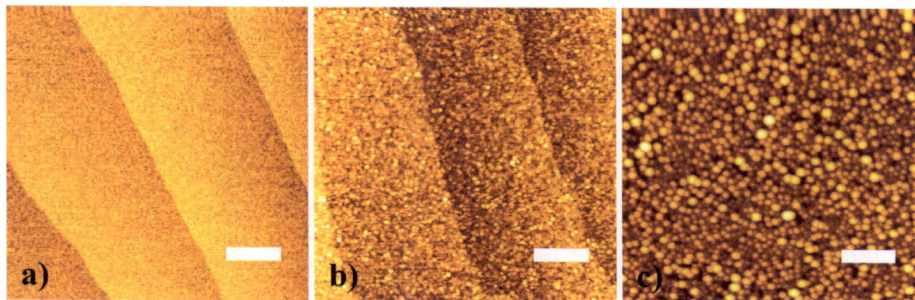

Figure 1: AFM images on a ($\bar{1}\bar{1}\bar{1}$) oriented GaAs substrate (scale bars: 200 nm). (a) GaAs buffer layer surface, (b) after deposition of Au at 200 °C, (c) after annealing the Au deposit at 500°C.

We performed a similar study for Au deposited on ($\bar{1}\bar{1}\bar{1}$) InP surface [14]. $Au_{(1-x)}In_x$ particles are formed and *in situ* RHEED monitoring indicates a solid to liquid transition above 425 °C while the liquid to solid transition takes place near 360 °C. Contrary to the former GaAs case, these temperatures are below the eutectic point of bulk Au-In alloys (454.3 °C) [13]. The melting point hysteresis can be due to supercooling and superheating phenomena that have been shown to be quite significant for nanocrystals [15]. Moreover, the equilibrium phase diagram of nanometric particles can differ from the macroscopic one [16]. These effects arise from the increase of surface to volume ratio and they are sensitive to the overpressure of the constituents, which influences the surface energies. Therefore, in a relatively large range of temperature, the state of the catalyst particle is uncertain and depends on the sample history. This gaves rise to questioning results: at the same temperature, InAs NWs could be elaborated by vapor-liquid-solid (VLS) [14,17] or vapor-solid-solid growth (VSS) [18]. Soon after this apparent controversy, the synthesis of Au-assisted Ge NWs was performed in a TEM microscope for real time monitoring [19]. It was directly observed that the NWs could grow below the macroscopic eutectic temperature, the catalyst particle being either solid or liquid for the same temperature. It was thus demonstrated that both VLS and VSS processes can operate under the same conditions with, however, a much slower growth rate for VSS.

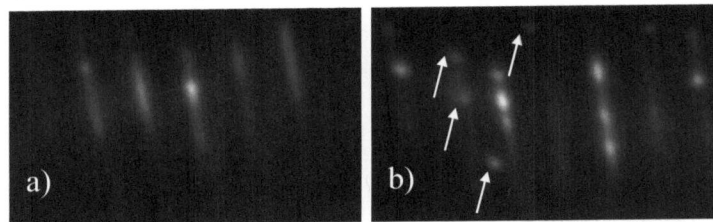

Figure 2: Reflection high energy electron diffraction patterns of Au deposited on a clean ($\bar{1}\bar{1}\bar{1}$) GaAs surface. **(a)** at 500 °C, after annealing; **(b)** at 300 °C, after cooling. Additional diffraction spots originating from the solid Au-Ga particles are observed. Reprinted with permission from Ref. [12]. *Nanotechnology*, Vol. 17, 4025. Copyright 2006 Institute of Physics.

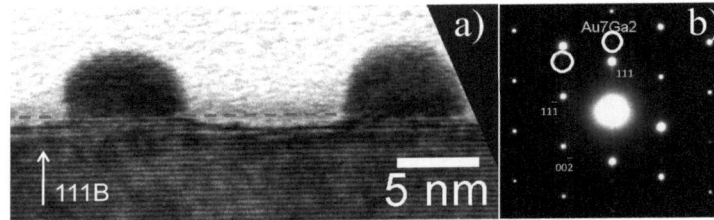

Figure 3: (a) TEM lattice image of solid Au_7Ga_2 catalyst droplets formed on a ($\bar{1}\bar{1}\bar{1}$) GaAs substrate after Au deposition and subsequent annealing. Dashes mark the top of the substrate surface before its local dissolution in the droplets. **(b)** Selected area TED revealing spots originating from the Au_7Ga_2 phase (two of them are highlighted) superimposed on the cubic GaAs substrate pattern.

Temperature Ranges for III-V NW Growth

We examine in which temperature ranges III-V NWs can be obtained on surfaces activated with Au. First, we detail the case of GaAs with a dedicated series of samples [20]. After the catalyst droplets are formed, the substrate is heated at the desired temperature and supplying Ga and As4 performs growth. When growth is performed at 300 °C, *i.e.* below the particle solidification point, we observe a rough surface without NWs, suggesting that catalyst particles are buried (Fig. **4a**). At 370 °C, *i.e.* in the range corresponding to the hysteresis between solidification and melting of the metallic particles, NWs are formed, but their morphology is very irregular and most of them present a crystallite at their top (Fig. **4b**). This morphology is possibly due to instabilities of the catalyst particles, which can easily switch from liquid to solid state at this temperature. Growth at 450 °C produces more regular NWs with a tapered end below which the NW diameter is significantly larger than that of the catalyst drop. This suggests that lateral growth occurred on the sidewall facets, concomitantly with the axial elongation of the NW (Fig. **4c**). At 550 °C, the lateral growth is not observed and cylindrical NWs are obtained (Fig. **4d**). Above 620 °C, the particles segregate at the GaAs surface with almost no NW formation (Fig. **4e**). Taking into account the RHEED observations reported in section II.1, the radical change of morphology occurring near 400 °C indicates that, in our conditions, regular NWs are formed only when the catalyst particles are liquid. We reached a similar conclusion for InP NWs. It is thus the VLS mechanism, which operates when regular NWs formed by Au-assisted MBE. For InP, InAs, GaP and GaAs binaries, this regime is obtained in the temperature ranges indicated in Fig. **5**. All the NWs presented in the following were elaborated within these ranges. The upper and lower limits of temperature may slightly vary with the incident fluxes and with the catalyst particle sizes, but those of Fig. **5** are meaningful enough to be discussed. Interestingly, the temperature ranges are quite similar for GaP and GaAs on one hand (420 – 620 °C), and for InP and InAs on the other hand (350 – 420 °C). By contrast, the overlap is very small for compounds of non-common group-III atoms. We have seen that the catalyst particles consist of liquid droplets of Au alloyed essentially with the group-III constituents. This is a strong indication that the temperature suitable for NW growth is mainly influenced by the chemical composition of the catalyst. In consequence, when axial heterostructures are sought, it seems much preferable to associate compounds of common group-III atoms, which have compatible growth temperatures. As a matter of fact, NWs containing InAs/InP [21, 22] and GaP/GaAs [23, 24] heterostructures show sharp interfaces in opposition to those made with InAs/GaAs heterostructures for instance.

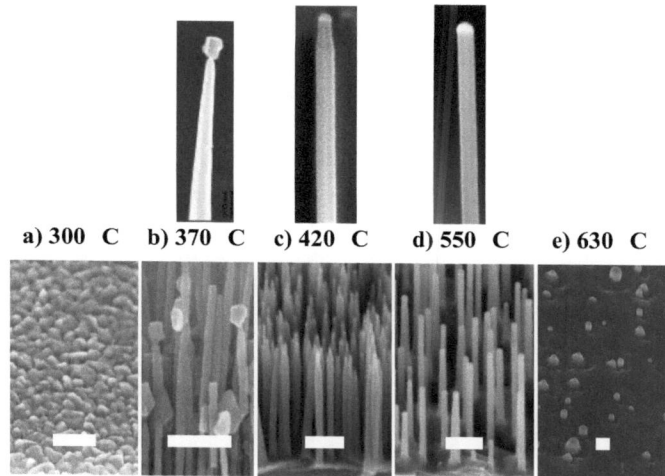

Figure 4: Morphologies of GaAs NWs elaborated at different growth temperatures (scale bars 200 nm).

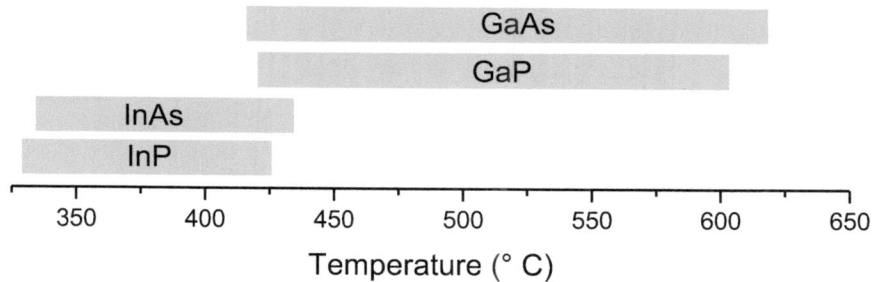

Figure 5: Ranges of substrate temperatures for which NWs of different binary III-V alloys can grow by Au catalyst-assisted MBE.

Catalyst Composition After Growth

We also analyzed the catalyst particles after growth of GaAs or InP NWs. Like other authors [25], we found that their composition depends much on the growth history and in particular on how growth was terminated and how the sample was cooled [26]. Let us consider the case of GaAs. When the Ga and As fluxes are stopped simultaneously, we observe the Au_7Ga_2 or AuGa crystalline phases (Fig. **6b** and **6c**). For growth performed at low As/Ga ratio, we identify the $AuGa_2$ compound surrounded by liquid Ga (Fig. **6d**). The average Ga content of this latter catalyst particle is estimated to more than 90%.

Referring to the macroscopic Au-Ga phase diagram, these compositions strengthen our conclusion that the catalyst particles are liquid during growth. However, when growth is terminated by stopping the Ga flux while maintaining the As flux and cooling the sample slowly, we observe almost pure gold (Fig. **6a**). In this case, most of the Ga atoms of the droplet have been used to continue forming NW layers with As supplied by the beam. This is supported by the constriction of the NW diameter (Fig. **6a**) which is observed in such samples [25, 26]: while the metallic droplet is depleted from its Ga atoms, its volume decreases and the wire diameter, which follows approximately that of the droplet, is also reduced. Thus, the Ga atoms contained in the NW constriction give a lower limit to the Ga content in the droplet before growth termination. The case of Fig. **6a** leads to a Ga content higher than 25%.

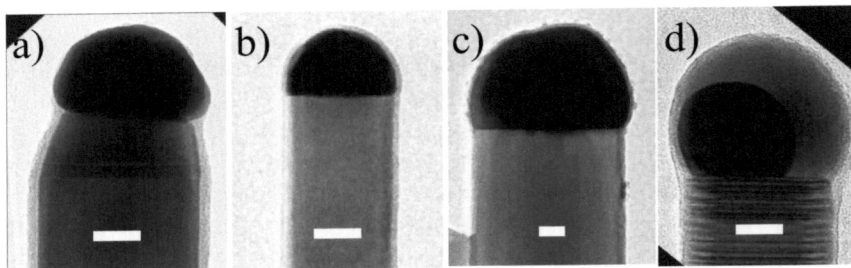

Figure 6: TEM views of catalyst particles after GaAs NW growth. At room temperature, various catalyst compositions are measured depending on growth conditions and growth termination. Ga content increases from a) to d), scale bars are 20nm. **a)** Pure Au after growth termination under As flux and slow cooling; **b)** Au_7Ga_2 for long wires and abrupt growth termination without As flux; **c)** AuGa for short wires and abrupt growth termination without As flux; **d)** $AuGa_2$ + liquid Ga for wires grown with low As/Ga flux ratio. Adapted with permission from Ref. [26], Applied Physics Letters 2005, *87*, 203101. Copyright 2005, American Institute of Physics.

CRYSTAL PHASE

Observation of Wurtzite Phase

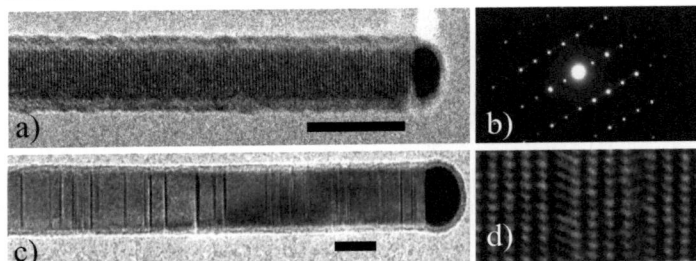

Figure 7: **(a)** GaAs NW segment with perfect WZ structure and **(b)** corresponding diffraction pattern. **(c)** GaAs NW segment of WZ structure with many stacking faults. **(d)** Detail of ZB layers in a dominantly WZ stacking (HRTEM image).

A surprising feature of the NWs is that they often adopt a crystalline structure, which is different from that of their bulk counterpart. The crystalline structure of all N-free III-V materials in bulk form is the cubic zinc-blende (ZB) structure. Most often, the NWs of the same III-V materials adopt the hexagonal wurtzite (WZ) structure independently of the growth technique [5, 25, 26, 27, 28]. Fig. **7a** shows for instance a portion of a GaAs NW with a pure WZ structure. The two compact structures, ZB and WZ, differ by their stacking sequence along the $[\bar{1}1\bar{1}]$ growth axis. The III-V monolayers (MLs) occupy three possible lateral positions A, B or C. ZB stacking is the repetition of ABC sequences whereas WZ repeats AB sequences. Although the WZ structure often dominates, the NWs usually contain stacking faults and short sequences of ZB structure (Fig. **7c**). The coexistence of these two crystalline phases can lead to original carrier confinement schemes in chemically homogeneous materials [29,30]. Although these "crystal phase heterostructures" might be attractive, their precise control is not achieved yet and today, controlling the phase purity remains one of the main challenges of III-V NW fabrication [31-35]. It is therefore essential to understand why WZ forms in NWs. This cannot be completely explained by considering fully formed NWs. Several ab initio calculations have shown that the total energy of NWs can indeed be lower in the WZ structure than in the ZB structure because of a lower specific energy of some specific sidewall facets [36-38] or of their vertical edges [39]. These contributions can compensate the difference of cohesive energy, which is favorable to the ZB structure. However, this compensation is verified only for very thin NWs: the calculated critical radii under which WZ might replace ZB are only of a few nm, whereas experimentally WZ is observed in NWs up to radii of several 100 nm. This disagreement disappears if one considers a similar energy balance at the nucleation stage. This argument is developed in section III.3 where we consider the specific case of GaAs [40]. Before, we briefly examine how nucleation proceeds in VLS.

Nucleation at the Triple Line

For NWs grown by VLS, the catalyst particle acts as a finite reservoir of variable composition. Before growth, this reservoir is in equilibrium with the substrate and the supersaturation is zero. When vapor fluxes are turned on, the supersaturation increases and the material can crystallize as small nuclei at the liquid/solid interface. It is generally admitted that the nuclei are 1 ML height and that one and only one nucleation event is required for growing each new ML for small enough NW radii. The initial nucleus extends and the ML is completed in a time much shorter than the mean time between two successive nucleation events. This *mononuclear* and layer-by-layer regime has been directly observed for group-IV NWs [41]. In the following, we always consider nuclei of height $h = 1$ ML. According to standard nucleation theory, the nucleation rate is limited by a nucleation barrier which writes $\Delta G_c = g\Omega h \frac{\Gamma_\ell}{\Delta\mu}$ and which must be overcome to form a nucleus of critical size. Γ_ℓ is the specific energy of its lateral interface with its immediate environment, $\Delta\mu$ is the difference of chemical potential between liquid and solid per pair of atoms III and V, Ω is the volume per III-V pair in the solid compound and g is a dimensionless factor related to the geometry of the nucleus. In the nano-droplet sitting on the substrate or on the top of a NW, the triple phase line that borders the liquid/solid interface is a preferential location for nucleation [42]. Indeed, at this boundary, the effective specific energy Γ_ℓ is significantly reduced. Assume that a fraction η of the peripheral edge of the nucleus is at the triple phase line and therefore in contact with the vapor phase. The complementary part $(1 - \eta)$ is in contact with the liquid. In that case, Γ_ℓ combines two contributions corresponding to the nucleus/vapor interface and to the nucleus/liquid interface. In addition, a negative term appears because the fraction η of the nucleus replaces the piece of pre-existing liquid-vapor interface which disappears (Fig. **8**). When nucleation occurs anywhere else than the triple phase line, the negative contribution is absent since $\eta = 0$. If we assume that nucleus/vapor and nucleus/liquid interface energies are comparable, the negative contribution strongly favors the nucleation at the triple line. This simple explanation is more rigorously argued in Ref. [42].

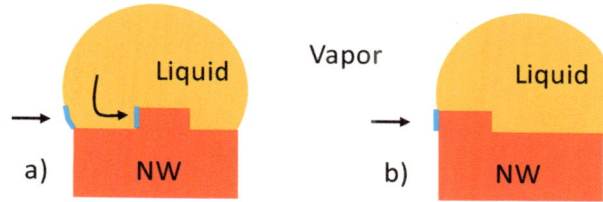

Figure 8: Two distinct locations for nucleation at the liquid NW interface. **(a)** nucleus surrounded by the liquid phase; **(b)** nucleus at the triple phase line. Adapted with permission from Ref. [42]. Copyright 2007 The American Physical Society.

Crystal Phase Determined at Nucleation

We now examine how ZB and WZ nucleations differentiate from each other [42]. Intrinsically, ZB or WZ structure cannot be assigned to a 1 ML high nucleus. Such a nucleus, with a given number of constituting atoms and a given edge configuration, can anchor to the same initial surface, either in ZB or WZ position. Assume that the NW growth just starts. Below the droplet we have the cubic substrate with its three last MLs stacked as ABC. The first NW nucleus cannot adopt the lateral position C which is occupied by the underneath layer. If the nucleus takes the position A, a new ZB ML is formed (the lateral position A, B or C of each new nucleus will fix that of the corresponding individual ML). If the B position is adopted, a WZ segment is initiated. Since this phase presents a loss of cohesive energy δE per pair of atoms III and V as compared to its ZB counterpart, $\Delta\mu$ must be replaced by $\Delta\mu - \delta E$ in the nucleation barrier of a WZ nucleus. Consider a nucleus completely surrounded by the liquid phase ($\eta = 0$). In that case, Γ_ℓ should not differ much for ZB or WZ position: the nucleus edge would be in the liquid with an interfacial energy weakly dependent on the nucleus position. On the other hand, nucleation at the triple line allows one to distinguish more clearly two different specific edge energies, Γ_ℓ^{ZB} and Γ_ℓ^{WZ}, associated with ZB and WZ positioning, respectively. Indeed, the fraction η of the nucleus edge, which sits at the triple line, can be seen as an extension of the NW sidewall facets. Those facets have different surface energies for ZB or WZ stacking. Thus we can write two distinct nucleation barriers:

$$\Delta G_c^{ZB} = g\Omega h \frac{(\Gamma_\ell^{ZB})^2}{\Delta\mu} \tag{1a}$$

$$\Delta G_c^{WZ} = g\Omega h \frac{(\Gamma_\ell^{WZ})^2}{\Delta\mu - \delta E} \tag{1b}$$

for ZB and WZ positions respectively. These expressions reveal two necessary conditions to favor the WZ positioning of the critical nucleus. The first and straightforward one is $\Gamma_\ell^{WZ} < \Gamma_\ell^{ZB}$. Calculations of specific energies of extended facets of WZ and ZB NWs [36,38] suggest that this condition is verified. The second condition is that the supersaturation must exceed a critical value. This writes:

$$\Delta\mu \geq \frac{(\Gamma_\ell^{ZB})^2}{(\Gamma_\ell^{ZB})^2 - (\Gamma_\ell^{WZ})^2} \delta E \tag{2}$$

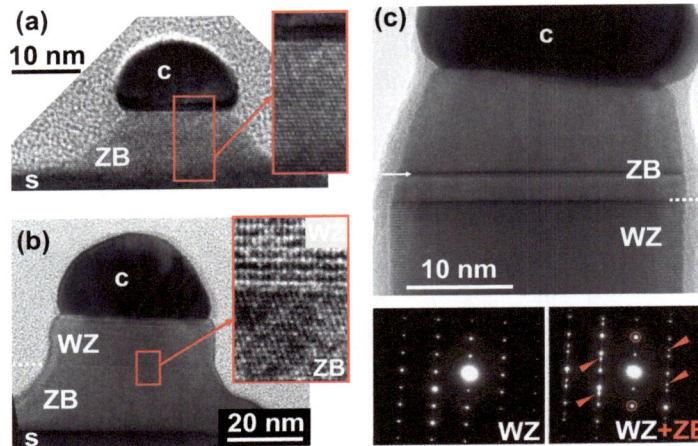

Figure 9: (a) TEM image of a very short GaAs NW with HREM close-up showing the pure ZB structure (s: substrate). **(b)** A short GaAs NW soon after the ZB-☐WZ transition, with HREM image of the transition zone. Reprinted with permission from Ref. [42]. Copyright 2007 The American Physical Society. **(c)** When growth is terminated under As only, the final tapered section of the GaAs NW (above dashed line) is pure ZB (with a stacking fault, pointed by arrow): the TED pattern from the main body of the NW (left) shows only WZ whereas the pattern from an area including also the terminal section (right) shows a superposition of WZ and ZB (arrowed spots).

This condition can be understood simply by considering the size of the critical nucleus whose characteristic dimension scales as $\Gamma_\ell/\Delta\mu$. Small $\Delta\mu$ results in a large critical nucleus, for which the energy of the peripheral edge is not sufficient to compensate the loss of cohesive energy δE of WZ positioning. Thus this analysis predicts that WZ phase cannot appear at too low $\Delta\mu$. This prediction is in full agreement with our experimental observations concerning two particular stages of the growth. The first one corresponds to the very beginning of growth. We grew GaAs NWs for only a few tens of seconds to observe the very first layers formed by VLS. TEM cross sectional views of the NWs epitaxially grown on their substrate are shown in Fig. **9**. ZB is unambiguously identified as the sole phase formed in the initial stage (Fig. **9a**). When the NW height reaches about 30 nm, growth switches abruptly to WZ stacking (Fig. **9b**). The second case where ZB forms systematically, is when we terminate MBE growth by switching off the Ga flux while maintaining an As flux during cooling. Over a few tens of nm, the final section of the NW is of ZB structure (Fig. **9c**). A similar observation was reported previously in the case of chemical beam epitaxy [25]. Thus, ZB segments form systematically in the two transient growth regimes during which $\Delta\mu$ starts from or decreases to zero. In these two particular stages, $\Delta\mu$ must be below the critical supersaturation. On the contrary, when the drop is rich enough in Ga and As, $\Delta\mu$ is above the critical value and WZ phase prevails as seen in the upper part of the NW in Fig. **9b** or in the lower part of the NW in Fig. **9c**.

This rule being highlighted, it does not give a foolproof means to control the crystal phase of the NWs. Whereas the lowest nucleation barrier (ΔG_c^{ZB} or ΔG_c^{WZ}) defines the dominant crystal phase, the probability to nucleate a ML of the alternate phase does not fall to zero [43]. Pure crystal phase can be obtained only when the difference between the two barrier heights is sufficiently large compared to $k_B T_s$, where T_s is the substrate temperature and k_B the Boltzmann constant. Second, controlling independently $\Delta\mu$ and Γ_ℓ *via* the growth parameters seems quite impossible. Generally, changing a growth parameter affects both the supersaturation and the surface energies. The global impact on the crystal phase is therefore hard to predict. Third, in the high supersaturation regime where the critical nucleus size is very small, the area available for nucleation at the triple line becomes much lower than the area out of it. This affects the probabilities of forming WZ or ZB phases in the opposite manner to the effect described above. Hence, the crystal phase issue is far from being perfectly understood and many efforts are dedicated nowadays to improve the control of the crystalline structure during NW growth [31-35].

KINETICS OF NW ELONGATION

Another important challenge is to form NWs which include heterostructures of predictable dimensions. This kind of objective requires a precise knowledge of the growth kinetics which appear much more complex for NWs than for standard 2D layers. In particular, the NW elongation with time was observed to be sublinear [44] or superlinear [14,45]. In contrast with 2D layer growth, at least two different surface orientations coexist during NW growth: the substrate surface and the NW sidewalls. These distinct surfaces intercept and collect the vapor flows under different angles. The adatoms diffuse on these surfaces with different characteristic lengths.

Experimental Method

In this sub-section, we present a method that provides detailed information on the chronology of growth of individual NWs. Our method relies on small periodic modulations of the incident vapor fluxes, produced on purpose. The amplitude and the time-period of the flux modulation are fixed experimentally and chosen to produce small but detectable changes in the NW composition along the growth direction. The growth kinetics is determined by counting the number of monolayers formed during each time-period of the modulation. Obviously, for III-V compounds, a ternary alloy at least is necessary to obtain the composition modulation. We illustrate this method with $InP_{1-x}As_x$ NWs grown at 420 °C. As_4 and P_2 fluxes are evaporated from solid sources and a periodic modulation of the incident As_4:P_2 flux ratio is produced. The flux modulations may be obtained by directly modulating the sources. In the present study, we use an indirect method: the modulation is obtained at constant source fluxes,

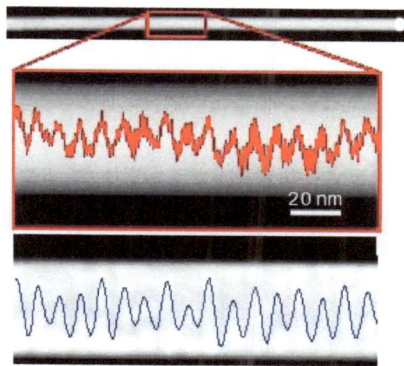

Figure 10: Modulation of composition in an $InP_{1-x}As_x$ nanowire segment grown with a modulated As_4:P_2 flux ratio. HAADF STEM image of the nanowire segment observed along the $<\bar{2}110>$ zone axis. (a) The HAADF signal is integrated over the whole diameter and the profile is plotted along the growth axis. (b) Fast-Fourier-transform filtered HAADF image and corresponding profile. From Ref. [49].

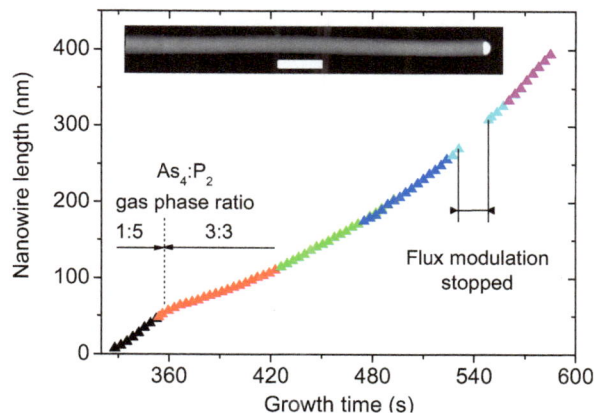

Figure 11: Chronology of growth of a single $InP_{1-x}As_x$ nanowire. The length of the nanowire shown in the inset (scale bar: 50 nm), is measured from its cleaved edge and plotted as function of the absolute growth time, as deduced from the composition oscillations along its axis. The different colors correspond to different overlapping HAADF STEM images. During the growth

experiment, the flux modulation was stopped for 12 time periods. The corresponding region, where composition oscillations are missing, serves as a time reference. At 355 s, the nominal As_4:P_2 flux ratio was changed abruptly from 1:5 to 3:3. From Ref. [49].

by exploiting the geometry of the MBE growth chamber, namely the fact that the molecular beams are not perfectly uniform on the wafer area. When the substrate is rotating, the NWs situated close to its periphery experience a small modulation of the incoming fluxes during each substrate rotation. Several important conditions must be met in order to transfer the flux modulation to an axial modulation of the NW composition. First, several monolayers must grow within one modulation time-period. We chose a time-period of 3.6 s corresponding to the growth of about 10 to 30 monolayers. The corresponding length (3-10 nm) is sufficiently small to probe finely the growth kinetics. Second, we must avoid a complete damping of the flux modulation upon transferring atoms into the liquid drop, which acts as a reservoir. In this respect, the high solubility of group-III atoms in the catalyst is not favorable. Modulating the group-V flux is expected to be more appropriate since P and As are much less soluble in the catalyst. On the other hand, the modulation must be small enough to minimize the resulting perturbation on NW growth kinetics. The NWs analyzed in the following have uniform diameter, close to the size of the catalyst particle and a fault-free WZ structure. This morphology indicates that nucleation at the NW sidewalls is negligible. Hence, the present experiments probe the sole axial growth. The $InP_{1-x}As_x$ NWs grown with modulated beams were removed from their substrate, deposited on a th*in silico*n nitride membrane and imaged by HAADF Cs-corrected STEM. This technique is very sensitive to composition variations [46]. The image of a NW segment of 45 nm diameter is presented in Fig. **10**. The faint contrast along the growth axis reveals that the composition of this NW has indeed been modulated. The modulation of the HAADF intensity profile integrated over the whole diameter is superimposed (Fig. **10a**). A fast-Fourier-transform filtered HAADF image with the corresponding intensity profile is also shown (Fig. **10b**).

The oscillation amplitude is about ± 0.7% of the total HAADF intensity. The corresponding variation Δx in the As concentration x was determined by EDXS to be about ±0.03 around the mean value of 0.66. An estimate based on Rutherford scattering indicates that the observed HAADF intensity oscillation amplitude is consistent with this composition excursion. The distance between two concentration extrema equals the increase of NW length during one time-period. Therefore, it is straightforward to determine the instantaneous growth rate at each oscillation. The complete chronology of the NW growth is thus accessible. In the present experiments, since the foot of the NW gets buried by the layer growing between the NWs and since the NWs get cleaved before TEM examination, the bottom part of the NW is not observed. However, counting the oscillations from a reference mark deliberately introduced in the growth sequence allows one to date the exact time at which each oscillation was formed. This procedure was applied to a 21 nm diameter NW. The positions of the oscillation maxima were measured from the cleaved end to the catalyst particle, using several overlapping HAADF images. The dependence with time of the NW height counted from the cleaved end is presented in Fig. **11**. In this growth experiment, we stopped the flux modulation for 12 time-periods. Consequently a short segment of the NW does not present regular composition oscillations. As mentioned above, the corresponding region, indicated in Fig. **11** may serve as a time reference. Second, at some point during growth, the nominal As_4:P_2 ratio was changed abruptly from 1:5 to 3:3 while the total average group-V flux was kept constant. Thanks to the former time reference, this event can be accurately dated, as shown in Fig. **11**. The impact of this large change of flux ratio is strong and immediate: as seen from the abrupt change of slope in Fig. **11**, the growth rate was suddenly reduced by a factor 2. This behavior contrasts much with the case of conventional MBE growth of 2D layers, for which the growth rate is fixed by the sole group-III elements.

Kinetic Models

As predicted by several models [14, 45], the measured height of the NW displays a non-linear behavior as a function of time. These models consider three material flows contributing to NW growth: (*A*) direct impingement on the drop, (*B*) direct impingement on the sidewalls followed by diffusion to the drop, (*C*) impingement on the substrate surface and diffusion to the drop along the sidewalls. The growth rate is then determined by solving a balance equation. For III-V alloys, this procedure can be applied to each component. In MBE, the problem is simpler for the group-III species that are elemental in the vapor or physisorbed states. Moreover, in a large range of growth temperature, their desorption rate from the substrate is negligible as compared to the incoming flux.

We analyze the experimental data of Fig. **11** with a model similar to that developed by Dubrovskii *et al.* [47], but adapted to our experimental conditions. The balance equation is solved for In atoms and their reevaporation from the InP(As) surfaces at 420 °C is neglected. We consider J, the beam flux of In atoms incoming onto the sample (3.96 $nm^{-2}s^{-1}$) with an angle α with respect to the substrate normal (32.5°). We take R as the NW radius (10.5 nm for NW of Fig. **11**) and L^2 as the mean substrate surface area per NW (the reverse of NW surface density). The catalyst droplet has a contact angle β with respect to its horizontal base plane. The sample under study has a low NW density ($L = 300$ nm), hence there is no shadowing between NWs.

The total number of group-III atoms impinging on the sample area L^2 per time unit is T=L^2Jcosα, The part of these atoms which is intercepted by the catalyst droplet on top of the NW is A=$\alpha\pi R^2$J where α is a geometrical factor which depends on angles α and β [48]. We consider that this part A is fully incorporated into the NW. We estimated that β must be close to 120° during growth, whereas at room temperature, after growth, this angle is about 100° (Fig. **11**). The difference comes from the drop inflation that is necessary to reach the supersaturation in the liquid during growth [49]. The two other contributions B and C stem from the collection of adatoms from the NW sidewalls and from the substrate surface, respectively, followed by their migration to the drop. B and C are evaluated by solving the diffusion equations with adequate boundary conditions and their exact expressions can be found in Ref. [47]. They are reformulated in Ref. [49] with a limited number of unknown parameters: λ_s and λ_f, the adatom diffusion lengths on the substrate surface and on the sidewalls, respectively, $\Delta\mu_{sl} = \mu_s - \mu_l$ and $\Delta\mu_{fl} = \mu_f - \mu_l$ where μ_l, μ_s and μ_f are the chemical potentials per III-V pair in the liquid droplet and at the substrate surface far from the NW; μ_f is the chemical potential on the sidewall at the maximum possible adatom coverage of the latter (limited by desorption only). B and C also depend on $H_w(t)$, the apparent NW height emerging at time t from the 2D layer which grows between the NWs. The NW elongation with time results from the sum of the three contributions:

$$\frac{dH_T(t)}{dt} = \frac{\Omega}{\pi R^2}(A + B + C) \tag{3}$$

where $H_T(t)$ is the total NW length measured from a fixed reference, namely the initial substrate-NW interface. We consider that the atoms, which do not participate in the NW elongation, are incorporated at the substrate surface and form the 2D layer that hence grows at the rate:

$$\frac{dH_{2D}(t)}{dt} = \frac{\Omega}{L^2 - \pi R^2}(T - A - B - C) \approx J\Omega \cos\alpha - \frac{\pi R^2}{L^2}\frac{dH_T(t)}{dt} \tag{4}$$

where, in the second equality, we have kept only the terms of lowest order in (R/L). Equation (4) describes the fact that NWs grow to the detriment of 2D growth. This 2D layer buries progressively the bottom of the NW. Therefore, the height of the emerging part of the NW is $H_W(t) = H_T(t) - H_{2D}(t)$ and using equation (4), its variation with time is:

$$\frac{dH_W(t)}{dt} = \frac{dH_T(t)}{dt}\left(1 + \frac{\pi R^2}{L^2}\right) - J\Omega \cos\alpha \tag{5}$$

Combining equations (3) and (5) gives a differential equation which is solved to fit our experimental data (Fig. **11**). For each growth conditions (low and high As₄:P₂ ratio), we have four adjustable parameters: $\Delta\mu_{sl}$, $\Delta\mu_{fl}$, λ_s and λ_f. λ_f has little influence on the calculation, provided that it is larger than the final NW length. We take $\lambda_f = 2$ μm for both growth stages to reduce the number of free parameters, a reasonable value since in our conditions, NWs up to this length can be obtained without radial growth. Concerning λ_s, we find that very small values (a few nm) are necessary to fit the experiment. Our best fit is obtained with $\lambda_s = 2$ nm. This very short length is surprising at first sight, but it can be understood as a limitation of adatom diffusion by atomic steps. Indeed, the epilayer between the NWs has a strong surface roughness and the widths of terraces between the atomic steps can be as small as a few nm. In these conditions, changing the group-V flux is expected to have little impact on the substrate surface diffusion. For that reason, we use also the same value of λ_s for both growth stages. On the other hand, the abrupt change of As₄:P₂ flux ratio

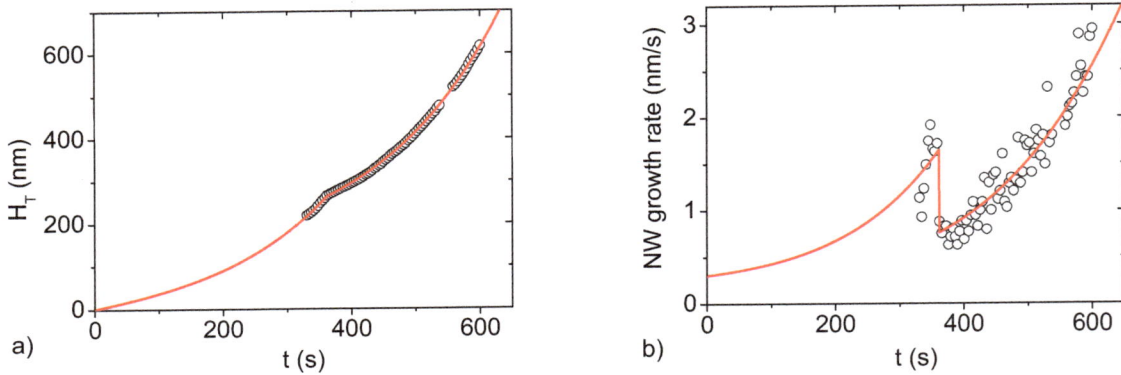

Figure 12: Experimental NW elongation (a) and corresponding axial growth rate (b) fitted with Dubrovskii's model [47] and a reduced number of free parameters. The adatom diffusion lengths on the nanowire sidewalls and on the substrate surface are $\lambda f = 2$ µm and $\lambda s = 2$ nm respectively. The differences of chemical potential between sidewalls adatoms or substrate adatoms and liquid droplet are $\Delta\mu fl$ and $\Delta\mu sl$, respectively. Different values are used for each As4/P2 ratio: $\Delta\mu fl = 270$ meV and $\Delta\mu sl = 18.2$ meV before, and $\Delta\mu fl = 160$ meV and $\Delta\mu sl = -145$ meV after t=355 s. From Ref. [49].

is expected to modify significantly the composition and the chemical potential of the catalyst droplet. Accordingly, we adjust two sets of chemical potential differences, valid before and after this change. We find $\Delta\mu_{fl} = 270$ meV and $\Delta\mu_{sl} = 18.2$ meV for the first part of growth, indicating net flows of diffusing adatoms from the sidewall facets and from the substrate surface to the droplet. The contribution from the substrate surface is very small due to the combination of low λ_s and moderate $\Delta\mu_{sl}$. Very quickly, the contribution from the sidewalls dominates. After the change of As$_4$:P$_2$ flux ratio, $\Delta\mu_{fl}$ must be reduced to 160 meV which indicates that the flow from the sidewalls to the droplet is less efficient when P$_2$ is replaced by As$_4$. At the same time, $\Delta\mu_{sl}$ takes a negative value of -145 meV showing that a significant part of the atoms intercepted by the NW now migrates to the substrate surface, to the detriment of NW growth. This behavior illustrates the crucial role of group-V fluxes which can, *via* their interaction with the other constituents of the drop, affect the chemical potential of In in the latter [50] and redirect part of the diffusion flow of In adatoms. With these parameters, Dubrovskii's model describes fairly well the experimental dependence $H_T(t)$, and the abrupt change in NW growth rate is well reproduced, as shown in Fig. **12**.

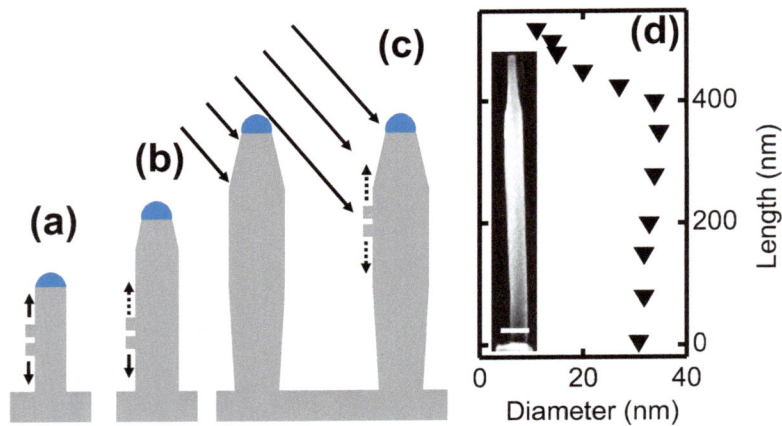

Figure 13: Schematic diagram of the morphology of NWs: **(a)** $H_W < \lambda_f$; **(b)** $H_W > \lambda_f$; **(c)** $H_W < \lambda_f$ and shadowing by neighboring NWs. The blocks shown on the NW sidewall represent group-III adatoms which can diffuse either towards the Au particle or towards the substrate surface, as shown by the directed arrows. The solid and dashed arrows indicate respectively that the adatoms can or cannot reach the Au particle or the substrate surface. **(d)** Effect of shadowing on NW morphology: diameter of the NW shown in the SEM view (scale bar is 50 nm) as a function of its length. Reprinted from Ref. [51], *Journal of Crystal Growth* 312, 2073–2077, C. Sartel *et al.*, Effect of arsenic species on the kinetics of GaAs nanowires growth by molecular beam epitaxy, copyright (2010), with permission from Elsevier.

We derive important conclusions from these results. In our experiments, NW growth is mainly sustained by contributions A and B. As the NW aspect ratio becomes significant, contribution A tends to be marginal as compared to B. The latter is responsible for the non-linear variation of $H_T(t)$: as the NW becomes longer, it collects more material on its sidewalls and consequently grows faster. During the first growth stage (As_4:P_2 ratio $\equiv$ 1:5), almost all the In atoms intercepted by the NW sidewall facets are transferred to the catalyst droplet. During the second stage (As_4:P_2 ratio $\equiv$ 3:3), this transfer is less efficient as a consequence of the change of As_4:P_2 ratio which affects the drop chemistry. The concentrations of As and P in the drop are modified and their interactions with the other constituents of the liquid phase have a significant impact on the chemical potentials in the liquid [50]. Since the As and P concentrations in the drop are low, they can be changed rapidly. Very few MLs are grown during the corresponding transient regime which therefore appears as a growth rate discontinuity (see Fig. **12b**). Contribution C is close to zero or even negative. This conclusion goes against the prevailing idea that collection of substrate adatoms is necessary to explain NW growth.

LATERAL GROWTH

The differences of chemical potential, which are derived in the previous sub-section, indicate that the highest chemical potential of the growing system, apart from that of the vapor phase, is always located on the sidewall facets and hence equal to μ_f. The NWs have hexagonal sections and the facets are $\{\bar{2}110\}$ or $\{10\bar{1}0\}$ planes. The exact altitude $\hat{z}$ of μ_f, where the group-III adatom concentration is the highest, varies with time. In the example treated in section IV.2, μ_f is located at the bottom of the NW ($\hat{z} = 0$) in the first growth stage. Then, $\hat{z}$ evolves slowly and for $H_W = 500$ nm, we have estimated that $\hat{z} = 114$ nm [49]. As the NW grows, the concentration of group-III adatoms on the sidewalls increases until the NW height gets of the order of λ_f. Then, nucleation on the sidewall facets is possible and likely occurs near $\hat{z}$ from which radial growth can develop by monolayer step flows that propagate parallel to the NW axis [45]. As long as the sidewalls receive the vapour fluxes uniformly over the NW height, the NW shape tapers at the catalyst side and the shell thickness is uniform elsewhere (Fig. **13b**). If the fluxes become shadowed by other neighboring NWs, tapering may also develop in the lower part of the NW (Fig. **13c**). This radial growth can be exploited to form core-shell heterostructures. To induce controllably the radial growth at arbitrary H_W, one can play with the growth parameters to reduce λ_f at the desired instant. An efficient way to do this is to switch the group III flux to provide species of lower λ_f. For instance, AlGaAs shells easily form around cylindrical GaAs cores because, for identical growth conditions, Al adatoms have shorter λ_f as compared to Ga adatoms. In that case, a radial heterostructure can be obtained without changing the other growth parameters (see Fig. **14a-b**). Other material

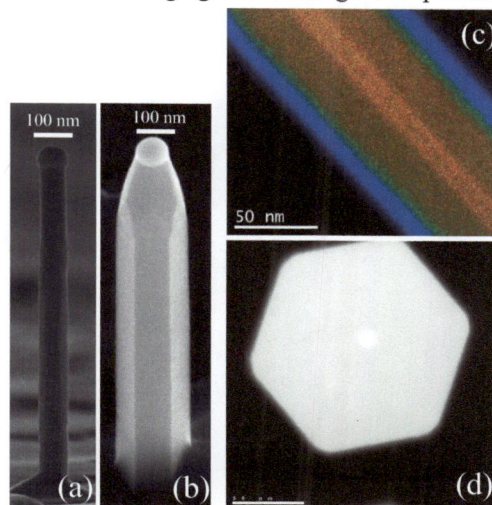

Figure 14: Formation of core-shell heterostructures. **(a)** SEM view of a cylindrical GaAs NW, **(b)** on which a AlGaAs shell can be formed in a later stage, at the same growth temperature. (c) and (d) HAADF STEM views of a InGaAs core/InP shell heterostructure grown with two temperature steps. **(c)** Longitudinal view, **(d)** cross-section view.

combinations, like InAs/InP, are less favorable in this respect. For this latter material system, one can reduce T_s to shorten the diffusion length of In adatoms. An example of InAsP/InP radial heterostructure grown with two temperature steps is shown in Fig. **14c-d**: a 10 nm InAsP core was grown at 420°C, then the temperature was reduced to 380°C to form a thick InP shell around it. This process leads to very sharp interfaces between core and shell. In general, the radial growth is not perfectly selective: as the shell forms, axial growth continues, possibly at reduced rate. In GaAs NW growth, using As_2 or As_4 species has also a strong impact on the diffusion length of Ga adatoms. Under As_2, radial growth is much more efficient than under As_4 flux. This is evidenced by observing the morphology of NWs grown with As_4 or As_2. With As_4, the NWs have a regular cylindrical rod-like shape with a diameter very close to that of their catalyst droplet. With As_2, the NWs have tapered shapes and they are markedly wider than the catalyst drop at their top [51]. In addition, for temperature lower than 550°C, the axial growth rate is significantly reduced for growth under As_2 (Fig. **15**). These morphological characteristics are consistent with the fact that the part of Ga adatoms which, under As_2, participate to lateral growth on the NW sidewalls are lost for the NW elongation. Thus, to form core-shell heterostructures of arsenides, it is also efficient to swap As_4 for As_2 after the core is grown.

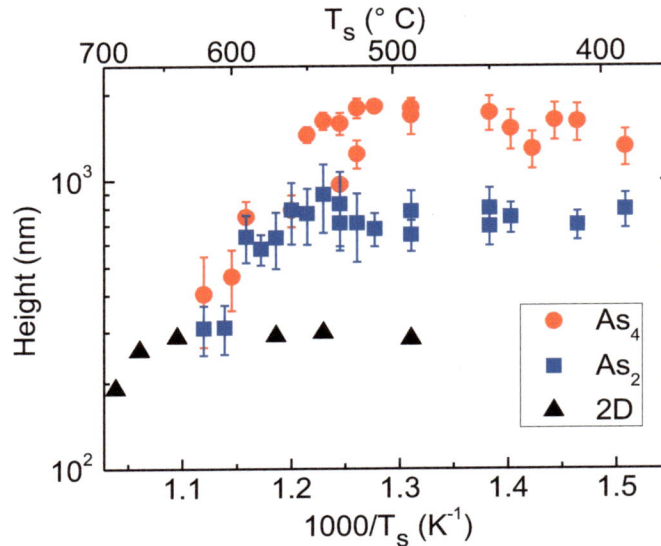

Figure 15: Height of GaAs nanowires after 25 min of growth on GaAs (111)B substrate using As_2 (squares) or As_4 (circles) fluxes, as a function of inverse substrate temperature. The temperature dependence of the height of a two-dimensional (2D) layer is shown for comparison (triangles). Reprinted from Ref. [51], *Journal of Crystal Growth* 312, 2073–2077, C. Sartel *et al.*, Effect of arsenic species on the kinetics of GaAs nanowires growth by molecular beam epitaxy, copyright (2010), with permission from Elsevier.

BURIED NANOWIRES

How to Bury Nanowires

As seen in section IV, concurrently to the VLS growth, a two-dimensional (2D) layer grows on the substrate surface between the NWs. The competition between these two growth modes is sensitive to the substrate temperature. Fig. **15**

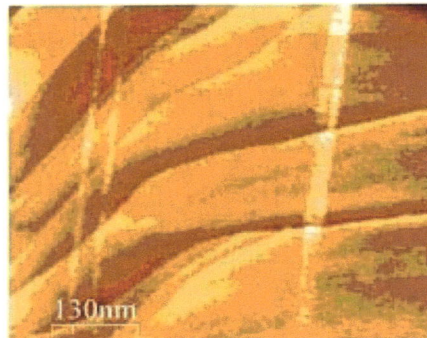

Figure 16: AFM image of a cleaved facet of a sample containing buried GaAs/AlGaAs core/shell NWs. The cross-sections of AlGaAs shells oxidize in air and thus present some topological contrast. When the cleavage crosses NWs near their periphery, the core is absent (NW on the right) while it is clearly visible for NWs cleaved near their median plan (NW on the left).

compares GaAs NW heights with the thickness of a 2D layer grown under the same conditions, but on a non-activated (catalyst-free) GaAs substrate. On the bare substrate, the growth rate has a constant value of 0.2 nm/s up to 650 °C. Above this temperature, Ga desorption becomes significant and the growth rate starts to decrease. NWs behave differently: their growth rate decreases steeply above 530-550°C, a temperature range well below the Ga desorption threshold. These distinct behaviors have allowed us to bury NWs, previously grown near 530°C, by an epitaxial layer grown subsequently near 640°C. Embedding NWs was first proposed for investigations by scanning tunneling microscopy and demonstrated with MOCVD growth schemes [52]. Our work extends this possibility to the MBE growth technique. A successful example of GaAs/AlGaAs core/shell NWs buried into epitaxial GaAs is shown in Fig. **16**.

Impact on the Crystalline Structure

The impact of the burying process on the crystalline structure of the NWs is of particular interest. The burying layer naturally adopts the cubic ZB structure of the GaAs substrate where its epitaxial growth takes place. In contrast, the dominant structure of free-standing (non-embedded) NWs is the hexagonal WZ, as seen in section III.1. Most of these NWs contain many stacking faults (Fig. **17a**). Remarkably, after being buried, the NWs present a perfect ZB structure (Fig. **17b-c**). The observation of a partially buried NW indicates that the phase transformation occurs progressively, at the same rate as the NW gets buried: the part of the NW emerging from the overgrowth retains the WZ structure whereas the buried part has switched to ZB and the interface between the two crystalline phases is very abrupt (Fig. **17d**). We proposed a mechanism to explain how the transformation proceeds [53]. Let us start from the situation shown in Fig. **17d**, where the burying process has already transformed the lower part of the NW

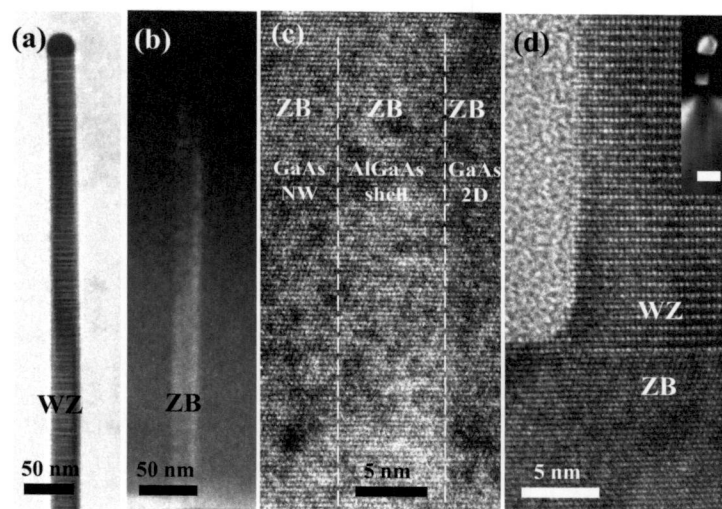

Figure 17: TEM images of GaAs nanowires with [111] B growth axis upwards. **(a)** bright field image of a non-buried nanowire; **(b)** 002 dark field image of a completely buried nanowire; **(c)** HRTEM image of the interface region between GaAs core, AlGaAs shell and burying GaAs; **(d)** HRTEM image of the interface region between the buried (bottom) and unburied (top) portions of the nanowire. The insert shows a partially buried NW at lower magnification (scale bar: 50 nm). Reprinted with permission from Ref. [53]. Copyright 2008 American Chemical Society.

to the ZB stacking. This buried part is in lateral registry with the stacking of the burying layer. When a new burying ML forms at the substrate surface, it reaches the side of the NW by step flow growth (Fig.**18a**, left). The lateral positioning of the As atoms in the burying ML and in the corresponding ML of the NW are not in registry, as opposed to the underlying Ga atoms whose positions are fixed by the ZB structure common to the overgrowth and to the already transformed NW (Fig. **18b**). The simplest way to correct this fault is to shift rigidly the whole emerging part of the NW situated above (and including) this As plane. This puts two MLs of the NW in the correct ZB position (Fig. **18a**, right). When the burying layer reaches the level above these two MLs, the same process repeats, and so on every two MLs. At each step, three equivalent translation vectors are permitted, namely $\frac{1}{6}[\bar{2}11]$, $\frac{1}{6}[1\bar{2}1]$ and $\frac{1}{6}[11\bar{2}]$ (Fig. **18b**). If the three translation vectors (adding to zero) are allowed with equal probabilities, the net tilt of the NW with respect to the substrate remains nearly null after complete burying. This is consistent with our observations (Figs. **17b** and **17d**). The process can also be described in terms of dislocations. If not in registry, the top burying layer and the corresponding NW layer create a linear crystalline defect at their boundary when they get in contact. With respect to the ZB structure, this defect may be considered as a Shockley partial dislocation [54], composed of segments lying along the edges of the NW (in the ZB indexing, the $\langle 0\bar{1}1\rangle$ directions of its hexagonal periphery) and bordering the faulted As plane. The Burgers vector of the partial dislocation is one of the three translations mentioned above. As soon as the dislocation forms along at least two sides of the hexagon (Fig. **19-1**), any of its parts can glide inward without increase of length (Fig. **19-2**) and therefore, without any energy increase. This gliding motion proceeds by shifting the NW ML into ZB position with respect to the underlying and already buried ZB portion of the NW. The translation of the whole ML fully eliminates dislocations and stacking faults. Glas calculated that eliminating the fault corresponds to a decrease of energy of about 28 mJ.m^{-2} [55]. Hence, the total energy of the defects decreases continuously during each elementary step of the transformation (Fig. **19**). Of course, when a NW ML is already in the correct ZB position, no dislocation is created and hence there is no driving force for translating the upper part of the NW. This happens in particular when there is a stacking defect or a ZB segment in the yet unburied WZ NW and explains why all such pre-existing defects (Fig. **17(a)**) get eliminated (Fig. **17(c)**). According to this simple analysis, there is no energy barrier to the transformation other than the Peierls barrier. Therefore, the translation of the whole ML may occur very soon after the burying layer reaches it, in agreement with the fact that the interface between the transformed and untransformed parts of the NW is abrupt and horizontal (Fig. **17d**). We have examined a second type of process which could also operate the WZ→ZB transformation [53]. It was found less likely, because it would not lead systematically to the abrupt ZB-WZ interface which is observed experimentally.

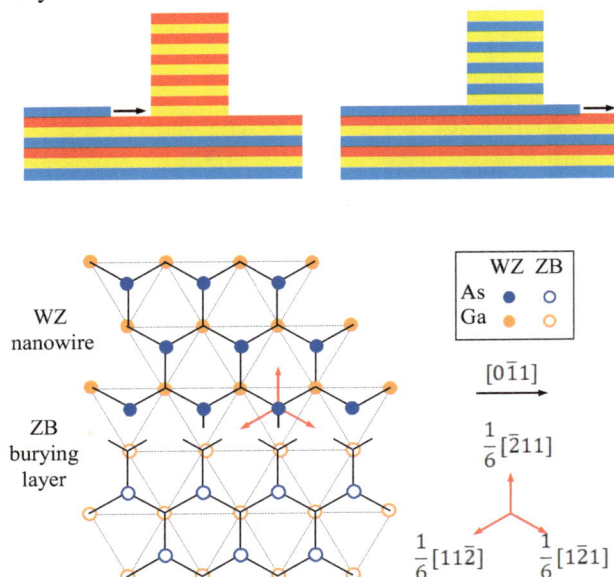

Figure 18: Top ML of the ZB burying layer meeting the corresponding ML of the NW in an out-of-registry condition. **(a)** Side view before (left) and after (right) the phase transformation. Each color denotes a different ML lateral position. **(b)** Top view. The partial dislocation created at the boundary can be annihilated by any of three translations of the NW (red arrows), indexed in the ZB lattice. Reprinted with permission from Ref. [53]. Copyright 2008 American Chemical Society.

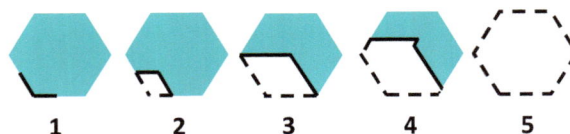

Figure 19: Some stages of the transformation of a nanowire ML from WZ stacking position into ZB position (top view). The faulted (untransformed) area is coloured. Solid lines indicate the progression of the linear defect initially formed (dashed line) at the boundary between burying ML and nanowire sidewall. The length ratio of solid to dashed lines equals 1 up to stage 3 and then decreases. Regarding the stacking fault, its area constantly decreases. Reprinted with permission from Ref. [53]. Copyright 2008 American Chemical Society.

Thus, burying is an efficient means to get perfect ZB GaAs NWs surrounded by their AlGaAs shells and embedded in the overgrown GaAs matrix. These objects are very promising to investigate the transport properties of 1D electronic system. They have no more free surfaces from which instabilities or degradation related to oxidation and surface states may occur. With a suitable design, the shell can be doped for transferring carriers into the core which is exempt of crystalline defects.

NUCLEATION STATISTICS

The modulated beam growth method presented in section IV.1 also permits the nucleation statistics to be investigated [56]. We have mentioned that, for thin NWs, VLS growth occurs in a mononuclear regime where each ML needs one and only one nucleation event to be formed. The number of MLs contained in each compositional oscillation is thus exactly equal to the number of nucleation events in the time interval τ of their growth. In our experiment, the modulation is periodic *i.e.* τ is constant. The histogram of the number m of nucleations taking place during τ thus gives the nucleation statistics. Fig. **20a** shows the lengths of successive oscillations (determined with an accuracy of $\pm 1ML$) contained in a NW segment of radius 25 nm and Fig. **20b** a nucleation histogram. The average growth rate increases with time, as already discussed in section IV.2. For the relatively short growth sequence studied here, we do not take into account this small variation in our statistical analysis (note that this variation tends to broaden the experimental distribution around the mean number of MLs, $\bar{m}$). If we assume that nucleation events occur independently of each other, the nucleation probability should be time-independent and the statistics should be Poissonian. The distribution of the oscillation lengths is compared with a homogeneous Poisson distribution normalized to same total number of oscillations in Fig. **20b**. The nucleation statistics are markedly sub-Poissonian. The experimental standard deviation, $\sigma = 1.75$, is less than half the Poisson standard deviation, $\sqrt{\bar{m}} = 4.31$. This proves that nucleation events are not independent. This is a consequence of the finite volume of the nanodroplet which acts as a reservoir of constituents. After each successful nucleation, the reservoir provides the group-III and group V-atoms which are necessary to complete the ML. Assume that the ML completion occurs in a time negligible as compared with the mean time between two successive nucleations [41, 57]. This causes an abrupt decrease of the numbers of group-III and group-V atoms in the liquid, $\delta N_{III} = \delta N_V = -\pi R^2 h/\Omega$. Since Au is conserved, the atomic concentrations of group-III and group-V constituents in the drop, c_i ($i = III, V$), and thus the difference of chemical potential $\Delta\mu$ between liquid and solid also decrease. In consequence, the nucleation probability P which is dominated by factor $exp[-\Delta G_c/(k_B T_s)]$, where ΔG_c is the $\Delta\mu$-dependent nucleation barrier defined in section III.2, is reduced. c_i, $\Delta\mu$ and P do not recover their pre-nucleation values before the drop is refilled. Conversely, overfilling of the drop due to a temporary lack of nucleation, leads to an increase of the nucleation probability. Nucleation is thus less likely after a nucleation event than before, nucleation events are temporally anti-correlated and hence not independent, so that their statistics become sub-Poissonian. Assume that the refilling rate is equal to the average growth rate for both constituents. Then, at any time, $\delta N_{III} = \delta N_V = \delta N$. At typical NW radius, $\delta N/\overline{N} \ll 1$, so that $\delta c_i = c_i - \overline{c_i} \approx (1 - 2\overline{c_i}) \delta N/\overline{N}$ where overbars indicate the average reference state of the drop. Using a first order development in c_{III} and c_V of $\Delta\mu$, one obtains an approximate expression of the nucleation probability [56]:

$$P \approx \overline{P} \, exp\left\{\left(B_{III}\frac{1-2\bar{c}_{III}}{\bar{c}_{III}} + B_V\frac{1-2\bar{c}_V}{\bar{c}_V}\right)\frac{\delta N}{\overline{N}}\right\} \tag{6}$$

where the dimensionless factors expressed as:

$$B_i = gh\Gamma_\ell^2 c_i \left(\frac{\partial\Delta\mu}{\partial c_i}\right)/k_B T_s \overline{\Delta\mu}^2 \tag{7}$$

contain parameters which are not precisely known, but can be estimated. According to Glas's thermodynamical calculations [50], B_{III} and B_V are of the same order. As seen in section II.3, the drop consists essentially of gold alloyed with group-III elements. $\overline{c_{III}}$ can be close to 0.5 whereas $\bar{c}_V \ll 1$, due to the low solubility of group-V elements in the catalyst [10,14,58]. Hence, the term $B_i\frac{1-2\bar{c}_i}{\bar{c}_i}$ is much larger for group-V than for group-III elements. Therefore, the group-V atoms influence the probability of nucleation much more than the group-III atoms. Moreover, the thinner the NW, the larger the effect, since $\delta N/\overline{N}$ scales as R^{-1}. We simulated growth self-consistently from Eq. (6), considering only the effect of the variations of c_V on the nucleation probability. Mean concentration $\bar{c}_V$ was taken in the range of a few percent. The experimental distribution of elongations per time period can be satisfactorily reproduced (Fig. **20(b)**), with $B_V = 3.8$ and $c_V = 0.03$ for instance. This B_V value is reasonable and may be realized with plausible values of Γ_ℓ (0.32 J.m^{-2}), $\Delta\mu$ (0.25 eV per III-V pair) and $\bar{c}_V \frac{\partial \Delta\mu}{\partial c_V}$ ($k_B T_s$).

Hence, the sub-Poissonian character of the nucleation statistics is perfectly explained by the negative feedback of the variation of the group-V concentration in the liquid on the nucleation rate. Different drop refilling scenarios would lead to different quantitative estimates but would not alter this conclusion, provided the refilling of the drop is slower than its depletion upon the growth of one ML.

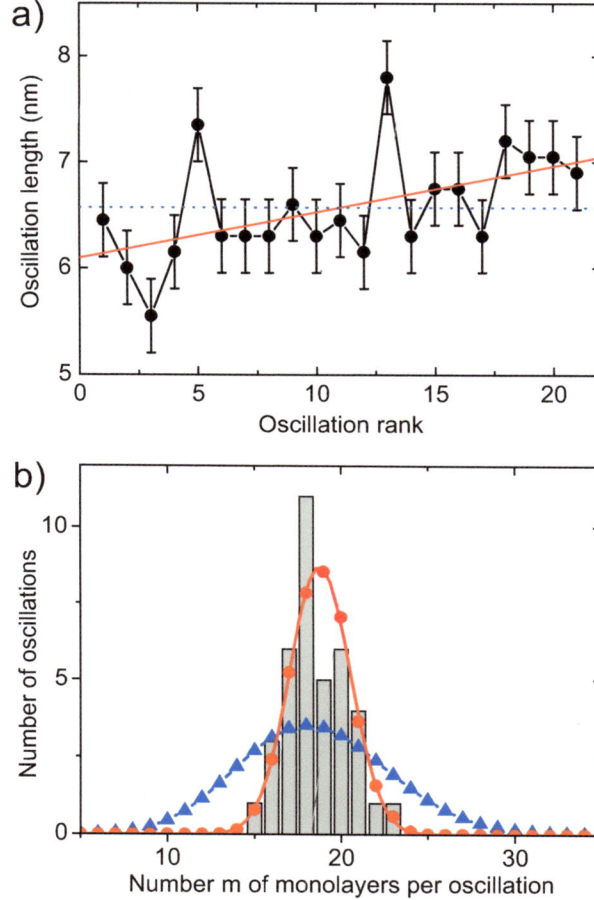

Figure 20: Experimental determination of the nucleation statistics. **(a)** NW elongations occurring in consecutive and equal time periods. The dashed line gives the mean elongation; the solid line is a linear regression. **(b)** Histogram of the numbers of nucleation events per time period. The two curves represent distributions calculated for Poissonian statistics (triangles) or self-regulated statistics (disks). Reprinted with permission from Ref. [56]. Copyright 2010 The American Physical Society.

A remarkable property of this self-regulation mechanism is its efficiency against fluctuations of the NW length after a large number of nucleations. This is evidenced by the regular dependence of the experimental NW elongation with time presented in section IV (Fig. **11**). The kinetic model used to fit the $H_T(t)$ dependence (Fig. **12**) ignores the fluctuations of the nucleation events. Hence, an upper limit of the experimental deviation due to these fluctuations is the quantity

$$\tilde{\sigma} = \left[\frac{\sum_{i=1}^{n}(O_i - E_i)^2}{n} \right]^{1/2}$$

(8)

where $\{O_i\}$ are the n observed values of $H_T(t)$, and $\{E_i\}$ the corresponding values of our best fit obtained with the kinetic model. $\tilde{\sigma}$ is evaluated to 5.8 MLs for the 630 nm long NW of Fig. **12**. Poissonian nucleation statistics would lead to a much larger standard deviation of 43 MLs. On the other hand, our model of self-regulated growth predicts that $H_T(t)$ should not deviate much from the fitting curve without nucleation statistics (Fig. **21**). We find that the deviation should become $\bar{m}$-independent after a few tens of nucleations. Using the values of B_V and c_V above-mentioned, we calculate that the standard deviation tends toward 1.2 ML for the 21 nm diameter NW considered here [56]. Because the kinetic model with a stationary droplet cannot describe perfectly the experiment, $\tilde{\sigma}$ is above this limit, but the large discrepancy with the Poissonian deviation for a 630 nm long NW confirms the existence of the self-regulation mechanism.

The self-regulated growth evidenced in this section is very beneficial to the control of NW length. Ensembles of NWs of identical diameter should present very small fluctuations of length. However, this mechanism operates only if one NW constituent has a low solubility in the catalyst particle, as is the case here for group-V atoms. The situation for Au-catalysed Si NWs, where the single constituent is highly soluble in the catalyst, is likely to be much less favourable in this respect. We thus confirm that the concentration of group-V species in the catalyst is low. This explains why sharp interfaces are easier to obtain by switching anions rather than cations (see section II.2).

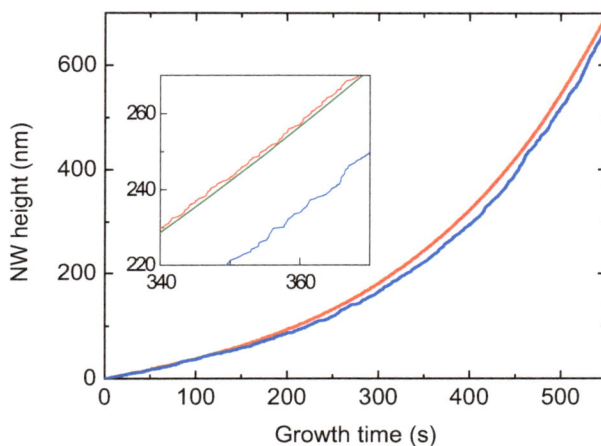

Figure 21: Simulation of NW elongation during growth. Green line: kinetic model with no statistics; blue line: kinetic model with Poissonian nucleation statistics; red line: kinetic model with self-regulated statistics. Close-up in insert: self-regulated growth deviates negligibly from the kinetic model with no statistics contrary to growth with Poissonian nucleation statistics.

CONCLUSIONS

NWs obtained by catalyst-assisted growth are fascinating objects which offer a great flexibility for the fabrication of original heterostructures, for instance the core-shell and buried nanowires presented here. However, their growth is complex and its control represents a real challenge. We have elucidated important mechanisms that operate in Au-assisted molecular beam epitaxy, some of them being valid in more general conditions. This method turns to be a vapor-liquid-solid growth for the III-V compounds investigated here. Nano-droplets of Au-based liquid catalyst are formed and act as reservoirs supersaturated with the NW constituents. The chemical potentials of III and V atoms in the drop govern the condensation of III-V nuclei forming the NW on one hand, and drive the incoming flows of atoms from vapor and adsorbed phases on the other hand. The specificity of the III-V NW growth is illustrated by the distinct roles of group-III and group-V atoms. The group-III atoms fill the catalyst drop *via* surface diffusion, the major contribution coming from the beam flux collected by the NW sidewalls. In this respect, the geometry of the growth chamber influences the growth kinetics. The group V atoms have a low concentration in the nano-sized drop and little variations of this concentration have a drastic impact on the nucleation rate. Their beneficial effect is to regulate the NW growth. The crystalline structure, dominantly WZ, is determined at the nucleation and depends on the supersaturation in the drop and on the surface energies in play. Those parameters are difficult to control independently. Nevertheless, fine tuning of III and V fluxes and temperature allows one to get crystalline structures free of stacking faults, for instance the InAsP NWs of WZ structure presented in this chapter. An alternative method is the epitaxial burying which transforms NWs with faulted WZ stacking into pure ZB NWs embedded in the epitaxial matrix. Other important issues, which are not addressed in this chapter, concern the doping and the control of NW growth along other crystalline directions, the [001] axis more particularly. Investigations on these topics are on-going worldwide.

ACKNOWLEDGEMENTS

We thank Christophe David, Nicolas Péré-Laperne and Marie-Noëlle Mérat-Combes for the AFM measurements shown in Figs. **1** and **16** and Damien Lucot for the electron beam lithography used for samples of Fig. **14**. We thank Vladimir Dubrovskii and George Cirlin (Academic University of the RAS, St Petersburg, Russia) for enlightening discussions and for George Cirlin's contribution to growth experiments. We thank Helge Weman, Bjørn-Ove Fimland, Dasa L. Dheeraj and Hailong Zhou (Norwegian University of Technology and Science, Trondheim, Norway) for fruitful discussions and for Dasa Dheeraj's and Hailong Zhou's participations to growth experiments. This work has received a financial support from the Agence Nationale de la Recherche (projects FILEMON 35 ANR-05-NANO-016 and BONAFO ANR-08-NANO-031) and from SANDiE European network of excellence.

REFERENCES

[1]   Deshpande, V. V.; Bockrath, M.; Glazman, L. I. ; Yacoby, A. Electron liquids and solids in one dimension. *Nature*, **2010**, *464*, 209-216.

[2]   Givargizov, E. I. Fundamental aspects of VLS growth. *J. Cryst. Growth*, **1975**, *31*, 20-30.

[3]   Petroff, P. M.; Gossard A. C.; Wiegmann W. Structure of AlAs-GaAs interfaces grown on (100) vicinal surfaces by molecular beam epitaxy. *Appl. Phys. Lett.* **1984**, *45*, 620-622

[4]   Bhat, R.; Kapon, E.; Simhony, S.; Colas, E.; Hwang, D.M.; Stoffel, N.G.; Koza, M.A. Quantum wire lasers by OMCVD growth on nonplanar substrates. *J. Cryst. Growth*, **1991**, *107*, 716-723

[5]   Koguchi, M.; Kakibayashi, H.; Yasawa, M.; Hiruma, K.; Katsuyama, T. Crystal structure change of GaAs and InAs whiskers from zinc-blende to wurtzite type. *Jpn. J. Appl. Phys.*, **1992**, *31*, 2061-2065.

[6]   Wagner R. S.; Ellis W. C. Vapor-liquid-solid mechanism of single crystal growth. *Appl. Phys. Lett.*, **1964**, *4*, 89-91.

[7]   Bryans, R. G.; James, D. W. F. The morphology of silicon whiskers grown by the vapor liquid solid process. *Micron,* **1969**, *2*, 268-281

[8]   Givargizov, E.I.; Sheftal, N.N. Morphology of silicon whiskers grown by the VLS-technique. *J. Cryst. Growth*, **1971**, *9*, 326-329.

[9]   Trentler, T. J.; Hickman, K. M.; Goel, S. C.; Viano, A. M.; Gibbons, P. C.; Buhro, W. E. Solution-Liquid-Solid Growth of crystalline III-V semiconductors: an analogy to vapor-liquid-solid growth. *Science*, **1995**, *270*, 1791-1794.

[10] Duan, X.F.; Huang, Y.; Cui, Y.; Wang, J.F.; Lieber, C.M. Indium phosphide nanowires as building blocks for nanoscale electronic and optoelectronic devices. *Nature*, **2001**, *409*, 66-69.

[11] Bjork, M. T.; Ohlsson, B. J.; Sass, T.; Persson, A. I.; Thelander, C.; Magnusson, M. H.; Deppert, K.; Wallenberg, L. R.; Samuelson, L. Indium phosphide nanowires as building blocks for nanoscale electronic and optoelectronic devices. *Appl. Phys. Lett.*, **2002**, *80*, 1058-1060.

[12] Tchernycheva, M.; Harmand, J.-C.; Patriarche, G.; Travers, L.; Cirlin, G. E. Temperatures conditions for GaAs nanowire formation by Au-assisted molecular beam epitaxy. *Nanotechnology*, **2006**, 17, 4025-.

[13] Massalski T. B. *Binary alloy phase diagrams*, 1st ed.; American Society for Metals, Metals Park: Ohio, **1986**.

[14] Tchernycheva M.; Travers L.; Patriarche G.; Glas F.; Harmand J.-C.; Cirlin G. E.; Dubrovskii V. G. Au-assisted molecular beam epitaxy of InAs nanowires: growth and theoretical analysis. *J. Appl. Phys.*, **2007**, *102*, 094313.

[15] Xu Q.; Sharp I. D.; Yuan C. W.; Yi D. O.; Liao C. Y.; Glaeser A. M.; Minor A. M.; Beeman J. W.; Ridgway M. C.; Kluth P.; Ager J. W.; Chrzan D. C.; Haller E. E. Large melting-point hysteresis of Ge nanocrystals embedded in SiO_2. *Phys. Rev. Lett.,* **2006**, *97*, 155701.

[16] Buffat, P.; Borel, J.-P. Size effect on the melting temperature of gold particles, *Phys. Rev. A*, **1976**, *13*, 2287-2298.

[17] Park, H.D.; Gaillot, A.C.; Prokes, S. M.; Cammarata, R. C. Observation of size dependent liquidus depression in the growth of InAs nanowires. *J. Cryst. Growth* **2006**, *296*, 159–164.

[18] Dick, K. A. ; Deppert, K. ; Mårtensson,T. ; Mandl, B. ; Samuelson, L. ; Seifert, W. Failure of the vapor–liquid–solid mechanism in Au-assisted MOVPE growth of InAs nanowires. *Nano Lett.*, **2005**, *5*, 761-764.

[19]   Kodambaka, S.; Tersoff, J.; Reuter, M. C.; Ross F. M. Germanium nanowire growth below the eutectic temperature. *Science*, **2007**, *316*, 729-732.

[20]   Harmand, J.-C.; Tchernycheva, M.; Patriarche, G.; Travers, L.; Glas, F.; Cirlin, G. E. GaAs nanowires formed by Au-assisted molecular beam epitaxy: Effect of growth temperature. *J. Cryst. Growth*, **2007**, *301*, 853-856.

[21]   Thelander, C.; Martensson, T.; Bjork, M.T.; Ohlsson, B.J.; Larsson, M.W.; Wallenberg, L. R.; Samuelson, L. Single-electron transistors in heterostructure nanowires. *Appl. Phys. Lett.,* **2003**, *83*, 2052-2054.

[22]   Tchernycheva, M.; Cirlin, G. E. ; Patriarche, G.; Travers, L.; Zwiller, V.; Perinetti, U.; Harmand, J. C. Growth and characterization of InP nanowires with InAsP insertions. *Nano Lett.*, **2007**, *7*, 1500-1504.

[23]   Mohseni, P. K.; Maunders, C.; Botton, G. A.; LaPierre, R. R. GaP/GaAsP/GaP core-multishell nanowire heterostructures on (111) silicon. *Nanotechnology*, **2007**, *18*, 445304.

[24]   Verheijen, M.A.; Immink, G.; de Smet, T.; Borgstrom, M. T.; Bakkers, E. P. A. M. Growth kinetics of heterostructured GaP-GaAs nanowires. *J. Am. Chem. Soc.*, **2006**, *128*, 1353-1359.

[25]   Persson, A. I.; Larsson, M. W.; Stenstrom, S.; Ohlsson, B. J.; Samuelson, L.; Wallenberg, L. R. Solid-phase diffusion mechanism for GaAs nanowire growth. *Nat Mater.*, **2004**, *3*, 678–681.

[26]   Harmand, J.-C.; Patriarche, G.; Pere-Laperne, N.; Mérat-Combes, M.-N.; Travers,L.; Glas, F. Analysis of vapor-liquid-solid mechanism for Au-assisted GaAs nanowire growth. *Appl. Phys. Lett.*, **2005**, *87*, 203101.

[27]   Mohan, P.; Motohisa, J.; Fukui, T. Controlled growth of highly uniform, axial/radial direction-defined, individually addressable InP nanowire arrays. *Nanotechnology*, **2005**, *16*, 2903-2907.

[28]   Soshnikov, I. P.; Cirlin, G. E.; Tonkikh, A. A.; Samsonenko, Y. B.; Dubovskii, V. G.; Ustinov, V. M.; Gorbenko O. M.; Litvinov, D.; Gerthsen, D. Atomic structure of MBE-grown GaAs nanowhiskers. *Phys. Sol. State*, **2005**, *47*, 2121-2126.

[29]   Spirkoska, D.; Arbiol, J.; Gustafsson, A.; Conesa-Boj, S.; Glas, F.; Zardo, I. ; Heigoldt, M.; Gass, M. H.; Bleloch, A. L.; Estrade, S.; Kaniber, M.; Rossler, J.; Peiro, F.; Morante, J. R.; Abstreiter, G.; Samuelson; L.; Fontcuberta i Morral, A. Structural and optical properties of high quality zinc-blende/wurtzite GaAs nanowire heterostructures. *Phys. Rev. B*, **2009**, *80*, 245325.

[30]   Akopian, N.; Patriarche, G.; Liu, L.; Harmand J.-C.; Zwiller, V. Crystal phase quantum dots. *Nano Lett.*, **2010**, *10*, 1198–1201.

[31]   Joyce, H. J.; Gao,Q.; Tan, H.H.; Jagadish, C.; Kim, Y.; Fickenscher, M. A.; Perera, S.; Hoang, T.B.; Smith, L. M.; Jackson, H. E.; Yarrison-Rice, J. M.; Zhang, X.; Zou, J. High purity GaAs nanowires free of planar defects: growth and characterization. *Adv Funct Mater*, **2008**, *18*, 3794 – 3800.

[32]   Dheeraj, D. L.; Patriarche, G.; Zhou, H.; Hoang, T.B.; Moses, A. F.; Grønsberg, S.; van Helvoort, A. T. J; Fimland, B.-O.; Weman, H. Growth and characterization of wurtzite GaAs nanowires with defect-free zinc-blende GaAsSb inserts. *Nano Lett.*, **2008**, *8*, 4459.

[33]   Shtrikman, H.; Popovitz-Biro, R.; Kretinin, A.; Houben, L.; Heiblum, M.; Bukała, M.; Galicka, M.; Buczko, R.; Kacman, P. Method for suppression of stacking faults in wurtzite III-V nanowires. *Nano Lett.*, **2009**, *9*, 1506-1510.

[34]   Caroff, P.; Dick, K. A.; Johansson, J.; Messing, M. E.; Deppert, K.; Samuelson, L. Controlled polytypic and twin-plane superlattices in III-V nanowires. *Nature Nanotechnol*, **2009**, *4*, 50-55.

[35]   Algra, R. E.; Verheijen, M. A.; Borgström, M. T.; Feiner, L.-F.; Immink, G.; van Enckevort, W. J. P.; Vlieg, E.; Bakkers, E. P. A. M. Twinning superlattices in InP nanowires. *Nature*, **2008**, *456*, 369-372.

[36]   Leitsmann, R.; Bechstedt, F. Surface influence on stability and structure of hexagon-shaped III-V semiconductor nanorods. *J. Appl. Phys.*, **2007**, *102*, 063528.

[37]   Galicka, M.; Bukala, M.; Buczko, R.; Kacman, P. Modelling the structure of GaAs and InAs nanowires. *J. Phys.: Condens. Matter*, **2008**, *20*, 454226.

[38]   Hilner, E.; Håkanson, U.; Fröberg L. E.; Karlsson M.; Kratzer P.; Lundgren E.; Samuelson L.; Mikkelsen, A. Direct atomic scale imaging of III-V nanowire surfaces. *Nano Lett.*, **2008**, *8*, 3978-3982.

[39]   Akiyama, T.; Sano, K.; Nakamura, K.; Ito, T. An empirical potential approach to wurtzite–zinc-blende polytypism in group III–V semiconductor nanowires. *Jpn. J. Appl. Phys.*, **2006**, *45*, L275-L278.

[40]   Glas, F.; Patriarche, G.; Harmand, J.-C. Growth, structure and phase transitions of epitaxial nanowires of III-V semiconductors. 16th International Conference on Microscopy of Semiconducting Materials. *Journal of Physics: Conference Series* **2010**, *209*, 012002.

[41]   Wen, C.-Y.; Reuter M. C.; Bruley, J.; Tersoff, J. ; Kodambaka, S.; Stach, E. A.; Ross, F. M. Formation of compositionally abrupt axial heterojunctions in Si/Ge nanowires. *Science*, **2009**, *326*, 1247.

[42]   Glas, F.; Harmand, J.-C.; Patriarche G. Why does wurtzite form in nanowires of III-V zinc blende semiconductors? *Phys. Rev. Lett.*, **2007**, *99*, 146101.

[43]   Dubrovskii, V. G.; Sibirev, N. V. ; Harmand, J.-C.; Glas, F. Growth kinetics and crystal structure of semiconductor nanowires. *Phys. Rev. B*, **2008**, *78*, 235301.

[44] Dayeh, S. A.; Yu, E. T.; Wang, D. D. Surface diffusion and substrate–nanowire adatom exchange in InAs nanowire growth, *Nano Lett.*, **2009**, *9*, 1967-1972

[45] Plante, M. C.; LaPierre, R. R. Analytical description of the metal-assisted growth of III-V nanowires: axial and radial growths. *J. Appl. Phys.*, **2009**, *105*, 114304.

[46] Carlino, E.; Grillo, V. Atomic-resolution quantitative composition analysis using scanning transmission electron microscopy Z-contrast experiments. *Phys. Rev. B*, **2005**, *71*, 235303.

[47] Dubrovskii, V. G.; Sibirev, N. V.; Cirlin, G. E.; Bouravleuv, A. D.; Samsonenko, Y. B.; Dheeraj, D. L.; Zhou, H. L. ; Sartel, C.; Harmand, J. -C.; Patriarche, G. ; Glas, F. Role of nonlinear effects in nanowire growth and crystal phase. *Phys. Rev. B*, **2009**, *80*, 205305.

[48] Glas, F. Vapor fluxes on the apical droplet during nanowire growth by molecular beam epitaxy. *Phys. Stat. Sol. (b)*, **2010**, *247*, 254-258.

[49] Harmand, J.-C.; Glas, F.; Patriarche, G. Chronology of nanowire growth. Submitted to *Phys. Rev. B*.

[50] F. Glas, submitted to *J. Appl. Phys.*

[51] Sartel, C.; Dheeraj, D.L.; Jabeen, F.; Harmand, J.-C. Effect of arsenic species on the kinetics of GaAs nanowires growth by molecular beam epitaxy. *J. Cryst. Growth*, **2010**, *312*, 2073-2077.

[52] Mikkelsen, A.; Sköld, N.; Ouattara, L.; Lundgren, E. Nanowire growth and nanowire doping studied by cross-sectional scanning tunneling microscopy. *Nanotechnology*, **2006**, *17*, S362-S368.

[53] Patriarche, G.; Glas, F.; Tchernycheva, M.; Sartel, C.; Largeau, L.; Harmand J.-C.; Cirlin G. E. Wurtzite to Zinc Blende Phase Transition in GaAs Nanowires Induced by Epitaxial Burying. *Nano Lett.*, **2008**, *8*, 1638–1643

[54] Bollmann, W. *Crystal defects and crystalline interfaces*; Springer: Berlin, 1970.

[55] Glas, F. A simple calculation of energy changes upon stacking fault formation or local crystalline phase transition in semiconductors. *J. Appl. Phys.*, **2008**, *104*, 093520.

[56] Glas, F.; Harmand, J.-C.; Patriarche, G. Nucleation antibunching in catalyst-assisted nanowire growth. *Phys. Rev. Lett.*, **2010**, *104*, 135501.

[57] Kashchiev, D. Dependence of the growth rate of nanowires on the nanowire diameter. *Cryst. Growth and Design*, **2006**, *6*, 1154-1156.

[58] Chatillon, C.; Hodaj, F.; Pisch, A. Thermodynamics of GaAs nanowire MBE growth with gold droplets. *J. Cryst. Growth*, **2009**, *311*, 3598-3608.

CHAPTER 5

III-Antimonide Nanowires

Helge Weman* and Dheeraj L. Dasa

Department of Electronics and Telecommunications, Norwegian University of Science and Technology (NTNU), NO-7491 Trondheim, Norway

Abstract: We review the recent progress in the growth, structural, optical and electrical properties of GaAs/GaAsSb, GaAs/GaSb, and InAs/InSb nanowires (NWs) grown by the vapor-liquid-solid (VLS) mechanism. The structural characterization of GaAsSb NWs reveals that they adopt zinc blende (ZB) crystal phase, whereas the same growth conditions (except the Sb flux) produce GaAs NWs with wurtzite (WZ) crystal phase. With increasing mole fraction of Sb, the ZB GaAsSb NW forms a lower density of twinning planes. Low-temperature photo-luminescence (PL) characterization of axially heterostructured GaAs/GaAsSb NWs shows that the linear polarization of the PL emission from the ZB GaAsSb inserts is opposite (parallel to NW axis) to the WZ GaAs segments (perpendicular to NW axis) due to a difference in the optical selection rules between ZB and WZ crystals. GaSb NWs have been grown on top of GaAs NWs by the Au-assisted VLS method. The two most striking differences between the GaSb and GaAs NWs are: 1) The diameter of the GaSb NWs is significantly larger than that of the GaAs NWs. 2) Whereas the GaAs NWs have WZ crystal phase with stacking faults, the GaSb NWs exhibit a defect-free ZB crystal phase. Undoped GaSb NWs have so far been determined to be *p*-type. Low-temperature PL measurements on GaSb NWs reveal a PL peak near 0.8 eV, with the energy position dependent on the V/III ratio. InSb NWs grown on top of InAs NWs show similar behavior in terms of diameter and crystal phase change as the GaAs/GaSb NWs.

Keywords: GaAsSb, GaSb, InSb, GaAs, antimonides, nanowire, zinc blende, wurtzite.

INTRODUCTION

III-antimonide materials such as GaSb ($E_g \sim 0.72$ eV), InSb ($E_g \sim 0.17$ eV), and AlSb ($E_g \sim 1.6$ eV) are useful semiconductor materials in their binary form as well as in alloys with Al, Ga, In, As, and P. By modulating the composition of such III-antimonides they can be used in various near-, mid- and long-wave infrared (IR) detectors and laser devices [1]. In fact, the detection of longer wavelengths, 8 - 14 µm, is possible with intersubband absorption in Sb based superlattices [2], and in dilute nitride antimonides. This reduces the need for the mercury- and cadmium-containing toxic materials that are currently used to accomplish these long wavelengths. There is also large interest in Sb compounds due the high carrier mobility, large exciton Bohr radii, and large spin-orbit coupling.

GaSb is often referred to as an intermediate-gap semiconductor, *i.e.*, its bandgap of 0.72 eV is neither as wide as in GaAs and InP nor as narrow as in InAs and InSb. GaSb based ternary and quaternary heterostructures offers a wide range of band gaps available from 0.8 - 4.4 µm that are interesting for different applications in optical communication, infrared imaging, and gas sensing. The unique properties of GaSb based materials such as high hole mobility (highest hole mobility of all III-V compounds) and high hole ionization coefficients, have led to significant improvements in avalanche photo detectors with high signal-to-noise ratio [3], and laser diodes with low threshold currents [1,4], respectively. Since the GaAs/GaAsSb hetero-junction can exhibit type II band-alignment, it can also be useful for devices like optical memories and solar cells [5]. It is also known that the band-alignment can be altered between type II and type I depending on the amount of strain across the lattice mismatched GaAs/GaAsSb hetero-junction [6]. A comprehensive review of the material and structural properties of GaSb was reported by Dutta *et al.* [7].

InSb with the most narrow band gap (0.17 eV), the largest bulk electron mobility (78 000 cm^2/Vs), and largest electron g-factor (51) of any III-V material have lead to significant improvement in e.g. the performance of high-speed electronic devices [8]. InSb also has a high thermoelectric figure of merit which can make it useful in thermoelectric power applications [9]. InSb is also a widely used material for mid-IR detector applications.

*Address correspondence to Helge Weman: Department of Electronics and Telecommunications, Norwegian University of Science and Technology (NTNU), NO-7491 Trondheim, Norway; Tel: +47-735 94409; E-mail: helge.weman@iet.ntnu.no

Jianye Li, Deli Wang and Ray R. LaPierre (Eds)

In spite of several useful properties for device application, the development of III-Sb NWs have lagged behind other semiconductor NWs. One reason to this is their large lattice mismatch to GaAs (7.8 % lattice mismatch with GaSb) and InAs (7 % lattice mismatch with InSb). Such problems can be avoided in the growth of NWs, which provides the provision to relax in the radial direction without forming any dislocations. Recent attempts have shown that high-quality GaAs/GaSb NWs can indeed be fabricated despite the large lattice-mismatch [10]. The low ionicity of GaSb is believed to be part of the reason in the realization of the defect-free ZB GaSb NWs by metal-organic vapor phase epitaxy (MOVPE) [11]. So far, even less work has been made on the growth of ternary GaAsSb NWs. We have demonstrated the growth of GaAsSb NWs, and axially heterostructured GaAs/GaAsSb NWs on (111)B oriented GaAs substrates by Au-assisted molecular beam epitaxy (MBE) [12-14]. Recently, the growth of axially hetero-structured GaAs/GaAsSb NWs grown on Si(111) substrates by Ga-assisted MBE was also demonstrated [15].

Axially hetero-structured InAs/InSb and InSb NWs have been grown by Au-assisted MOVPE [16,17] and chemical beam epitaxy (CBE) [18], as well as catalyst-free techniques [19].

In the next three sections we review the recent progress in the growth, structural, optical and electrical properties of GaAs/GaAsSb (section 2), GaAs/GaSb (section 3), and InAs/InSb NWs (section 4) grown by the vapor-liquid-solid (VLS) mechanism.

GaAs/GaAsSb NANOWIRES

The growth of GaAs/GaAsSb NWs has so far been demonstrated by Au-assisted [12-14] and self-catalyzed [15] MBE. Typical tilted-view scanning electron microscopy (SEM) images of the as-grown GaAsSb NWs (~ 25 % Sb) grown on GaAs(111)B substrates by Au-assisted MBE, and GaAs/GaAsSb heterostructured NWs grown on Si(111) substrates by self-catalyzed MBE are shown in Figs. **1(a)** and **1(b)**, respectively. The NWs grow normal to the surface with the Au or Ga nano particle seen at the top of the NWs. Diameters of the NWs are around ~ 50 - 100 nm and nearly uniform along the NW. The NW length is observed to decrease with increase in NW diameter [12]. This dependence of the NW length on the diameter confirms that the growth is to a large extent fed by diffusion of adatoms from the surface of the substrate to the Au droplet at top of the NW [12].

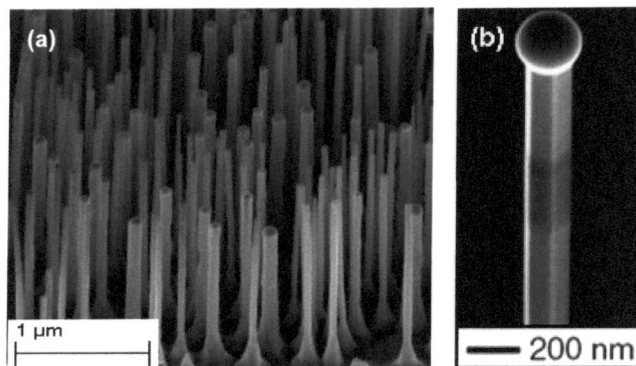

Figure 1: Tilted-view SEM images of (a) as-grown GaAsSb NWs (25 % Sb) grown on a GaAs(111)B substrate by Au-assisted MBE, and (b) GaAs/GaAsSb axial heterostructured NW grown on a Si(111) substrate by self-catalyzed MBE. **(a)** Reproduced from Dheeraj *et al.* [12]. **(b)** Reprinted with permission from [15], Applied Physics Letters, *96*, 121901. Copyright 2010 American Institute of Physics.

We have studied the GaAsSb NW growth under different Sb fluxes. The Sb content of the GaAsSb NW was investigated by energy dispersive X-ray (EDX) spectroscopy. The Sb mole fraction was determined from the loss of As signal in the GaAsSb NW compared to that of a pure GaAs NW. The average Sb mole fraction in the GaAsSb NWs grown under 2, 4, 6 and 8 × 10^{-7} Torr Sb flux is approximately 5, 15, 25 and 35 %, respectively. However, a large distribution in the Sb mole-fraction between different GaAsSb NWs within the same sample, and change in the Sb mole fraction along the length of the NW has been observed. Most of the GaAsSb NWs exhibit a decrease in the Sb mole fraction with its length. We believe this is mainly caused by either an increase in the substrate temperature or an increase in the growth rate of the NWs with its length.

In Fig. **2**, we show transmission electron microscopy (TEM) images of a typical GaAsSb NW (~ 25 % Sb) [12]. The most striking feature is that the GaAsSb NWs adopt zinc blende (ZB) structure, whereas the same growth conditions (except the Sb flux) produce GaAs NWs with wurtzite (WZ) structure.

Fig. **2(a)** shows a [110] zone axis high-resolution TEM image revealing the ZB structure of the GaAsSb NW, with a twinning plane (indicated by a black arrow) perpendicular to the [111]B growth axis. The crystal planes above the twinning plane (orientation A) is a rotation by 60° relative to the crystal planes below the twinning plane (orientation B) about the growth axis. The selected area electron diffraction pattern in Fig. **2(b)** shows additional spots related to the presence of both crystal orientations. Using the $1\bar{1}1$ diffraction spot marked by a circle in Fig. **2(b)**, we obtained the dark field TEM image as shown in Fig. **2(c)**, where crystal orientation A appears bright and the crystal orientation B dark. By investigating the full length of several GaAsSb NWs (~ 25 % Sb), single-oriented ZB segments as long as 500 nm were observed [12]. It is believed that the formation of a ZB crystal structure in the GaAsSb NW is due to either an increase of the material-dependent critical supersaturation value or a decrease of the supersaturation in the Au particle (or both) [12].

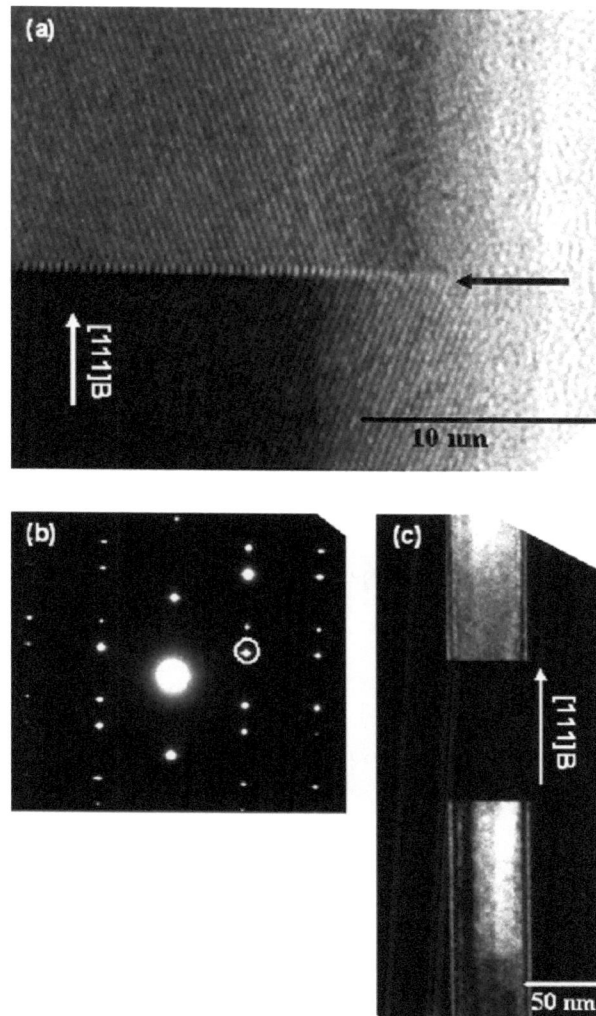

Figure 2: (a) High-resolution TEM image and **(b)** selected area electron diffraction pattern showing a ZB structure with twins. **(c)** Dark-field image obtained by using the $1\bar{1}1$ diffraction spot, marked by a circle in (b). Reproduced from Dheeraj *et al.* [12].

By using the $1\bar{1}1$ diffraction spot from the TEM images of GaAsSb NWs with different Sb composition, we obtained the dark field image as shown in Fig. **3**, where twinned segments appear bright and un-twinned dark.

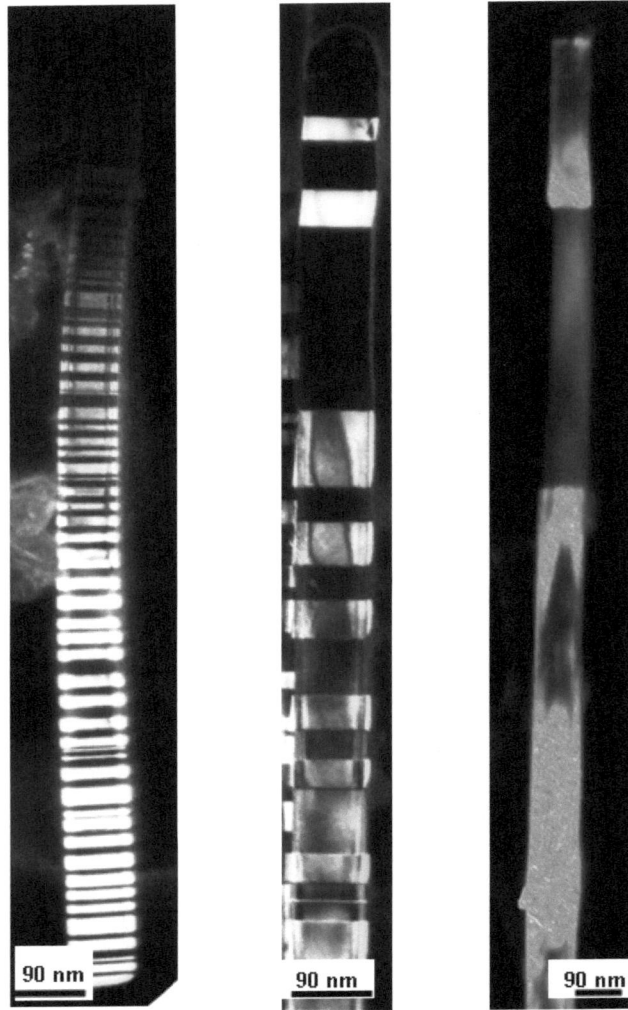

Figure 3: Dark-field TEM image obtained by using the 1 $\bar{1}$ 1 diffraction spot as in Fig. **2(c)**, for GaAsSb NWs with different Sb fluxes. The average Sb composition was estimated to be 2 %, 10 % and 25 % in the NW, from left to right, respectively. Unpublished work by Dheeraj *et al.*

The twinning frequency is observed to decrease with increase in Sb composition. Further, WZ GaAs NWs with short ZB GaAsSb inserts have also been grown. It was found that the transition from WZ GaAs to ZB GaAsSb was abrupt while the transition back from ZB GaAsSb to WZ GaAs often exhibit twin defects, a 4H polytype and extended stacking faults [14]. A TEM image of such NW is shown in Fig. **4**.

The GaAs NW segment before the GaAsSb insert exhibits a pure WZ phase with almost no stacking faults. As can be seen in Fig. **4(a)** and the high-resolution TEM image of Fig. **4(c)**, the transition (marked as T1) from the GaAs WZ phase to the GaAsSb ZB phase insert is very abrupt. The second transition, from the GaAsSb ZB insert to the upper GaAs NW segment (marked as T2), is shown by the high-resolution TEM in Fig. **4(b)**. In most of the NWs, the defect-free GaAsSb ZB insert was directly followed by a few nanometers of a twinned cubic GaAs phase (ZB/3C). Above this microtwin, a GaAs 4H polytype phase is formed, before the GaAs NW segment returns to a WZ phase. The appearance of a $1/4$ (0002) reflection in fast Fourier transforms of the high-resolution TEM images of the 4H phase (not shown here) confirms a GaAs 4H polytype phase with the stacking sequence of ABCBABCB….

The 4H polytype phase is well known in bulk form in wide band gap materials such as SiC and AlN. The 4H polytype of SiC is studied in detail due to its superior electron mobility [20]. We believe that the 4H polytype is the intermediate phase between the ZB and WZ phases. The gradual change in the supersaturation conditions from conditions favouring the ZB crystal phase to WZ crystal phase, could lead to the formation of intermediate phases such as cubic twins and a 4H polytype phase and delay the formation of the WZ phase [21]. It is reasonable to believe that the supersaturation is gradually increasing after the shut-off of the Sb flux and that this induces the 4H polytype crystal phase. However, no Sb was found above the insert or in the gold particle by EDX [14]. In fact, we have recently shown that adapting a growth interruption after the growth of the GaAsSb insert eliminates the formation of intermediate phases [22].

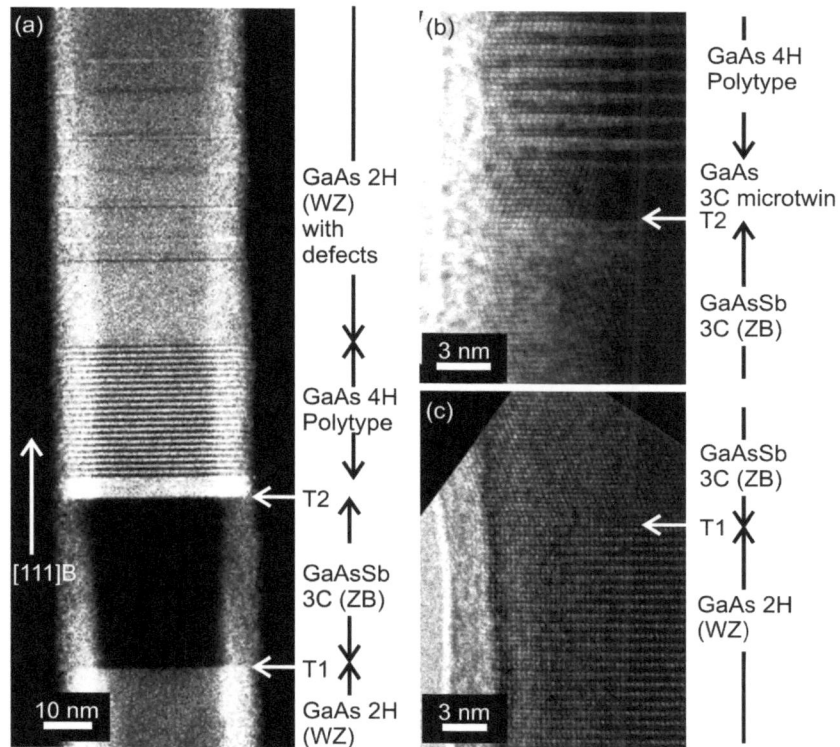

Figure 4: Dark-field **(a)** and high-resolution **(b, c)** TEM images of a typical WZ GaAs NW with a ZB GaAsSb insert. Reprinted with permission from Ref. [14]. Copyright 2008 American Chemical Society.

The WZ GaAs NWs with short ZB GaAsSb inserts with Sb mole fraction of 25 % is believed to exhibit type II band alignment [22]. However, due to the high surface area to volume ratio of NWs, surface states deteriorate the optical properties of these GaAs-based NWs. To avoid the problems with surface states and to enhance the optical properties, we have thus enclosed these heterostructured NWs with radial AlGaAs shells [23]. Such WZ GaAs/AlGaAs core-shell NWs with a single ZB GaAsSb insert have been characterized by low temperature micro-photoluminescence (μ-PL) [22].

A typical PL emission spectrum from a single WZ GaAs/AlGaAs core-shell NW with a ZB GaAsSb insert is shown in Fig. **5(a)**. The PL emission peak related to the GaAsSb insert is observed at ~ 1.27 eV. The ZB segment in the GaAs barrier at the top interface of the GaAsSb insert causes a type II band alignment. This results in the spatial separation of electrons and holes in the adjacent ZB GaAs and ZB GaAsSb layers, respectively. Fig. **5(b)** shows the power dependent spectra from the GaAsSb related peak at 1.27 eV.

With an increase of the excitation power with a factor of ten the PL emission is broadened and the PL peak is blue shifted by ~ 100 meV. The large blue shift can be attributed to band filling and band bending effect as the excitation intensity increases [24,25]. The spatially separated electrons and holes create an electric field at the GaAsSb/GaAs interface that results in band bending as the excitation density increases. The broadening of the PL peak at higher excitation intensities is probably due to the electron state filling in the ZB GaAs segment.

(a)

(b)

Energy (eV)

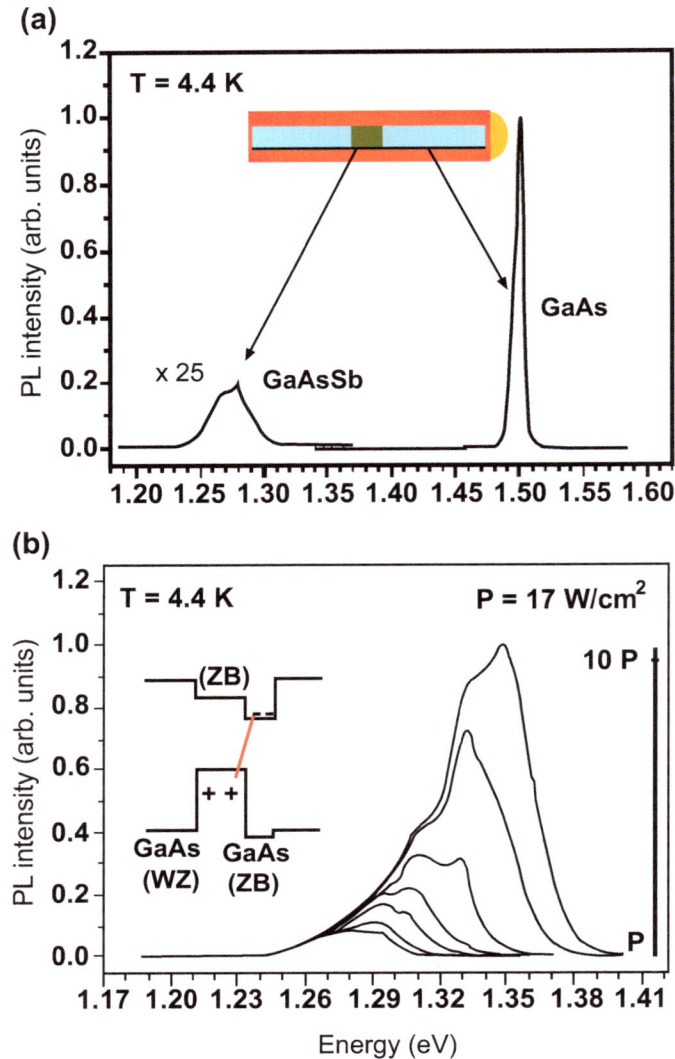

Figure 5: **(a)** μ-PL spectrum from a single WZ GaAs NW with a single ZB GaAsSb core. Inset shows a schematic image of the NW. **(b)** Power dependent spectra from the GaAsSb PL peak. Inset shows schematically the type II band alignment when the upper GaAs barrier contains a few nm thin ZB GaAs segment. Reproduced from Moses *et al.* [22].

Fig. **6(a)** shows the linear polarization dependence of the PL emission from a single WZ GaAs/AlGaAs core-shell NW with a ZB GaAsSb insert [26]. Zero degree angle is chosen to be along the long axis of the NW. The PL emissions from the GaAsSb core insert and the GaAs NW core barriers are here observed at energies ~ 1.28 eV and 1.49 eV, respectively. It can be seen that the PL emission from the ZB GaAsSb insert is strongly enhanced when it is measured parallel to the long axis of the NW, whereas the emission from the WZ GaAs NW barriers is strongly enhanced when measured perpendicular. In Fig. **6(b)** we show the PL spectra taken at 0 and 90 degrees polarization, respectively. It is clear that the PL emissions from the ZB GaAsSb insert and WZ GaAs NW barriers have orthogonal directions of linear polarization. In a WZ NW where the crystal *c*-axis is oriented along the long axis of the NW ([0001]), optical selection rules require that the emission is only allowed if the electric dipole moment is perpendicular to the crystal *c*-axis [27,28]. For a ZB NW, the optical selection rules play no role in the linear polarization of the emitted photons, and the polarization is thus completely dominated by the effect due to the dielectric mismatch as shown in earlier works [28-30].

The method of analyzing the linear polarization of interband transitions from NWs with mixed crystal phases is a useful optical characterization method to identify from which crystal phase the optical transitions originate.

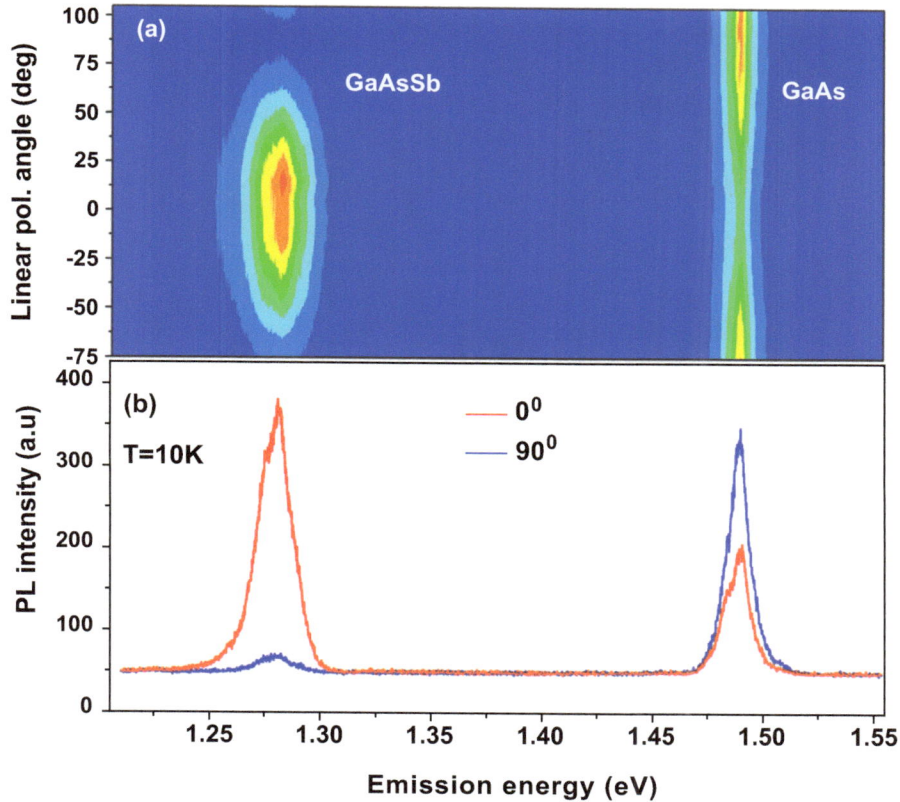

Figure 6: Linear polarization dependence of the PL emissions from a WZ GaAs NW with a ZB GaAsSb insert measured at 10 K. **(a)** False-color (red indicates high and blue low intensity) map shows the dependence of the PL intensity as a function of emission energy and the linear analyzer angle. Zero degree corresponds to the long axis of the NW. **(b)** PL spectra measured parallel (0°) and perpendicular (90°) to the NW. Reprinted with permission from Ref. [26]. Copyright 2010 American Chemical Society.

GaSb NANOWIRES

Growth of axially heterostructured GaAs/GaSb NWs overcomes the problems of variations in the Sb composition from one NW to another, or within a single NW. The growth of such NWs has been demonstrated by Au-assisted metal organic vapor phase epitaxy (MOVPE) [10,11,31]. Typical SEM images of such GaAs/GaSb NWs, grown at different substrate temperatures and Sb fluxes is shown in Fig. 7. [10].

The growth rate of GaSb NWs was reported to increase exponentially with substrate temperatures up to 470 °C, above which the growth rate saturates, and even decreases, with further increase in the temperature. An increase in the growth rate of GaSb NWs has also been observed with an increase in the Sb flux until the growth is no longer group V-limited. These results suggest that the rate limiting step in the MOVPE growth of GaSb NWs at lower substrate temperatures to be the thermal decomposition of TMSb, which is lower compared to TMGa [32].

The most significant differences between the GaSb and GaAs NW segments are: 1) The diameter of the GaSb segments is significantly larger than that of the GaAs segments. 2) Whereas the GaAs segments have WZ crystal phase with stacking faults the GaSb segments exhibit ZB crystal phase with no stacking faults or twinning planes. 3) The growth rate of the GaSb segments is 50 to 100 times lower than that of the GaAs segments. 4) The GaSb segments have vertical $\{\bar{1}10\}$ facets, in contrast to the {112} side facets normally observed in ZB dominated GaAs NWs grown by MOVPE. Interestingly no tapering was observed in the GaSb NWs, even after significant radial growth.

The increase in the diameter of the GaSb NW (when grown on top of a GaAs NW) was initially attributed to the result of strain relaxation *via* lateral expansion [31].

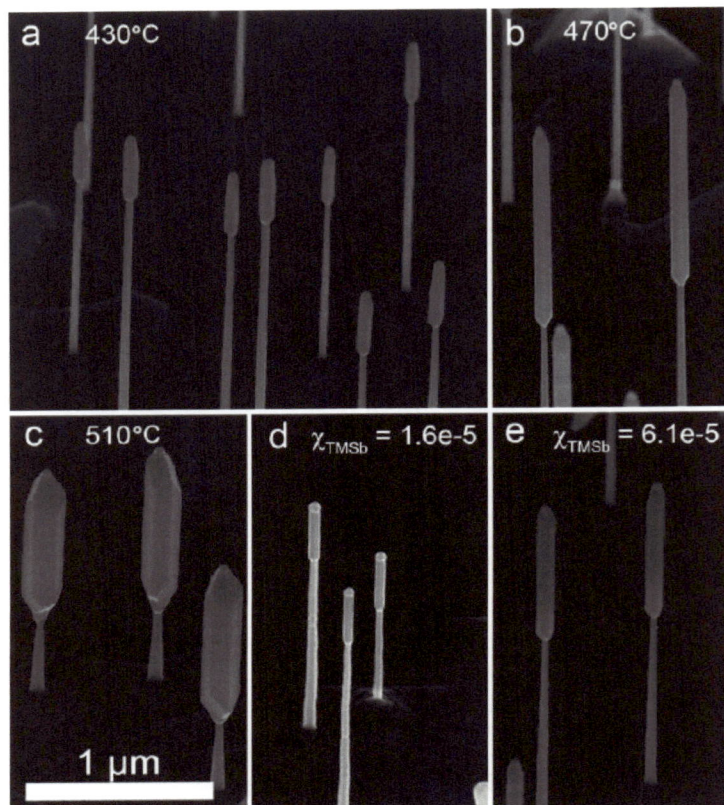

Figure 7: Tilted-view SEM images of axially heterostructured GaAs/GaSb NWs. The effect of growth temperature on the NW morphology is shown in **(a–c).** The effect of χ_{TMSb} on variation on NW morphology is shown in (d) and (e). All images have the same scale bar. Reprinted from [10], *Journal of Crystal Growth*, Vol. 310, 4115-4121. Jepppson *et al.*, GaAs/GaSb nanowire heterostructures grown by MOVPE. Copyright (2008), with permission from Elsevier.

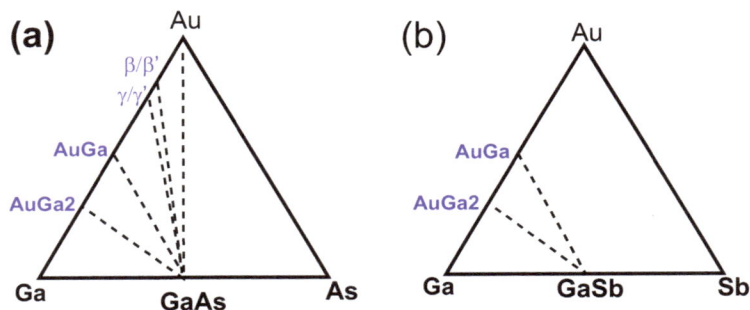

Figure 8: Schematic ternary phase diagrams of **(a)** the Au-Ga-As, and **(b)** the Au-Ga-Sb system. The pseudo-binary tie-lines between different compositions are shown by dotted lines.

However, EDX characterizations have revealed that the Au seed particles contain either 49 at % Ga or 64 at % Ga, when grown at higher or lower Sb flux, respectively, in comparison to between 10 and 15 at % Ga observed in GaAs NWs grown by the same MOVPE system [10]. This study confirms the difference between the Au-Ga-As and Au-Ga-Sb ternary phase diagrams. As shown in Fig. **8**, there exists only two pseudo-binary sections in this temperature range with tie-lines between GaSb and either AuGa or AuGa$_2$ (Fig. **8(b)**), while several such tie-lines exist between GaAs and different AuGa compounds (Fig. **8(a)**). In fact, the difference in the composition of the Au seed particle during the growth of GaAs and GaSb NWs explains the reason for the increase in the diameter of the GaSb NW during growth. The AuGa and AuGa$_2$ segments obtained during the growth of the GaSb NWs have orthorhombic B31 and cubic CaF$_2$ structure types, respectively. The unit cell volumes of these crystal structures are about two and three times larger, respectively, than for the Au-Ga crystal obtained during the growth of the GaAs NWs. Thus for a certain number of Au adatoms in the Au particle, the volume of AuGa and AuGa$_2$ will be approximately 1.28 and 1.5 times larger, respectively, than for a Au-Ga particle.

The high crystal quality of the ZB GaSb NWs is attributed to the low ionicity of the Ga-Sb bonds. GaSb has the smallest ionic bonds of all III-V materials [33], and is hence more stable in the ZB crystal structure than in the WZ crystal structure. Therefore, the twin planes, which originate due to a transition from ZB to WZ crystal structure, are not favorable in the GaSb NWs. As mentioned earlier, the high crystal quality of ZB GaAsSb NWs grown by Au-assisted MBE is attributed to the variation in supersaturation levels [12].

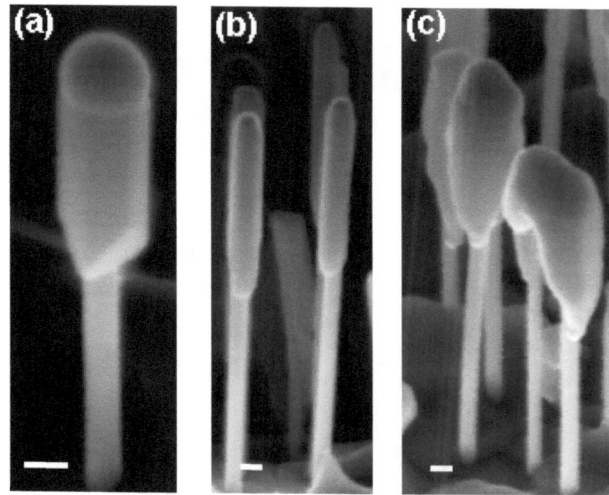

Figure 9: Tilted-view SEM images of axial hetero-structured GaAs/GaSb NWs grown by MBE. Typical morphology of NWs grown at substrate temperature of 450 °C with Sb flux of **(a)** 6×10^{-7}, **(b)** 8.6×10^{-7}, and **(c)** 9.6×10^{-7} Torr, respectively. Scale bar is 40 nm. Unpublished work by Dheeraj *et al.*

The lower growth rate of GaSb NWs has been attributed to the lower decomposition rate of TMSb [10]. In fact, it has been shown that by choosing higher TMGa and TMSb mol fractions with a V/III ratio of ~ 1, growth rates as high as 5 μm/min have been achieved [32].

The formation of NWs with { $\bar{1}$ 10} side facets have only been reported for materials such as GaSb, InSb, and Si, which adopt a defect-free ZB crystal phase, *i.e.* without any stacking faults or twin defects. However, at present very little is known about the relation between the NW side facets and the crystal phase [12].

To gain more insight into the mechanisms involved in the growth of GaSb NWs, we have made an attempt to grow heterostructured GaAs/GaSb NWs also by MBE. The temperature of the Ga effusion cell was preset to yield a nominal planar growth rate of 0.7 MLs^{-1}. The GaAs NW growth has been initiated by opening the shutter of the Ga effusion cell at a substrate temperature of 540 °C under an As$_4$ flux of 6 × 10^{-6} Torr. After the growth of GaAs NWs for 20 minutes, the substrate temperature was decreased to the desired temperature for the GaSb NW growth. The GaSb NWs were grown for 10 minutes with substrate temperatures between 400 °C and 540 °C, and a Sb$_2$ flux between 6 × 10^{-7} and 1.1 × 10^{-6} Torr. Typical tilted-view SEM images of axially heterostructured GaAs/GaSb NWs are shown in Fig. **9**. The different SEM images show the morphology of the GaSb NWs grown at a substrate temperature of 450 °C with a low (Fig. **9(a)**), optimum (Fig. **9(b)**) and high (Fig. **9(c)**) Sb flux. Similar to the observations made in GaSb NWs grown by MOVPE, the diameter of the GaSb NWs grown by MBE is larger than that of the GaAs NWs, with no observed tapering. However, in contrast to MOVPE, in MBE we observe comparable growth rates for GaSb and GaAs. NWs with morphology as shown in Fig. **9(a)**, have been observed in samples grown either at low Sb flux and low substrate temperature or at high Sb flux and high substrate temperatures. This indicates that the desorbtion of Sb increases with increase in substrate temperature, leading to Ga-rich conditions. Therefore, it is important to adjust the V/III ratio with a change in substrate temperature. The NWs with such morphology have been characterized by TEM and a typical NW is shown in Fig. **10**.

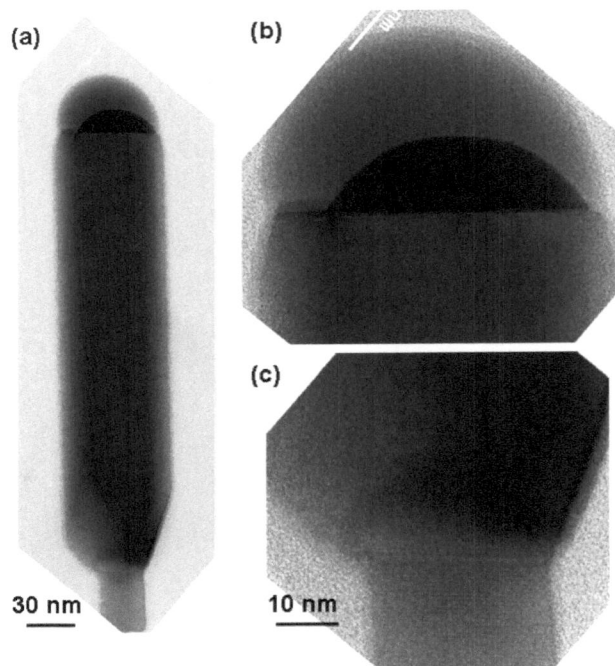

Figure 10: High-resolution TEM images of an axially heterostructured GaAs/GaSb NW grown with a low Sb flux (6×10^{-7} Torr) showing **(a)** an overview of the GaSb part of the NW, **(b)** the AuGa2 particle and amorphous Ga at the tip of GaSb NW, and **(c)** the abrupt interface of the WZ GaAs to ZB GaSb transition. Unpublished work by Dheeraj *et al.*

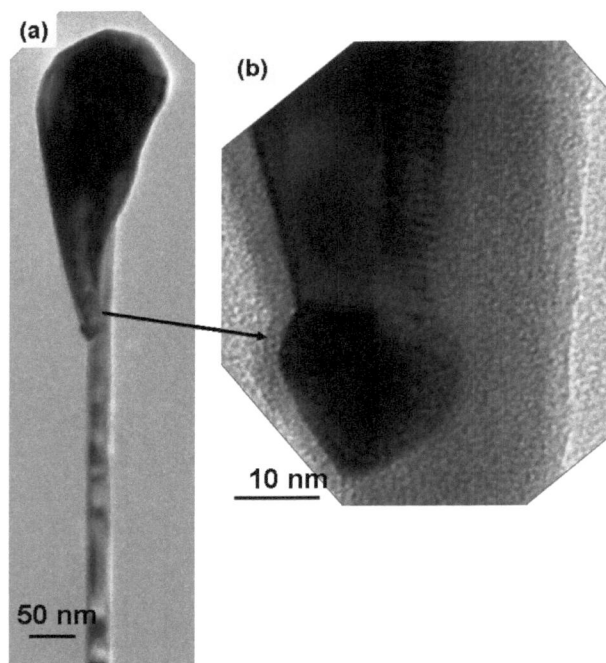

Figure 11: TEM images of an axially heterostructured GaAs/GaSb NW grown with a high Sb flux (9.6×10^{-7} Torr). **(a)** Shows an overview of the NW, and **(b)** at the GaAs/GaSb interface. Unpublished work by Dheeraj *et al.*

As can be seen from Figs. **10(a)** and **10(c)**, the GaAs NWs adapt WZ crystal phase while the GaSb NWs adapt a defect-free ZB crystal phase without any stacking faults or twin plane defects. Figs. **10(a)** and **10(b)** depict the presence of an Au nano particle with different composition at the NW tip. An EDX investigation revealed that the dark phase is $AuGa_2$ surrounded by an amorphous Ga phase in lighter contrast. Interestingly, no Sb was detected in the Au particle. It is plausible that Sb has already evaporated from the Au particle during cooling after the NW growth. We believe that the diameter of the amorphous Ga phase increases with increase in the diameter of the NW. Further, we have found that the diameter of $AuGa_2$ particle is about from 1.75 times to 2.25 times larger than that of the diameter of the GaAs NW, which is somewhat higher than the estimation by Jeppsson *et al.* [10]. We believe that this difference is due to the change in the contact angle of the Au particle with the NW surface.

NWs with morphology obtained in Sb-rich conditions, as shown in Fig. **9(c)**, indicates that the Au-assisted GaSb NW growth is hindered with an increase in the Sb flux. This is consistent with the observations made previously. Typical TEM images of such NWs is shown in Fig. **11**.

As shown in Fig. **11(b)**, the Au catalyst particle has solidified leading to the cessation of VLS growth of the GaSb NW. However, the GaSb NW continues to grow by the vapor-solid (VS) mechanism, and soon after looses its one-dimensional morphology.

Low temperature (5 K) μ-PL of single GaSb NW grown by MOVPE was reported by Jeppson *et al.* [11]. The PL peak was relatively narrow (8 meV) peaking at energy in the range from 0.785 eV to 0.805 eV. The origin of the PL emission is uncertain due to the large variations in peak position between different NWs [11]. More recently it has been shown that the PL peak energy as well the intensity from GaSb NWs is strongly dependent on the V/III ratio, as shown in Fig. **12** [32]. By making peak fits to the relatively broad PL peaks, transitions associated with various native defects were identified. At a V/III ratio of 1, higher-energy transitions were found to dominate the spectrum and the linewidth was reduced, indicating that the GaSb NWs were of higher optical quality [32].

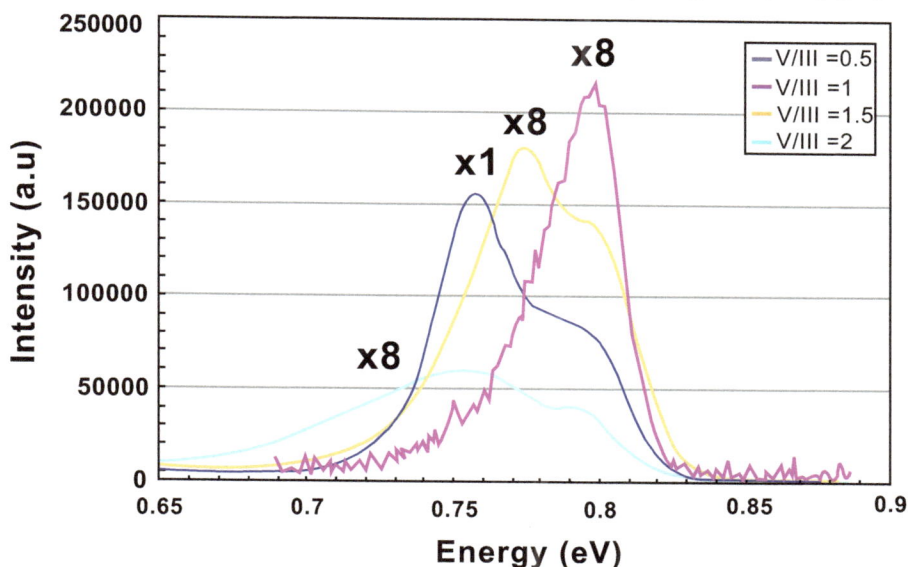

Figure 12: Low temperature (~ 5 K) PL spectra of GaSb NWs grown at V/III ratios of 0.5, 1, 1.5, and 2. Reprinted with permission from Ref. [32]. Copyright 2010 TMS.

The GaSb NWs grown at a V/III ratio of 1 were found to have the highest optical quality. Apart from the PL measurement of these GaSb NWs, the transport properties were also reported [11,32]. Typical room temperature *I–V* characteristics of the GaSb NWs grown by Jeppson *et al.* [11], are shown in Fig. **13**. The insets of Fig. **13** show the transfer characteristics and a SEM image of the contacted GaSb NW. The results indicate that intentionally undoped GaSb NWs have p-type conduction with a resistivity of ~ 0.2 cm. Similar results were observed in the GaSb NWs grown by Burke *et al.* [32].

Previously the unintentional p-type doping in bulk GaSb has been attributed with an excess of Ga or a deficiency of Sb [34]. One possible defect was speculated to be a native complex center, where a Ga atom occupies a Sb site due to a Sb vacancy and leaves a Ga vacancy behind. Due to the relatively low growth temperatures used for the GaSb NWs, it is also possible that carbon incorporation may also contribute to the p-type conductivity [32].

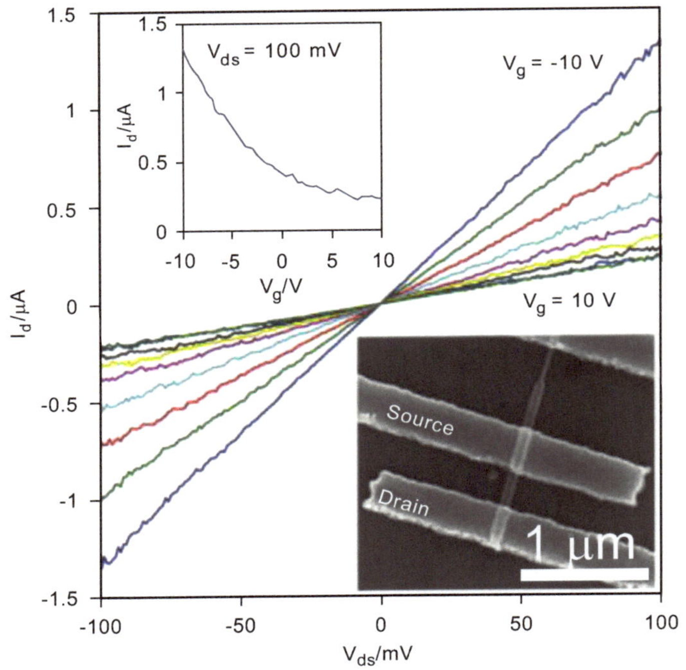

Figure 13: *I-V* characteristics of a back-gated GaSb NW at RT. The inset at the top shows the transfer characteristics indicating that the undoped GaSb NWs are p-type. The inset at the bottom shows a SEM image of the back-gated GaSb NW. Reprinted from [11], *Journal of Crystal Growth*, Vol. 310, 5119-5122. Jeppsson *et al.*, Characterization of GaSb nanowires grown by MOVPE. Copyright (2008), with permission from Elsevier.

InSb NANOWIRES

InSb NWs have been grown by Au-assisted MOVPE [16,17] and chemical beam epitaxy (CBE) [18], as well as catalyst-free techniques [19]. Figs. **14(a)** and **14(b)**, show typical SEM and TEM images of InAs/InSb NWs grown on InAs(111)B substrates by MOVPE [16]. The NWs grow vertically aligned on the substrate along the [111]B direction, and have a two-segment structure with a narrow base-segment (InAs segment) and a wide upper-segment (InSb segment). The InAs segments were found to be tapered, whereas the InSb segments were non-tapered [16].

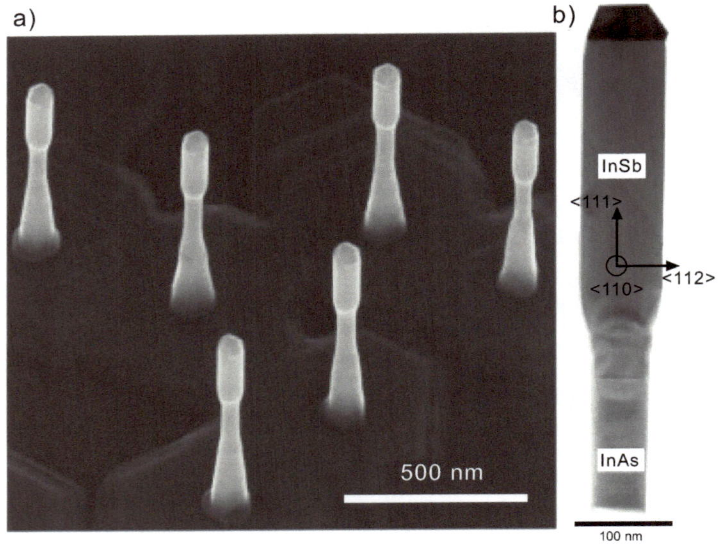

Figure 14: (a) Tilted-view SEM image of InAs/InSb NWs grown on a InAs(111)B substrate by Au-Assisted MOVPE, and **(b)** bright-field TEM image of a InAs/InSb NW from the same sample. Reprinted with permission from Ref. [16]. Copyright 2008 Wiley-VCH.

The observed differences between InAs and InSb NWs are similar to those observed for GaAs and GaSb NWs. For example, the diameter of the InSb NWs is larger than that of the InAs NWs, and whereas the InAs NW has WZ crystal phase (as GaAs NWs) the InSb NWs exhibit ZB crystal phase (as GaSb NWs) with no stacking faults and twinning planes. The formation of defect-free ZB InSb NWs is attributed to the low ionicity of the In-Sb bond.

The increased diameter of the InSb NW grown on top of an InAs NW has been determined to be due to the change of the In composition in the Au particle, *i.e.* the InAs NWs can grow from a supersaturated AuIn γ-phase particle (with ~ 30 % In) whereas the InSb NWs grow from the AuIn$_2$ phase. The AuIn$_2$-InSb system is known to be a pseudo binary eutectic system with up to 12 % InSb soluble in AuIn$_2$. In fact, InSb can be precipitated from a supersaturated particle (leaving AuIn$_2$) or from a supersaturated solid AuIn$_2$ particle. The state of the Au particle during the growth of InSb NWs is still not known [16]. Both solid and liquid particles may be involved even at the same growth temperature, as has been reported for Ge NWs [35].

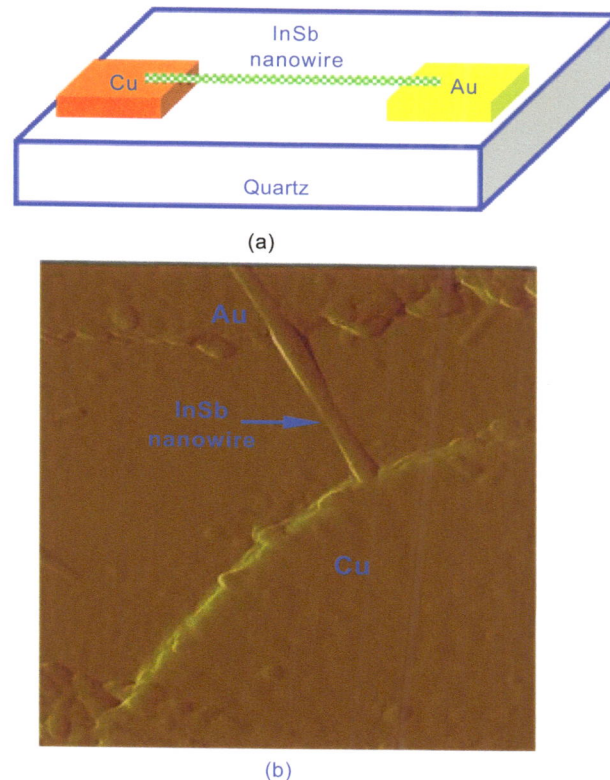

Figure 15: (a) Schematic picture of an InSb NW photodetector with contacts to Au and Cu electrodes on a quartz substrate. **(b)** AFM image of the contacted InSb NW photodetector. Reprinted with permission from Ref. [36]. Copyright 2009 IEEE.

Since InSb has very small band gap energy (0.17 eV at room temperature) this makes it difficult to measure its PL properties. This has only been scarcely done in bulk InSb, using Fourier transform PL and so far not for InSb NWs. The optical properties of InSb NWs can however be studied by IR absorption, which is important for NW based detector applications. Recently, an IR detector based on a single InSb NW was demonstrated [36]. The detector consists of an InSb NW connected to an Au and a Cu electrode. A schematic picture and an AFM image of the contacted InSb NW detector, is shown in Fig. **15**.

The InSb NW photodetector was demonstrated to detect IR signal at room temperature with a high quantum efficiency and low dark current [36].

The disadvantages of using narrow band gap semiconductor photodetectors, are that they normally generate very high dark currents. This may be overcome by using InSb NWs, due to the size shrinking and suppressed phonon scattering in NWs [36].

Electrical characterization was carried out on on the InAs/InSb NWs grown by Caroff *et al.* [16]. An SEM image of a contacted InSb NW segment is shown in Fig. **16(a)**. *I–V* characteristics are presented in Figs. **16(b)** and **16(c)**. The linear *I-V* curves indicate good ohmic behavior in the undoped InSb NW. The resistances in the InSb NW were measured to be 15 kΩ and 7 kΩ at 300 K and 4.2 K, respectively. From the back-gate dependence (Fig. **16(c)**), the undoped InSb NWs were shown to have n-type behavior [16].

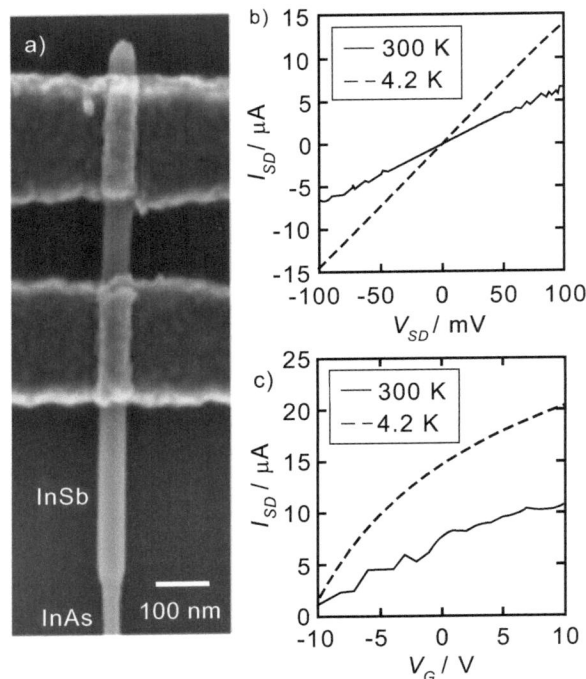

Figure 16: (a) SEM image of a contacted InSb NW. **(b)** *I–V* curves for an InSb NW at 300 K and 4.2 K. **(c)** Source–drain current trough the InSb NW plotted as a function of the backgate voltage for V_{s-d} =100 mV at 300 K and 4.2 K. Reprinted with permission from Ref. [16]. Copyright 2008 Wiley-VCH.

Magneto-transport measurements on InSb NW quantum dots were reported by Nilsson *et al.* [28]. They deduced giant *g* factors on such InSb NW quantum dots, with absolute values up to ~ 70.These measurements indicate that InSb NW also can have interesting properties for applications in spintronics.

SUMMARY

In this review, we have discussed the recent progress in the growth, structural, optical and electrical properties of III-antimonide NWs grown by the VLS mechanism.

GaAs/GaAsSb NWs grown by Au-assisted MBE reveals that the GaAsSb segment has ZB crystal phase, whereas the GaAs segments has WZ crystal phase. With increasing mole fraction of Sb, the ZB GaAsSb NW forms a lower density of twinning planes, being almost twinning-free at ~ 25 % Sb composition. Low-temperature PL of GaAs/GaAsSb NWs shows that the linear polarization of the PL emission from the ZB GaAsSb inserts is parallel to the NW axis whereas the WZ GaAs segments are perpendicular. The linear polarization aniso-tropy is explained by a difference in the optical selection rules between ZB and WZ crystals.

GaAs/GaSb NWs have been grown by Au-assisted VLS method by MOVPE and MBE. With both growth techniques, the diameter of the GaSb NWs is significantly larger than that of the GaAs NWs, and the GaAs NWs have WZ crystal phase and the GaSb NWs a defect-free ZB crystal phase. Un-intentionally doped GaSb NWs are found to be *p*-type. InSb NWs grown on top of InAs NWs show similar behavior in terms of diameter and crystal phase change as the GaAs/GaSb NWs. Transport studies of InSb NWs reveal that they are *n*-type, with a giant *g*-factor.

The results demonstrate that high quality ZB III-Sb NWs can be grown by the VLS method, and that these NWs have a high potential in future electronic and optoelectronic nano-device applications.

ACKNOWLEDGEMENTS

We want to thank our collaborators and group members at NTNU for their contributions to the work described in this review. Part of this work was supported by the 'NANOMAT' program (Grant No. 182091) of the Research Council of Norway.

REFERENCES

[1]   Rattunde, M.; Schmitz, J.; Mermelstein, C.; Kiefer, R.; Wagner, J., III-Sb-based Type-I QW Diode Lasers. In *Mid-infrared Semiconductor Optoelectronics*, **2006**; pp 131-157.

[2]   Xie, H.; Piao, J.; Katz, J.; Wang, W. I., Intersubband absorption in $Ga_{1-x}Al_xSb/AlSb$ superlattices for infrared detection. *J. Appl. Phys.* **1991**, *70*, 3152-3156.

[3]   Hildebrand, O.; Kuebart, W.; Pilkuhn, M. H., Resonant enhancement of impact in $Ga_{1-x}Al_xSb$. *Appl. Phys. Lett.* **1980**, *37*, 801-803.

[4]   Yu, S. Q.; Ding, D.; Wang, J. B.; Samal, N.; Jin, X.; Cao, Y.; Johnson, S. R.; Zhang, Y. H., High performance GaAsSb/GaAs quantum well lasers. *J. Vac. Sci. Technol. B* **2007**, *25*, 1658-1663.

[5]   Wang, J. B.; Johnson, S. R.; Chaparro, S. A.; Ding, D.; Cao, Y.; Sadofyev, Y. G.; Zhang, Y. H.; Gupta, J. A.; Guo, C. Z., Band edge alignment of pseudomorphic GaAs1-y Sby on GaAs. *Phys. Rev. B* **2004**, *70*, 195339.

[6]   Liu, G.; Chuang, S.-L.; Park, S.-H., Optical gain of strained GaAsSb/GaAs quantum-well lasers: A self-consistent approach. *J. Appl. Phys.* **2000**, *88*, 5554-5561.

[7]   Dutta, P. S.; Bhat, H. L.; Kumar, V., The physics and technology of gallium antimonide: An emerging optoelectronic material. *J. Appl. Phys.* **1997**, *81*, 5821-5870.

[8]   Ashley, T.; Dean, A. B.; Elliott, C. T.; Pryce, G. J.; Johnson, A. D.; Willis, H., Uncooled high-speed InSb field-effect transistors. *Appl. Phys. Lett.* **1995**, *66*, 481-483.

[9]   Seol, J. H.; Moore, A. L.; Saha, S. K.; Zhou, F.; Shi, L.; Ye, Q. L.; Scheffler, R.; Mingo, N.; Yamada, T., Measurement and analysis of thermopower and electrical conductivity of an indium antimonide nanowire from a vapor-liquid-solid method. *J. Appl. Phys.* **2007**, *101*, 023706-6.

[10]   Jeppsson, M.; Dick, K. A.; Wagner, J. B.; Caroff, P.; Deppert, K.; Samuelson, L.; Wernersson, L.-E., GaAs/GaSb nanowire heterostructures grown by MOVPE. *J. Crys. Growth* **2008**, *310*, 4115-4121.

[11]   Jeppsson, M.; Dick, K. A.; Nilsson, H. A.; Sköld, N.; Wagner, J. B.; Caroff, P.; Wernersson, L.-E., Characterization of GaSb nanowires grown by MOVPE. *J. Crys. Growth* **2008**, *310*, 5119-5122.

[12]   Dheeraj, D. L.; Patriarche, G.; Largeau, L.; Zhou, H.; van Helvoort, A. T. J.; Glas, F.; Harmand, J. C.; Fimland, B. O.; Weman, H., Zinc blende GaAsSb nanowires grown by molecular beam epitaxy. *Nanotechnology* **2008**, *19*, 275605.

[13]   Dheeraj, D. L.; Patriarche, G.; Zhou, H.; Harmand, J. C.; Weman, H.; Fimland, B. O., Growth and structural characterization of GaAs/GaAsSb axial heterostructured nanowires. *J. Crys. Growth* **2009**, *311*, 1847-1850.

[14]   Dheeraj, D. L.; Patriarche, G.; Zhou, H.; Hoang, T. B.; Moses, A. F.; Grønsberg, S.; van Helvoort, A. T. J.; Fimland, B. O.; Weman, H., Growth and characterization of wurtzite GaAs nanowires with defect-free zinc blende GaAsSb inserts. *Nano Lett.* **2008**, *8*, 4459-4463.

[15]   Plissard, S.; Dick, K. A.; Wallart, X.; Caroff, P., Gold-free GaAs/GaAsSb heterostructure nanowires grown on silicon. *Appl. Phys. Lett.* **2010**, *96*, 121901.

[16]   Caroff, P.; Wagner, J. B.; Dick, K. A.; Nilsson, H. A.; Jeppsson, M.; Deppert, K.; Samuelson, L.; Wallenberg, L. R.; Wernersson, L.-E., High-quality InAs/InSb nanowire heterostructures grown by metal-organic vapor-phase epitaxy. *Small* **2008**, *4*, 878-882.

[17]   Nilsson, H. A.; Caroff, P.; Thelander, C.; Larsson, M.; Wagner, J. B.; Wernersson, L.-E.; Samuelson, L.; Xu, H. Q., Giant, Level-Dependent g Factors in InSb Nanowire Quantum Dots. *Nano Lett.* **2009**, *9*, 3151-3156.

[18]   Ercolani, D.; Rossi, F.; Li, A.; Roddaro, S.; Grillo, V.; Salviati, G.; Beltram, F.; Sorba, L., InAs/InSb nanowire heterostructures grown by chemical beam epitaxy. *Nanotechnology* **2009**, *20*, 505605.

[19]   Chandrashekhar, P.; Vaddiraju, S.; Kim, J. H.; Jacinski, J.; Chen, Z.; Sunkara, M. K., Self-nucleation and growth of group III-antimonide nanowires. *Semicond. Sci. Technol.* **2010**, *25*, 024014.

[20]   Meyer, B. K.; Hofmann, D. M.; Volm, D.; Chen, W. M.; Son, N. T.; Janzén, E., Optically detected cyclotron resonance investigations on 4H and 6H SiC: Band-structure and transport properties. *Phys. Rev. B* **2000**, *61*, 4844.

[21] Dubrovskii, V. G.; Sibirev, N. V., Growth thermodynamics of nanowires and its application to polytypism of zinc blende III-V nanowires. *Phys. Rev. B* **2008,** *77,* 035414.

[22] Moses, A. F.; Hoang, T. B.; Dheeraj, D. L.; Zhou, H. L.; Helvoort, A. T. J. v.; Fimland, B. O.; Weman, H., Micro-photoluminescence study of single GaAsSb/GaAs radial and axial heterostructured core-shell nanowires. *IOP Conference Series: Materials Science and Engineering* **2009,** *6,* 012001.

[23] Zhou, H. L.; Hoang, T. B.; Dheeraj, D. L.; van Helvoort, A. T. J.; Liu, L.; Harmand, J. C.; Fimland, B. O.; Weman, H., Wurtzite GaAs/AlGaAs core-shell nanowires grown by molecular beam epitaxy. *Nanotechnology* **2009,** *20,* 415701.

[24] Dinu, M.; Cunningham, J. E.; Quochi, F.; Shah, J., Optical properties of strained antimonide-based heterostructures. *J. Appl. Phys.* **2003,** *94,* 1506-1512.

[25] Chiu, Y. S.; Ya, M. H.; Su, W. S.; Chen, Y. F., Properties of photoluminescence in type-II GaAsSb/GaAs multiple quantum wells. *J. Appl. Phys.* **2002,** *92,* 5810-5813.

[26] Hoang, T. B.; Moses, A. F.; Ahtapodov, L.; Zhou, H.; Dheeraj, D. L.; Helvoort, A. T. J. v.; Fimland, B. O.; Weman, H., Engineering Parallel and Perpendicular Polarized Photoluminescence from a Single Semiconductor Nanowire by Crystal Phase Control. *Nano Lett.* **2010,** *10,* 2927-2933.

[27] Birman, J. L., Polarization of fluorescence in CdS and ZnS single crystals. *Phys. Rev. Lett.* **1959,** *2,* 157.

[28] Mishra, A.; Titova, L. V.; Hoang, T. B.; Jackson, H. E.; Smith, L. M.; Yarrison-Rice, J. M.; Kim, Y.; Joyce, H. J.; Gao, Q.; Tan, H. H.; Jagadish, C., Polarization and temperature dependence of photoluminescence from zincblende and wurtzite InP nanowires. *Appl. Phys. Lett.* **2007,** *91,* 263104-3.

[29] Wang, J.; Gudiksen, M. S.; Duan, X.; Cui, Y.; Lieber, C. M., Highly polarized photoluminescence and photodetection from single InP nanowires. *Science* **2001,** *293,* 1455-1457.

[30] Hoang, T. B.; Titova, L. V.; Yarrison-Rice, J. M.; Jackson, H. E.; Govorov, A. O.; Kim, Y.; Joyce, H. J.; Tan, H. H.; Jagadish, C.; Smith, L. M., Resonant excitation and imaging of nonequilibrium exciton spins in single core-shell GaAs-AlGaAs nanowires. *Nano Lett.* **2007,** *7,* 588-595.

[31] Guo, Y. N.; Zou, J.; Paladugu, M.; Wang, H.; Gao, Q.; Tan, H. H.; Jagadish, C., Structural characteristics of GaSb/GaAs nanowire heterostructures grown by metal-organic chemical vapor deposition. *Appl. Phys. Lett.* **2006,** *89,* 231917-3.

[32] Burke, R. A.; Weng, X.; Kuo, M.-W.; Song, Y.-W.; Itsuno, A. M.; Mayer, T. S.; Durbin, S. M.; Reeves, R. J.; Redwing, J. M., Growth and characterization of unintentionally doped GaSb nanowires. *J. Electron. Mater.* **2010,** *39,* 355.

[33] Abu-Farsakh, H.; Qteish, A., Ionicity scale based on the centers of maximally localized Wannier functions. *Phys. Rev. B* **2007,** *75,* 085201.

[34] Hu, W. G.; Wang, Z.; Su, B. F.; Dai, Y. Q.; Wang, S. J.; Zhao, Y. W., Gallium antisite defect and residual acceptors in undoped GaSb. *Phys. Lett. A* **2004,** *332,* 286-290.

[35] Wen, C. Y.; Reuter, M. C.; Bruley, J.; Tersoff, J.; Kodambaka, S.; Stach, E. A.; Ross, F. M., Formation of compositionally abrupt axial heterojunctions in *silico*n-germanium nanowires. *Science* **2009,** *326,* 1247-1250.

[36] Chen, H.; Sun, X.; King, W. C. L.; Meyyappan, M.; Xi, N. In *Infrared Detection Using an InSb Nanowire,* IEEE Nanotechnology Materials and Devices Conference, Traverse City, Michigan, USA., 2009; Traverse City, Michigan, USA., 2009; p 212.

CHAPTER 6

III-V Ternary Nanowires

Faustino Martelli*

Istituto per la Microelettronica e i Microsistemi del Consiglio Nazionale delle RicercheVia del Fosso del Cavaliere 100, 00133 Roma, Italy

Abstract: In this chapter I will give an overview of the existing knowledge on the growth and the main properties of III-V ternary nanowires. I will describe bare ternary as well as heterostructure nanowires, both axial and radial. A particular emphasis is given to the competing behaviors occurring during the growth of this class of nanowires, that is different ad-atom incorporation due to different properties of impinging species of the same group: diffusion lengths, solution of the different element in the catalyst nanoparticle and the thermal stability of chemical bonds. These competing processes have effects on alloy composition, composition uniformity and nanowire shape.

Keywords: Nanowires, ternary alloys, III-V compounds, crystal growth, arsenides, phosphides, antimonides, nitrides.

INTRODUCTION

Ternary and quaternary semiconductor materials have a great importance in the development of III-V-based devices as, e.g., semiconductor lasers, photodetectors, multijunction solar cells or high-electron-mobility transistors, because they allow the precise tuning of the most important structural and electronic parameters, like lattice constant, energy band-gap and carrier effective mass. They can then be used to design in the most precise way the performance needed by the device to be fabricated.

It is then easy to foresee that this wide class of materials will also play an important role in the development of future devices based on III-V semiconductor nanowires (NWs). III-V ternary NWs have indeed been grown [1] already during the first stages of the present, mature NW research. In the case of NWs the lack of constraints relative to lattice match between substrate and epitaxial layers, generally necessary for two-dimensional (2D) structures, will further widen the possible combinations of ternary alloys and substrates. On the other hand, we will see that some lattice-match constraint remains for the fabrication of heterostructure NWs.

To date, results on quaternary nanowires still lack and this review will then only cover ternary compounds. However, we will see that quaternary alloys are occasionally found in some NWs as the unexpected result of atomic diffusion in adjacent layers.

Presently, the many advantages given by the tunability of the structural and electronic properties of ternary nanowires are partially shadowed by the complex thermodynamics and kinetics that rule and compete during the growth process. As it has been explained in the previous chapters, a comprehensive growth model of binary III-V nanowires is still in development. The presence of three or four atomic species that contribute to the nanostructure formation is a further complication that adds variables to the solution of this specific problem for alloyed NWs. As we will indeed see, different thermodynamic and kinetic mechanisms compete during the growth of ternary NWs, both in metal-induced and in self-catalyzed or selective-area processes.

Nevertheless, or just for this reason, the growth of III-V ternary nanowires is a fascinating research topic that challenges researchers to find the proper interpretation that will allow them to achieve perfectly controlled growth of these complex nanowires.

*Address correspondence to Faustino Martelli: Istituto per la Microelettronica e i Microsistemi del Consiglio Nazionale delle RicercheVia del Fosso del Cavaliere 100, 00133 Roma, Italy. Tel.: +390645488708; fax: +390649934066; E-mail: *faustino.martelli@cnr.it*

Jianye Li, Deli Wang and Ray R. LaPierre (Eds)

This chapter will summarize the present knowledge on the growth and on the main properties of III-V ternary nanowires. I will show that ternary NWs have been obtained by several growth methods, exploiting or not the vapor-liquid-solid (VLS) model. Metal-induced, self-catalyzed and selective-area processes, although very different in important details, share not only the final product but also the presence of an important kinetic factor as the ad-atom competition in forming the nanowires.

Another important aspect of the NW technology is the formation of axial and/or radial heterostructures. The core-shell (c-s) structures allow an effective passivation of the NW surface with the consequent improvement of the electrical and optical properties of the NWs. As I will show, ternary compounds play an important role in these structures both as core- and shell-materials. Although this chapter is focused on the growth and properties of the ternary alloy NWs, I will briefly discuss the electronic properties of binary cores when their investigation is made possible by the presence of a ternary shell. Moreover, topics such as lattice structure, growth homogeneity or others that are common to all III-V semiconductor nanowires will be discussed, restricting the chapter to the description of the features found in ternary alloys. A more general discussion on these important topics of the NW technology will be found in other chapters of this book.

This chapter is organized as follows. In Section 2 some general information on the growth of ternary NWs will be first given. The following two sections will instead contain the review of the results obtained on III-III-V and on III-V-V ternary alloys, respectively. The choice to divide the ternary alloys in these two families on one hand follows a traditional way to distinguish them, and on the other hand is somehow driven by the observation that in the case of the III-III-V alloys the different diffusion lengths of the group-III ad-atoms lead to competing mechanisms that modify the final result in terms of NW morphology, alloy composition, and composition homogeneity. Such a competition is less important in the case of III-V-V alloys. Moreover in the case of the Au-induced growth, only the group-III elements actively interact with the catalyst metal particle that induces the growth because the group-V elements have low solubility in the metal particle. The use of different metals to induce the NW growth could make this assumption not true.

Catalysts different than Au have been used to grow ternary NWs and selective-area or self-catalyzing processes have been widely used to the same aim. All these aspects will be reported. For sake of simplicity, however, unless otherwise indicated, the growth of the NWs is intended as induced by Au nanoparticles.

GENERAL ASPECTS OF THE GROWTH OF TERNARY III-V NANOWIRES

As mentioned in the previous paragraph, the growth of ternary nanowires is complicated by competing mechanisms that influence the relative incorporation of the different impinging species of the same group.

The first competing mechanism, which only occurs in the metal-induced growth, is due to the different solutions of the different elements in the used metal. This has an effect on the relative supersaturation of the intermetallic compound, which forms before and during the growth, and the following release of the distinct species that may have different rates for different elements. The first aspect influences the relative presence of the species in the intermetallic compounds, while the second has effects on the relative growth rate. Both aspects have effects on the final composition of the alloy. Because of the low solubility of the group-V species in Au, when this metal is used to induce the growth, only competition among the group-III elements takes place.

The second process that gives rise to species competition is the ad-atom diffusion on the substrate and along the NW side-walls. This latter species competition also occurs in self-catalyzed and in selective-area growth.

Both mechanisms have influence on the average composition of the alloy and also on its homogeneity. Composition variation along and across the nanowires has been observed in several cases.

The most used techniques to grow ternary NWs are molecular beam epitaxy (MBE), along with its variants chemical beam epitaxy (CBE) and gas-source MBE (GSMBE), and metal-organic chemical vapor deposition (MOCVD, also called metal-organic vapor phase epitaxy, MOVPE). With these techniques, the growth generally takes place under group-V overpressure. This feature is responsible for the fact that only the different diffusion lengths of the group-

III elements play a role in the relative competition in species incorporation. The difference of bond energy that the group-V elements may have with the group-III species is taken into account when the partial pressures of the gases in the growth chambers are set. As we will see in several cases of III-V-V alloys, the switch among group-V elements gives rise to sharp interfaces, although some composition tail can be observed because of the time needed for gas phase transitions.

We will see that the importance of one or the other competing mechanism will depend on the species involved and on the growth technique. For this reason, their effects will be reported and discussed when describing the results on the specific alloys.

III-III-V ALLOYS

Based on the above discussion, the III-III-V alloys are probably the most challenging species combination to obtain nanowires with homogeneous composition. The reason lies just in the competing mechanisms described above. The variation from species to species of the diffusion length of group III atoms gives indeed rise to an incorporation competition that influences both radial and axial alloy composition, which is often different from the reference 2D composition used to set the growth parameters. This ad-atom incorporation competition also affects the composition homogeneity and the shape of the nanowires. Moreover, group III elements form intermetallic compounds with gold with different composition, a difference that affects the relative presence in the intermetallic nanoparticle (NP) that forms at the NW tip.

$Al_xGa_{1-x}As$

AlGaAs has a direct band-gap for any alloy composition that is larger than that of GaAs. Moreover, the two binary compounds forming this alloy have almost the same lattice constant. These features have made AlGaAs the most investigated III-V alloy and the GaAs/AlGaAs system the most investigated heterostructure among the epitaxial layers. It is therefore almost obvious to find AlGaAs among the most used alloys in NW fabrication. In the case of core-shell (c-s) structures, AlGaAs is the perfect shell material for GaAs cores because the shell can be grown lattice matched to the core.

Differences among the AlGaAs composition in the NWs and in the 2D epitaxial layer used as a reference, were already pointed out in the pioneering work by Wu and coworkers [2]. Their Au-catalyzed NWs, grown by MBE had an Al content $x=0.13$, as determined by photoluminescence (PL), although the value measured on the 2D reference layer was $x=0.24$. We notice here that PL generally measures the lowest-energy state of an ensemble and this means that the value measured in the NWs is the lowest x value given by the composition fluctuations. In any case, this was the first indication that the NW composition may differ from the designed one. In their paper Wu and coworkers attributed the observed difference to the VLS mechanism that is to the different incorporation/release rate in the Au nanoparticle for the two different elements.

PL is a technique that averages over the large number of illuminated NWs and along their extension. Energy dispersive X-ray spectroscopy (EDS) performed in transmission electron microscopy (TEM) along AlGaAs NWs grown by GSMBE [3] has shown that the Al content is inhomogeneous from wire to wire and also within single wires. Also in that work, the average Al content in the NWs was found to be lower than in the 2D reference layer ($x=0.37$). EDS determined x to be in the range of 0.26-0.39 at the center of the NW bases with an average measurement of 0.36, that is very close to the reference value. However, the Al content decreases going towards the NW tip, where values of $x=0.20-0.23$ were found. An inhomogeneity of the Al content was also found in the radial direction, with higher Al concentration near the outside of the NW.

Those observations were explained with the different diffusion length that is at least four times larger for Ga than for Al [4]. As the NW length increases, for Al atoms it becomes more difficult to reach the NW tip, reducing the Al incorporation in the NWs, a picture that nicely explains the experimental data. This picture gives also account for the more tapered morphology of the AlGaAs NWs compared to GaAs NWs grown in the same conditions [5].

The inhomogeneous radial distribution of the Al content has been exploited by the same group to fabricate tapered multiple quantum well AlGaAs NWs [6], see Fig. **1**, that show intense single-wire luminescence.

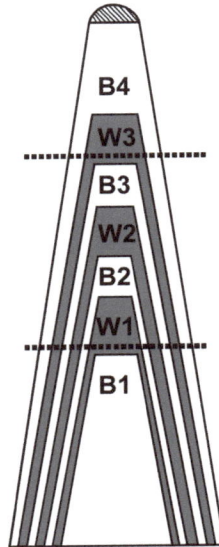

Figure 1: Cross section sketch of the nanowires multiple quantum well configuration proposed in reference 6: wells: W1, W2, and W3 (shaded regions) are separated by barriers B1-B4 (white regions). Reprinted with permission from Ref. [6]. Copyright 2008 American Chemical Society.

In contrast with what is observed about AlGaAs NWs grown by MBE, the growth of AlGaAs NWs by MOCVD results in NWs with Al content that increases along the wire going towards the NW tip [7]. In the case of the nanowires shown in Fig. **2**, the composition is quite large, x=0.36 right below the catalyst and decreases to x=0.18 at the NW base.

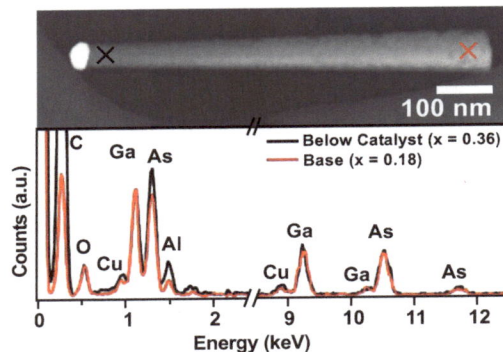

Figure 2: Scanning TEM image (top) and EDS spectra (bottom) collected at two points (indicated by markers) along AlGaAs NWs grown by MOCVD. The nanowire exhibits a compositional gradient along the NW length with higher Al composition at the NW final end (x=0.36) than at the base (x=0.18). Reprinted in part with permission from Ref. [7]. Copyright 2008 American Chemical Society.

To justify this major difference with MBE-grown NWs, the authors claim that in MOCVD, material release because of supersaturation of the intermetallic NP is a more important growth mechanism than ad-atom incorporation of species diffused from the substrate. This claim is supported by the observation that the AlGaAs shell has a lower Al content than the AlGaAs core. The shell is radially formed after a vapor-solid mechanism, while the core axially grows after the VLS model. The lower Al diffusion length is then responsible for the lower Al content in the shell. The NW base has a larger contribution from shell formation and thus a lower average Al content. The different decomposition rate for trimethylaluminum and trimethylgallium also plays a role in the final result. A scheme of the growth model is proposed in Fig. **3**.

The larger complexity of the MOCVD growth process is confirmed by the investigation on the effect of the V/III ratio, made in the same work [7]. The Al content in the core increases as the V/III ratio increases, while the Al content in the shell shows little variations remaining constant in the study. Again, the observed differences are attributed to the different growth mechanism that dominates in the two sections of the core-shell structure.

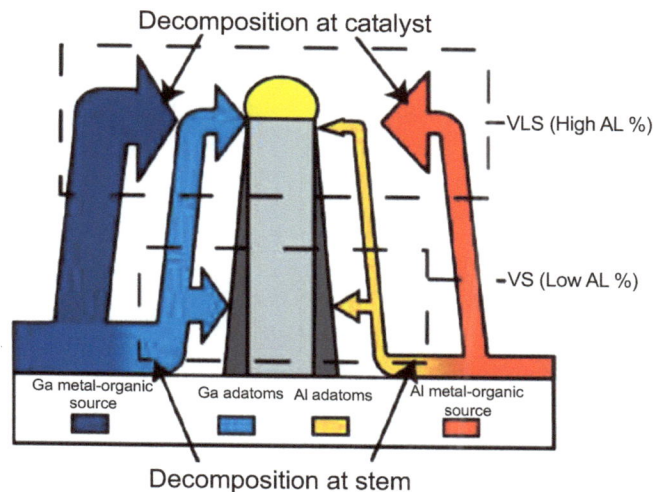

Figure 3: Sketch of the $Al_xGa_{1-x}As$ nanowire growth using MOCVD. The nanowire shell growth rate is proportional to the amount of ad-atoms created by decomposition of metalorganic (MO) precursors and inversely proportional to the diffusion length of the ad-atoms. At the stem, Al ad-atoms have shorter diffusion lengths but lower decomposition rate than Ga ad-atoms, resulting in the low x_{shell}. VLS is instead responsible for x_{core} and depends on the ad-atom diffusion and MO precursor decomposition at the catalyst. Ad-atom contribution tends to lower Al content in the core relative to the shell due to shorter Al diffusion length of Al with respect to Ga. On the contrary, relatively high concentration of the Al precursor and its higher decomposition rate at the catalyst results in higher x_{core} than x_{shell}. Reprinted with permission from Ref. [7]. Copyright 2008 American Chemical Society..

A much smaller Al content in the wires grown by MOCVD than in the reference epilayer was also reported by Hoang and coworkers [8]. Instead, the importance of growing at high temperatures ($T_g > 650$ °C) to achieve high Al contents has been pointed out in the work by Tambe *et al.* [9]. The important role of the growth temperature for the growth of ternary III-V NWs will later emerge also for other alloys.

The important role of the AuAlGa intermetallic compound in the MOCVD growth has been also pointed out by Ouattara and coworkers [10] for GaAs/AlGaAs NWs containing both axial and radial GaAs/AlGaAs interfaces. With the use of scanning tunnel microscopy (STM) the authors have shown that the GaAs-AlGaAs axial interface is monolayer sharp, while the reverse interface is diffuse, showing a gradual decline of the Al concentration over 50 nm. The slow decrease of Al content is attributed to the slow release of Al atoms from the AuAlGa reservoir.

The use of AlGaAs shells grown around a GaAs core has been exploited to realize optically efficient NWs [8, 11-16]. Indeed the GaAs PL efficiency strongly benefits from the passivating shell of the larger-gap alloy, and the exciton lifetime in twin-free structures of this type has been measured to be of the order of 1 ns [12], similar to the bulk-GaAs case. In reference 13 the AlGaAs shell is grown epitaxially on the GaAs core. In agreement with what was described above, the shorter diffusion length of Al gives rise to the tapering of the GaAs/AlGaAs NWs, tapering that is not observed if only the GaAs core is grown. Finally, Moewe and coworkers [14], have shown that GaAs/AlGaAs core-shell NWs can be also obtained *via* a self-catalyzed process induced by surface roughness. Also in this case the NWs have a strongly tapered shape, a feature that confirms the great importance of diffusion mechanisms, where in this case the VLS process was inactive.

The selective-area growth of GaAs/AlGaAs core-shell NWs has been demonstrated by a series of works of Hokkaido University [17-20], with improvements of the GaAs NW luminescence up to a factor 500 [19] due to the presence of the AlGaAs shell. The growth, by MOCVD, takes place on GaAs(111) or Si(111) substrates where openings have been properly patterned. The NWs realized with this method do not show the strong tapering observed in the self-assembled wires described above.

The improvements of the NW emission efficiency has finally led to the fabrication of vertically aligned light emitting diodes (LEDs) based on the GaAs/AlGaAs core-shell NWs grown on Si [20]. The growth of the AlGaAs shell is obtained by lowering the growth temperature by 50 °C (700 °C) with respect to that used for the vertical growth of the GaAs core (750 °C). The structure used for the device is made by a n-GaAs core followed by a n-AlGaAs/p-AlGaAs/p-GaAs multishell structure. A superlinear increase of the electroluminescence intensity with increasing junction current is observed, similar to a superluminescence behavior. A good thermal stability of the devices is observed, probably because of the optimal mechanical contact that the wires grown by this method have with the substrate.

An interesting aspect of Hokkaido's works is that with a selective etching of the GaAs core, the authors have succeeded in fabricating AlGaAs nanotubes [18].

Selective-area MOCVD was also used by Sladek and coworkers to grow GaAs/AlGaAs c-s NWs [21]. The scope of the AlGaAs shell is to provide a manner to obtain conductive GaAs NWs. An increase of the NW diameter at the top has been observed in this case, a rare but not isolated observation.

A feature that is worth pointing out is that wurtzite structure of the lattice has been found in the AlGaAs NWs independently of the growth method and technique. Both VLS and SA growths, the latter induced without the use of any metal NP, lead to the presence of extended wurtzite phases.

In$_x$Ga$_{1-x}$As

The ternary alloy InGaAs is widely used for optoelectronic devices in the near-infrared range like laser diodes and photodiodes. Moreover, the smaller electron mass makes it preferable to GaAs also for high-speed transistors. These features have made it one of the most investigated alloys among the epitaxial layers. Its importance is reproduced in the several papers that deal with the growth of InGaAs NWs. As for AlGaAs, InGaAs NWs have been obtained by VLS using different growth techniques as well as by self-catalyzed processes and by SA-MOCVD.

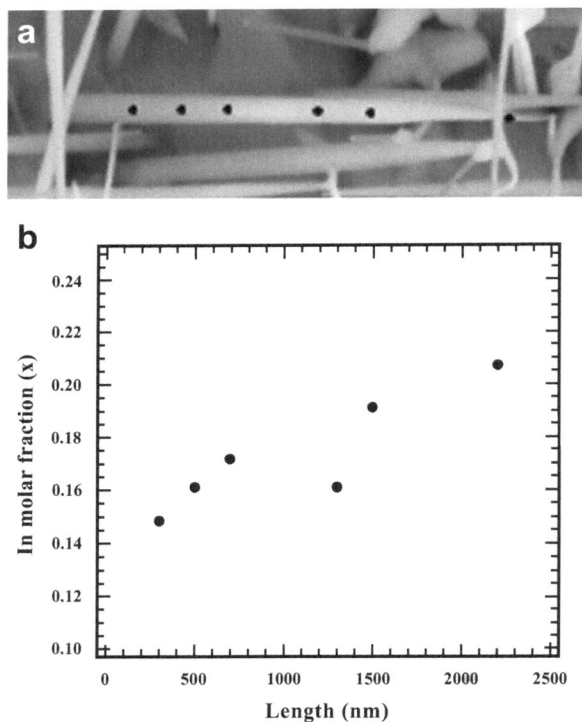

Figure 4: The upper part **(a)** of the figure shows an InGaAs wire grown by MBE at 460 °C. The marks indicate the locations where EDS spectra have been recorded. The lower part **(b)** depicts the In molar fraction in the wire along its axis as obtained from EDS. The figure is built in order to have marks in the image over the corresponding points in the graph. Reprinted from *Microelectronics Journal*, Vol. 40, F. Jabeen *et al.*, Growth of III– V semiconductor nanowires by molecular beam epitaxy, 442-445, Copyright (2009), with permission from Elsevier.

The importance of the ad-atom diffusion for the NW growth by MBE is confirmed by the observation that the In content in InGaAs NWs increases going from the bottom towards the tip of InGaAs NWs [22], see Fig. **4**. In diffusion length is higher than that of Ga [23]. Hence, In atoms have a higher probability than Ga atoms to reach the NW tip moving along the NW side-walls and hence the upper part of the NWs will have a higher In content than the bottom part. This finding and its explanation perfectly agree with what is observed for AlGaAs NWs grown by MBE and described above. In ref. [22], content homogeneity is obtained at relatively high growth temperature (550 °C), when incongruous In desorption compensates the higher In diffusion length.

The large difference of diffusion length between In and Ga is also invoked to explain the features observed in InGaAs NWs on GaAs(111)B substrates by MOCVD [24]. In this work it is shown that for increasing In content the height of the InGaAs NWs shows an increasing inverse dependence on the NW density. The longer height at low densities is explained with the larger diffusion area of impinging atoms that contribute to the NW growth. Moreover a systematic enhancement of tapering with increasing In content is observed along with a change of the shape of the NW base. The base changes indeed from a triangular shape for GaAs to a hexagonal shape for high In contents. This change has no correlation with the NW density and hence should not be due to the different diffusion length of In and Ga species.

Contrary to the case of MBE, the MOCVD-grown InGaAs NWs reported in reference 24 show a lower In content at the top of the NW with respect to the NW base. This content variation along the wires is responsible for the wide PL spectrum observed in those samples.

Regolin and coworkers [25] also used MOCVD to grow InGaAs NWs. They found that the average In composition in the wires was lower than in the reference layer used for calibration. This feature is common to the wires grown by MBE [22]. The difference in In content between reference layer and NWs is attributed to an active role of the Au nanoparticle (NP). The miscibility of Ga and In in Au is in principle different and the resulting In and Ga content in the InGaAu intermetallic compound that forms in the NP may be different from that given by the mere flux ratio. We will see later that the NP also plays an active role in the sharpness of heterostructure interfaces along the wires.

It is worth noticing here that in both cases of AlGaAs and InGaAs, MBE and MOCVD give reverse results about the gradient direction of the alloy composition. For MBE diffusion mechanisms appear to dominate, while for MOCVD, NP-related mechanisms are of major importance.

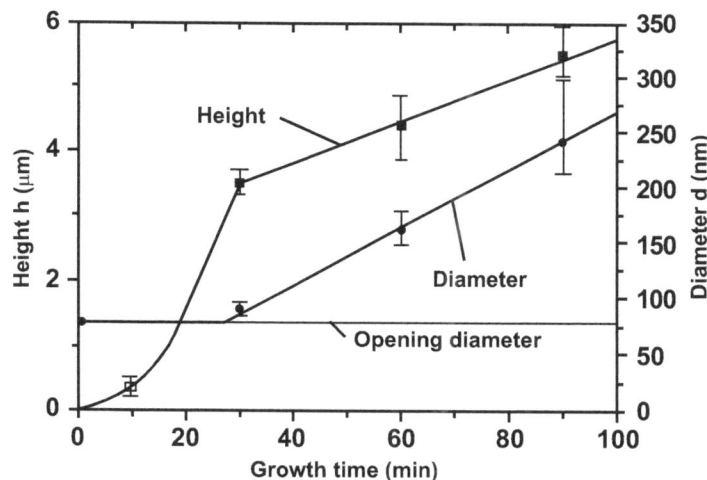

Figure 5: Time evolution of height (closed squares) and diameter (closed circles) of InGaAs nanowires. The opening diameter of the mask is 80 nm and the nominal x value is 0.52. Reprinted from *J. Cryst. Growth*, Vol. 310, T. Sato *et al.*, Growth of InGaAs nanowires by selective-area metalorganic vapor phase epitaxy, 2359-2364, Copyright (2008), with permission from Elsevier.

InGaAs NWs were successfully grown also with the selective-area method [26, 27] on GaAs(111)B and InP(111)B substrates using MOCVD. The growth takes place with a complex dependence on growth time, summarized in Fig. **5**. The results show nonlinear behaviors in the growth rate both in the vertical and lateral directions. Moreover, the growth rate depends on the wire diameter. The nonlinearities depend on the ratio of relative supply of the metalorganic precursors of In and Ga. The inhomogeneous enlarging of the NW above the opening diameter is due to lateral growth. All these features can be explained by a complex diffusion model that includes diffusion from masked areas to openings and diffusion from sidewalls to the nanowires tips. We refer the reader to the original paper [25] for a detailed discussion of the diffusion process that also affects the alloy content in the NWs. It is worth noting that in this case the authors find a larger In content in the wires than in the reference planar region, a different result than that described above for references [22] and [25].

InGaAs has been also used in core-shell or axial heterostructures. In this case, because InGaAs has a smaller band-gap than GaAs and InP, it is used as the active region in light emitting heterostructures.

InGaAs has been used as the core material in InGaAs/GaAs [28] and InGaAs/GaAs/InAlAs [29] c-s NWs. The passivation effect of the shell make it possible to observe room temperature PL in both c-s heterostructure NWs. In the first work [28], the presence of the GaAs shell increases the PL intensity by a factor 600 with respect to the bare InGaAs NWs obtained with the same growth conditions. Carrier confinement has been observed in the 30-40 nm wide InGaAs cores. In the second work [29] the emission wavelength in the c-s NWs has been tuned up to 2 μm by varying the InGaAs alloy composition. In this case, the bare InGaAs NWs did not show any PL. Also in these wires, a compositional variation along the growth axis has been observed. In analogy with what is found in references 24 and 26, the In content decreases towards the NW tip in these sample grown by MOCVD.

Reference 29 also reports the fabrication of axial InGaAs/InAlAs NWs. Five InGaAs inserts were grown in AlInAs NWs. EDS analysis reveals that at the InGaAs/AlInAs interface the Ga profile changed within 6 nm, while at the opposite interface, Al was detected in the whole InGaAs region and a quaternary AlInGaAs region was formed.

Regolin and coworkers [30] grew InGaAs/GaAs axial NW heterostructures by MOCVD. They found no sharp interface. On the contrary they found that the In content changes over a few hundred nanometers and attribute this behavior to a memory effect in the Au droplet, an effect that we will find also for the growth of other alloys. The growth of a GaAs section above the InGaAs region gave rise to the realization of an InGaAs/GaAs c-s structure, due to lateral growth of GaAs. No information on the PL characteristics of the InGaAs region was however reported.

Systematic investigations on the growth by MOCVD of GaAs/InGaAs/GaAs double heterostructures were performed by Bauer and coworkers [31]. A detailed analysis of the In content in the InGaAs segment and of the intermetallic compound formed in the NP demonstrates the slow In accumulation inside the droplet with the following smoothness of the heterostructure interfaces. Sharper interfaces were found using growth interruption. As expected by general knowledge on MOCVD growth, it was observed that an increase of the As pressure gives rise to a higher In content in the alloy segment.

InGaAs/GaAs axial heterostructures have been also obtained with a catalyst-free growth [32]. With this method a Ga NP forms at the beginning of the growth process and induces the NW growth. When In is switched on an InGa intermetallic compound forms. This process takes about 1000 seconds to complete and the final InGaAs composition is determined by the composition of the intermetallic NP. The highest In content obtained with this method has been $x=0.05$.

It is also worth mentioning that InGaAs/GaAs superlattice NWs were grown by MOCVD [33].

InGaAs/GaAs axial heterostructures were also realized by selective-area MOCVD [34-38] in which InGaAs quantum wells were grown between GaAs barriers. In the InGaAs segment of those NWs the In content increases with increasing diameter of the NWs. The diameter is varied *via* the size of the holes patterned in the substrate. Again, the diameter dependence of the In content can be explained with the different diffusion length of Ga and In species in the used growth conditions [34]. The In content also varies with the NW length, being lower in longer NWs.

In the SA InGaAs/GaAs wires the lattice structure is zinc blende with rotational twins, defects that are shown to affect the luminescence of the wires. The PL also allows the observation of weak carrier confinement in the InGaAs wells. Field-effect transistors [37] and Fabry-Pérot microcavities [38] were fabricated with this method. This latter device is made possible by the ordering of the SA growth given by the lithographic step that produces the openings in the substrate.

InGaAs was also used to form axial heterostructures with InAs [39] and InP [40]. In the former case InGaAs acts as the barrier material in the InAs NW. The growth of axial InAs/GaAs/InAs wires is made difficult because the interfacial energy between the Au-catalyst and InAs is higher than the interfacial energy between Au and GaAs [41-43]. This prevents axial growth of InAs on GaAs as the system minimizes its free energy. This issue concerns other material combinations used for NW growth as well. Using MBE the authors of references 41-43 have been able to grow InAs/In$_x$Ga$_{1-x}$As/InAs axial heterostructure NWs with Ga content up to 1-x=0.47. By scanning transmission electron microscopy they find that the In$_x$Ga$_{1-x}$As/InAs junctions are abrupt, while the InAs/In$_x$Ga$_{1-x}$As junctions are broadened. Studies on GaAs/InAs NWs [43] show that when the AuGa intermetallic compound is exposed to In flux, the NP quickly changes its composition by expelling Ga and absorbing In, where in the opposite case In tends to remain in the NP. These differences make the In$_x$Ga$_{1-x}$As/InAs junction sharper than the InAs/In$_x$Ga$_{1-x}$As junction. In those works, the In$_x$Ga$_{1-x}$As segments have a higher In content than the calibration layers.

InGaAs/InP heterostructure NWs were grown on InP(111)B substrates by gas-source MBE [40]. With respect to the heterostructures described above, this heterostructure involves the interchange of both group-III and group-V elements. The NWs consisted of an initial InP segment, an InGaAs middle segment, and a terminating InP segment. These NWs present a prominent bulge near the wire middle followed by a slight reduction in wire diameter, see Fig. **6**. The bulge is attributed to a change of the intermetallic NP size due to a transient supersaturation with group III material. Indeed, the bulge coincides predominantly with changes in group-III composition. In these samples the smaller diffusion length of Ga than that of In is responsible for Ga deficiency in the InGaAs segments with respect to the designed value. Moreover the differences in diffusion lengths are also responsible for the lateral growth of InGaAs on InP that gives rise to a core-shell structure. Because of some lingering P after the growth of the bottom InP segment or because of residual sidewall deposition during the growth of the top InP segment, the shell material is actually a quaternary InGaAsP alloy.

Figure 6: Typical shape of a InGaAs/InP heterostructure NW. The bulge at the center of the image corresponds to changes in the group III composition. Reprinted with permission from Ref. [40], *Nanotechnology*, 2007, Vol. 18, 385305. Copyright 2007 Institute of Physics.

In these samples the smaller diffusion length of Ga than that of In is responsible for Ga deficiency in the InGaAs segments with respect to the designed value. Moreover the differences in diffusion lengths are also responsible for the lateral growth of InGaAs on InP that gives rise to a core-shell structure.

AlInAs

AlInAs is a ternary alloy usually used as cladding layer in planar heterostructures used in telecommunication optoelectronics. As described above it can be used as cladding layer also in nanowires. The same work [29] that reports the InGaAs/AlInAs heterostructure NWs also describes the growth by MOCVD of bare AlInAs NWs. InGaAs and AlInAs NWs show significant structural differences both in shape and in lattice structure. While InGaAs NWs appear hexagonal from the top views, many AlInAs NWs present a triangular shape. The lattice structure of the two types of nanowires is different: while InGaAs NWs are mainly zinc blende structures with many twins and some wurtzite regions, AlInAs NWs are mainly wurtzite with some twins.

Al$_{1-x}$In$_x$P

AlInP is a ternary alloy that can be made lattice matched to GaAs. For this reason it is mainly used for AlInP/InGaP/GaAs heterostructures for visible optoelectronic devices. In nanowires this compound has been only used as shell material grown by MOCVD around an Au-catalyzed GaAs core [44]. The average shell composition was approximately Al$_{0.5}$In$_{0.5}$P. Because of the surface state passivation of the GaAs, the core PL is two orders of magnitude larger than that of the bare GaAs NWs grown in the same conditions. Similarly to what was described for

AlGaAs/GaAs c-s structures, the authors of reference 44 use a selective etching to remove the core and investigate the properties of the AlInP nanotube thus obtained. Multiple peaks between 1.9 and 2.1 eV in the shell spectrum indicate that the shell is not homogeneous. These energies, lower that that expected for the alloy composition, indicate that local band gap minima exist, due to the presence of In-rich domains. Also in this case, the different diffusion lengths of the group-III species, shorter for Al than for In, are responsible for the observed phase segregation. The counterparts of the In-rich domains are the Al-rich regions ($Al_{0.6}In_{0.4}P$) that form in the {112} directions where two {110} facets meet during growth. Spontaneous phase segregation was also suggested to take place in AlGaAs nanowires grown by solution-solid-liquid mechanism [45].

$In_xGa_{1-x}P$

InGaP is generally used as the active material in GaAs based heterostructures for optoelectronic emitting devices operating in the visible region. This ternary alloy has been used to grow both bare $In_xGa_{1-x}P$ [46] NWs and GaAs/InGaP c-s heterostructure NWs [47, 48].

In the former case [46], the InGaP NWs were grown on polycrystalline InP substrates without the use of Au as the catalyst. Instead, metallic In, liberated from the InP substrate by phosphor evaporation, works as the catalyst metal. The resulting NP at the top of the NWs is composed by an InGa intermetallic compound. The presence of such a NP at the NW top suggests a growth mechanism similar to VLS. Raman measurements indicate that the NW composition is quite homogeneous in samples grown at 700 °C, while NWs grown at 650 °C show a composition variation along the NWs. Moreover, Raman and PL indicates a higher Ga content in the NWs grown at the highest temperature. These observations agree with that also observed for the InGaAs NWs (see above), where the group-III species are the same.

Au-catalyzed GaAs-$In_xGa_{1-x}P$ (1-x=0.34, 0.48, 0.58, and 0.69) core-shell nanowires were grown by MOCVD [47] on GaAs substrates. The growth of the shell has the usual effect to improve by 2 to 3 orders of magnitude the PL intensity of the GaAs core. Because of the dependence of the lattice constant on the alloy content, InGaP is grown with different strain, depending on its composition. The growth of a tensile- or of a compressive-strained shell causes the formation of strain also in the GaAs core. Exploiting this mechanism, with the alloy composition range used in their work the authors were able to tune the GaAs PL wavelength over a range of 240 meV. They also show that the use of a lattice matched shell gives rise to a more intense PL [48] in samples grown on GaP(111)B than in those grown on Si(111) substrates. With the use of slightly modified core structure that includes the growth of a GaP segment beneath the active GaAs, the lattice matched GaAs/InGaP heterostructure c-s- NWs were used to realize light emitting diodes (LEDs). In analogy to what was observed in the material, also the NW LEDs on Si displayed lower PL intensity and electroluminescence (EL) efficiency as compared with those on GaP substrates.

In reference 48 it is also pointed out the importance of removing the Au nanoparticles from the top of the NWs for a better use of the devices. Let me comment that this is actually a general issue of many technologies that will use NWs: the absence of metal NPs, typical of the self-catalyzed methods, is not only important for the proper operation of the devices but also for the purity of the grown nanowires.

The different behaviors of Ga and In cause striking effects also in the growth of In_5Ga_5P NWs by gas-source MBE [49]. The diffusion length difference is not only the cause of the more or less pronounced tapering of the NWs (depending on the growth conditions) but also of the radial inhomogeneity of these NWs that in practice exhibit a core-shell structure with an In-rich core and a Ga-rich shell, similar to what is observed in AlGaAs NWs [6]. This radial inhomogeneity might be also due to other competitive behavior as a possible suppression of Ga solubility in Au when In is present. However, the finding of an increase at the radial center of the element with the largest diffusion length is observed not only in the cited alloys but, as we will see later, also in the case of other alloys. It then appears as a general behavior. It is worth noting that this radial inhomogeneity might be exploited to build carrier confining radial heterostructures. An axial analysis of the alloy composition further reveals that In incorporation strongly depends on the V/III flux ratio: as expected in MBE, indeed, In incorporation increases with respect to Ga incorporation as the V/III flux ratio increases. Under the growth conditions used in the experiment, the authors find wurtzite structures with zinc blende insertions (stacking faults). The density of stacking faults can be reduced using low growth rates and high V/III flux ratios.

$Al_xGa_{1-x}N$

The AlGaN alloy finds application in the development of optoelectronic devices in the ultraviolet region of the light spectrum. Its band gap varies from 3.4 eV for pure GaN to 6.2 eV for AlN, both being direct-gap materials. As all other nitrides, also this ternary compound has already been used in several works on nanowires, mainly as barrier material in both axial and radial heterostructures.

A peculiar properties of nitride nanowires is their relatively easiness to grow *via* self-catalyzed processes. This feature has been often related to the use of a plasma source to produce N species, but this is not a narrow condition to obtain self-catalyzed III-N nanowires, as the growth of AlGaN NWs by CVD [50] using NH_3 vapors demonstrates. Although self-catalytic, the authors suggest that the growth of these NWs is obtained *via* the VLS process induced by a Ga(Al) molten droplet that forms on the substrate surface. This conclusion is drawn by the observation of a particle on top of most of the NWs. Both electron diffraction and Raman scattering show that the obtained NWs are single-phase wurtzite AlGaN alloys.

$Al_xGa_{1-x}N$ ($0.2 \leq x \leq 0.6$) nanocolumns were first grown by plasma-assisted MBE on Si(111) substrates *via* a pure self-catalytic process [51] without contribution from metals neither of extrinsic nor of intrinsic origin. The change from a compact growth to a columnar growth is obtained upon increasing the V/III ratio. The Al content, varied by changing the ratio between the Al flux and the total group III-element flux, is generally higher than the nominal one derived by reference layers. This is due to the Ga desorption at the high growth temperature used (760 °C). Similarly to what is observed in AlGaAs NWs grown by MBE, also in the case of AlGaN grown by this technique, the Al content slightly decreases moving toward the NW tip. The growth of AlGaN nanocolumns was later developed by inserting 2-5 nm thick GaN quantum discs [52, 53]. The diameters of the columns were in the range of 20-150 nm. Strong luminescence at low temperatures allowed the observation of carrier confinement in the GaN discs. Carrier confinement, the dependence of the luminescence intensity on the disc thickness, and the luminescence spectral broadening and long lifetimes were shown to be affected by the inhomogeneous strain revealed in those structures. The authors indeed argue that the inhomogeneous strain gives rise to a quantum confined Stark effect (QCSE) that is responsible for the e-h spatial separation that can explain the luminescence features described above.

The same growth technique allowed the realization of columnar AlGaN/GaN nanocavities with AlN/GaN Bragg reflectors [54].

Detailed studies of the optical properties of AlGaN/GaN quantum discs were made possible by alternating AlGaN axial sections in GaN NWs [55]. This latter paper contains a long discussion on the optical properties of binary GaN that goes beyond the scope of the present chapter and the reader is therefore referred to the original work.

Self-catalyzed AlGaN NWs by plasma-assisted MBE were also obtained by other groups [56, 57]. In the first of those works [56], species diffusion competition has been observed in $Al_xGa_{1-x}N$ ($0.2<x<0.4$) NWs grown on Si(111). This competition is suggested to have an impact in the NW shape. Among the results reported in the second work [57], the observation that the optimal growth temperature depends on the Al content merits to be underlined.

$GaN/Al_{.75}Ga_{.25}N$ c-s heterostructure NWs were grown by chemical vapor transport on a Ni-coated sapphire substrate [58]. Ni, an element often used to induce the growth of carbon nanotubes, is supposed to induce the VLS growth of the NWs and has been also used to induce the growth of $Al_{0.1}Ga_{0.9}N$ NWs grown by MBE [59]. In the c-s NWs [58], carrier confinement in GaN cores with section diameters of 20 and 52 nm, is observed, as indicated by the blue shift of the PL energy with respect to the emission of bare GaN NWs. Surprisingly, from the noise/signal ratio of the PL spectra, the presence of a shell seems to not improve the PL intensity of the GaN shells. Nevertheless, stimulated emission has been obtained in these c-s structures, but not in bare GaN NWs with d<100 nm. In the apparent absence of a beneficial effect of the surface passivation on the PL emission, the better performance of the c-s structure in emitting stimulated light can be attributed to the photon confinement obtained in GaN because of the presence of the cladding AlGaN shell.

Core-shell GaN/AlN/AlGaN NWs with triangular section have been grown by MOCVD [60]. In these structures, made to investigate the confined electron gas in the GaN core, the AlGaN alloy represents the outer shell. The same group has also realized GaN/InGaN/GaN/AlGaN radial single quantum well (SQW) nanowires [61]. These structures realized to investigate the optical properties of the InGaN quantum well will be discussed in the paragraph dedicated to this latter alloy.

Interesting information about species competition in this alloy is given in ref. [62], where the Ni-induced growth of $Al_xGa_{1-x}N$ (0.2<x<0.9) by MOCVD is reported. A first observation of this work is that as the Al concentration is increased, an inverse dependence is observed between the length of nanowires and the density of nanowires. Samples with high x exhibit a clean background and a low density of long (4–5 μm) nanowires while samples with low x are characterized by short, rod-like (1–2 μm long) nanowires. Also for this alloy a large radial inhomogeneity in the NWs is observed, with a much lower Al content in the centre of the NW and a higher Al content on the NW sidewalls. The Ga content in the NWs has of course the opposite behavior. A similar behavior has been described above for AlGaAs and InGaP [49]. In all cases, the core is depleted with the element with the shortest diffusion length. As reported above for AlGaAs, InGaP and InGaAs, also in the nonequilibrium synthesis of AlGaN nanowires the presence of different group-III ad-atoms with a large difference in bond strength and consequently diffusion mobility create a unique interplay between kinetic and thermodynamic processes. As a result, the grown structure behaves like a core-shell structure with a GaN core and an Al-rich shell. This interpretation is supported by the GaN-like luminescence observed in NWs grown with different nominal x value. The reduced Al diffusivity is also reflected by the increasing NW tapering observed as x increases.

$In_xGa_{1-x}N$

The ternary alloy InGaN is assuming an increasing importance for the development of active optical devices operating in the visible and for the fabrication of multijunction photovoltaic cells. The development of this material has been delayed with respect to other alloys because of the uncertainty about the band-gap value of InN that has been reliably determined only recently [63]. In nanowires, InGaN finds application as the carrier confining material in radial heterostructures with GaN and AlGaN used as cladding layers. A first example can be found in GaN/$In_{0.2}Ga_{0.8}N$/GaN core-shell-shell NWs [64]. Ni-catalyzed, defect-free NWs of this type, with triangular cross section were grown by MOCVD. A strong EL emission from InGaN has been observed at 448 nm. EL was made possible by the n(Si) or p(Mg) doping of GaN core and GaN external shell, respectively. TEM on those structures shows that the InGaN shell grows uniformly with smooth surfaces and maintains the overall triangular cross section of the core nanowire. These features suggest that InGaN shell growth is epitaxial on the GaN core sidewalls.

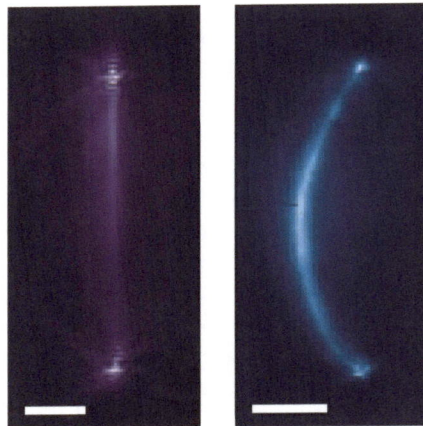

Figure 7: Photoluminescence images (false colour) from GaN/$In_{0.05}Ga_{0.95}N$ (left) and GaN/$In_{0.23}Ga_{0.77}N$ (right) MQW nanowires. Scale bars are 5 μm. Reprinted by permission from Macmillan Publishers Ltd: [Nature Materials] [67], copyright (2008).

The described structure has been later developed into multishell nanowire heterostructures [65, 66]. In order to achieve multicolor emission [65], two more shells were added to the structure discussed above, in order to have a five-shell n-GaN/$In_xGa_{1-x}N$/GaN/p-AlGaN/p-GaN heterostructure. The AlGaN layer improves both carrier and photon confinement in the InGaN active layers. The In content in the NWs was adjusted by varying the growth temperature. NW heterostructures with x=0.01, 0.1, 0.2, 0.25 and 0.35 were realized. In this way the emission wavelength was tuned in the 367-577 nm range. Lasing was finally achieved from 365 to 494 nm with InGaN/GaN multiple quantum well (MQW) structures radially grown on GaN NW sidewalls [67]. The nanowires were grown on a sapphire substrate. An interesting feature of these last structures is the large difference of growth rate observed in the different lateral facets of the NWs: higher on the two {1-101} facets, slower on the {0001} facet. The difference is probably due to the N-termination of the {0001} facets which is known to give rise to slower planar growth rates than Ga-terminated facets [68]. Fig. **7** shows PL images of two GaN/InGaN MQW NWs that show how the NW containing the largest amount of In is bent. The NW bending is probably due to the strain asymmetry that builds because of the different thicknesses of the MQW structures that grow on the different facets.

An axial approach has been instead used by two different groups to build $In_xGa_{1-x}N$ quantum dots (QDs) in catalyst-free GaN NWs grown by MBE on Si (111) [69, 70]. Green, yellow or amber light was emitted at room temperature from the dots realized in reference 69, the emission wavelength depending on alloy composition (0.15<x<0.25). Armitage and Tsubaki, on the other hand, succeeded in the realization of a white-emitting LED [70]. The result was obtained exploiting the two-colors emission from the InGaN inserts. With a proper choice of the density and diameter of the GaN NWs embedding the InGaN inserts and of the InGaN growth parameters it was possible to combine the two colors in order to have white light emission. In this work the authors rule out the presence of any QCSE in their strained structures, an effect that is instead invoked in AlGaN/GaN NWs [52]. Similarly to what was observed for InGaP [49] and AlGaN [62], also Armitage and Tsubaki find a depletion of the element with the longest diffusion length (In in this case) at the NW sidewalls.

The absence of the QCSE in InGaN/GaN NWs is also claimed by Bardoux and coworkers [71]. A further intersting result of that work is the measurement of a positive binding energy of 13 meV for the biexciton in the InGaN discs.

Bare InGaN NWs were grown by CVD [72-74]. In the first two works, the use of hydride CVD leads to a maximum In content of x=0.2. In the first of those works, where the substrate used was (0001)sapphire, CL was observed in wires in the wavelength region of 380-470 nm. In reference 73, the authors observed that the relative increase of In content increases the final NW length, keeping constant the amount of the group III species. This observation confirms once more the importance of the ad-atom diffusion length in the growth of III-III-V NWs. Also in this case, both cubic and hexagonal lattice structures were observed.

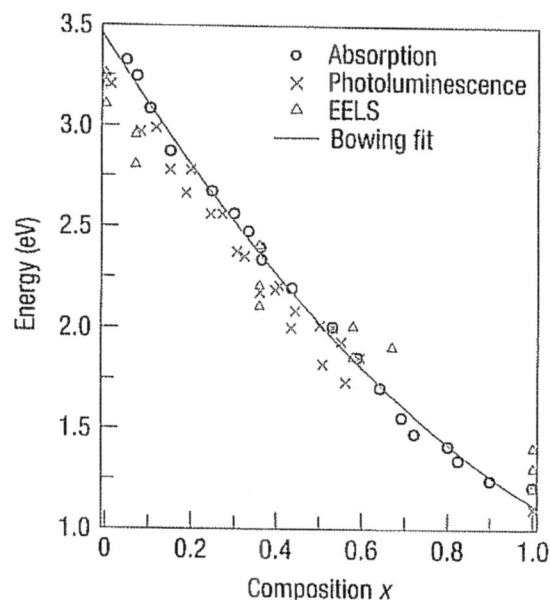

Figure 8: Band-gap energy of $In_xGa_{1-x}N$ nanowires vs alloy composition. The curve is a fit obtained using a bowing parameter of 1.01 eV. Reprinted by permission from Macmillan Publishers Ltd: [Nature Materials] [74], copyright (2007).

The complete tunability of the InGaN composition was instead obtained by MOCVD [74]. The result was obtained using halide CVD at low temperatures. Photoluminescence in the 400-700 nm wavelength region was reported for In contents up to x=0.6. For higher values the PL intensity drops. The band gap of the InGaN, see Fig. **8**, was determined over the whole alloy composition by the combination of PL, optical absorption and electron-energy loss spectroscopy. The linear and bowing parameters of the band-gap were determined by the fit of the combined data to be 1.12 eV and 1.01 eV, respectively.

$In_xAl_{1-x}N$

InAlN alloys have potential application in light emitting devices and solar cells because its bandgap varies from 0.6 eV for InN to 6 eV for AlN, namely from infrared to ultraviolet.

Very dense $In_xAl_{1-x}N$ (x= 0.71, 0,81, 0.92) nanocolumns were grown by plasma-assisted MBE on Si (111) surfaces [75, 76] starting from GaN three-dimensional seeds. These nanocolumns, having diameters ranging from 40 to 130 nm show room temperature luminescence at energy that blue shifts and broadens as the Al content increases. $InN/In_{0.77}Al_{0.23}N$ multiple QW structures with 100 wells embedded into the nanocolumns were also grown. Well widths were of 1, 2, and 3 nm. Energy shift of the luminescence of the different QW samples is a clear indication of carrier confinement in these nanostructures.

ScAlN

The band gap of ScN, 2.1 eV, makes the ScAlN alloy an alternative to InAlN for use in optoelectronic devices at wavelengths from the UV up to the red region. $Sc_{0.05}Al_{0.95}N$ NWs, with the alloy composition determined by EDS measurements, were grown by vapor phase epitaxy through an unintentional VLS process based on the formation of ScAl droplets on a ScN film [77]. The resulting NWs have wurtzite lattice structure, are about 1 µm long and 100 nm thick. CL reveals an emission peak at 2.4 eV that the authors attribute to e-h recombination in ScAlN without identification of the exact electronic state.

III-V-V ALLOYS

The techniques that are mostly used to grow III-V nanowires are MOCVD and MBE with its variants. In both processes, the growth rate of 2D epitaxial layers only depend on the group-III element flux and the growth takes place under large group-V overpressures. This means that in the growth of epitaxial III-V-V ternary alloys, the content of each group-V element in the material depends on the flux ratio of the two group-V species and on the bonding energy that each species has with the group-III element of the alloy. This latter parameter is automatically taken into account when the calibration of the concentration is performed: as a matter of fact, the relative content of the group-V elements in the materials is determined by the relative pressures of the impinging elements. This assumption can be assumed as valid also for the growth of NWs. Moreover, the miscibility of the group-V elements in Au is very poor, leading to the generally accepted opinion that the catalyst particle does not play any role in the determination of the relative content of the group-V species in III-V-V ternary NWs grown using Au as the catalyst. As we will see in the following, these features allow a better control of the growth of this type of ternary nanowires compared to the III-III-V case described in the previous section.

$GaAs_{1-x}P_x$

$GaAs_{1-x}P_x$ has the direct-indirect band-gap crossover for $x \approx 0.45$ [78]. This is therefore the highest P content that can be used in optically active sections of nanowires. This alloy also finds application as quantum barrier material in either axial or core-shell heterostructures embedding GaAs.

GaAsP nanowires were first grown by Duan and coworkers [1] with x=0.4. The NWs were single crystals with (111) growth direction. The measured alloy composition was very close to that designed in the laser-assisted catalytic technique used for the growth and no evidence for compositional modulation was observed. These two last observations were a first confirmation of the good compositional control that can be achieved in III-V-V alloy NWs.

Optically active GaAsP segments were embedded in GaP NWs grown by MOCVD on Si [79] or GaP [80] substrates. To grow the GaAsP segment, arsine was switched on at a certain time during the growth. The alloy content was controlled by adjusting the arsine-to-phosphine ratio.

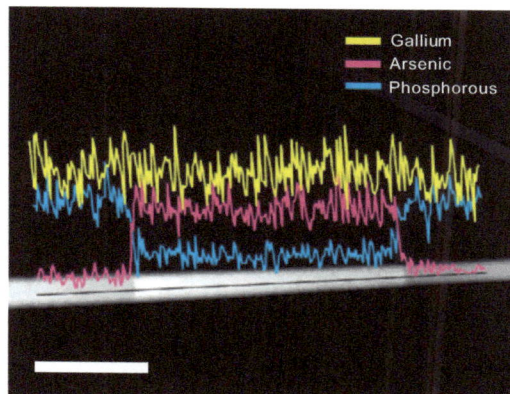

Figure 9: EDS line scan of a GaP nanowire with a GaAsP segment showing the sharp nature of the interface. Scale bar is 200 nm. Reprinted in part with permission from Ref. [79]. Copyright 2004 American Chemical Society.

Fig. **9** shows the X-ray energy dispersion measurements that confirm the expected sharpness of the GaAsP/GaP interfaces, obtained by switching on or off the arsine flux.

Bright PL is obtained at room temperature with these NWs. This result is probably made possible by the growth of a thin GaP shell over the GaAsP core due to lateral growth that occurs during the growth of the final GaP segment. Tuning of the emission wavelength is obtained by varying the alloy content. However, a limit has been found as luminescence at higher energy than expected was observed when GaAs sections were embedded in the wires. A background P-pressure during the growth of the GaAs segments may be the reason for this observation. Another possibility is As-to-P exchange reaction during the growth of the final GaP segment that leads to the incorporation of some phosphorus into the GaAs segment.

GaAsP/GaP NW heterostructures have been widely investigated by LaPierre and coworkers [81-84]. They have grown both radial or axial heterostructures by gas-source MBE on GaAs(111) [81] or Si(111) [82-84] substrates. The morphology of the NWs appears to be more regular and homogeneous on the latter substrates, maybe because of the smaller difference of the GaP lattice constant from that of Si than from that of GaAs. The growth on the sidewalls, observed in the first work [81] when axial heterostructures were designed, has been exploited in the later works to obtain core-(multi)shell structures similar in morphology to the GaAs/AlGaAs heterostructure NW scheme shown in Fig. **1**. Despite the consideration made above on the low solubility of the group-V elements in Au, consideration also made by the authors of those works, the transition in As and P occurs in the relatively smooth length of 7 nm. Stacking faults in the wurtzite NWs have been observed in the samples grown with a continuous growth, that is without growth interruption at the interfaces, of the GaAsP/GaP sequences. Single-crystalline, wurtzite NWs have been instead obtained by using growth interruption at the GaAsP/GaP interface [83].

These heterostructure NWs have been characterized by Raman spectroscopy and photoluminescence. Raman spectroscopy is a technique that allows a quick screening of the structural properties of NWs. Applied to alloys it also permits the determination of their composition. In the case of the $GaAs_{0.56}P_{0.44}/GaAs_{0.68}P_{0.32}/GaP$ heterostructures, the strain built in the core-shell heterostructure has been revealed by means of the energy shift of the LO phonon peaks. The $GaAs_xP_{1-x}$ alloy composition determined by the phonon frequencies ($x{\approx}0.57$-0.58) is in good agreement with the EDS spectra ($x{\approx}0.56$) [84]. In those heterostructure NWs carrier confinement has been supposed on the grounds of PL measurements. The PL measurements also show that the GaP shell guarantees the surface passivation and gives rise to an intense light emission from the GaAsP segment.

Finally, $GaAs/GaAs_{0.8}P_{0.2}$ core-shell NWs were grown by selective-area MOCVD to fabricate optically pumped lasers operating at low temperatures (4.2 K) [85]. Non tapered, regular parallelograms with hexagonal sections have been obtained with this growth technique. The NW diameter, between 200 and 500 nm, is determined by the opening holes in the substrates. The relatively large diameters are necessary to enhance reflectivity at both ends of the surface and bring sufficient optical confinement for lasing. The increase of PL intensity due to the presence of the GaAsP shell is of two orders of magnitude with respect to bare GaAs NWs grown in the same conditions. It is worth noting that the intensity increase reported for most of the core-shell structures discussed in this chapter is always of two to three orders of magnitude with respect to the bare NWs made of the core material. The PL of these NWs shows a narrowing of the line shape and a rapid increase of the emission intensity at an excitation power density of about 8 kW/cm^2. The emission wavelength is 816 nm. A quick gain saturation occurs in these optically pumped devices probably because of heating and/or non homogeneous excitation. An unexplained and unusual feature observed in these NWs is the blue shift of the emission wavelength with increasing excitation power. Contrary to what was observed in ref. 85, in bulk and 2D heterostructures, as the excitation intensity increases, the combined effects of Moss-Burstein shift (blue) and band-gap renormalization (red) due to many body effects always lead to a red shift of the emission wavelength [86].

InAs$_{1-x}$P$_x$

InAsP has a direct band gap over the entire compositional range that varies from 0.35 eV in InAs to 1.3 eV in InP at room temperature. Despite the interesting band gap range, in the telecommunication range, this alloy hardly finds application in epitaxial structures because its lattice parameter badly matches with that of typical substrates. As mentioned in the introduction this disadvantage plays a minor role in the growth of nanowires.

InAsP and InAsP/InAs heterostructure NWs have been topics of a few papers of Lund University [87-90]. InAs$_{1-x}$P$_x$ NWs were grown by CBE using different Au particle dimensions (30, 50, and 70 nm). The dimension of the Au NPs has no effect on the composition of the alloys, while higher growth temperatures favor P incorporation in the material, similarly to what occurs in the epitaxial growth of this alloy. Field-effect electrical measurements show that the InAsP NWs has a type-n conductivity. Temperature-dependent electrical transport measurements, performed on individual heterostructured wires to extract the conduction band offset of the InAs$_{1-x}$P$_x$/InAs interface, allow the determination of the value of the linear coefficient of the conduction band versus x (0.6 eV) and of its bowing parameter (0.2 eV) [87].

During the growth of the axial InAsP segments an InAsP shell also grows on the wire sidewalls, with the same composition of the InAsP axial segment. High-resolution TEM images reveal that these NWs present a wurtzite structure in both InAs and InAsP [88]. Low-temperature (5 K) photocurrent spectroscopy for different alloy compositions (x=0.14, 0.23, 0.29, 0.34, and 0.48) leads the authors to determine the band gap for wurtzite InAs and InP. These values are 0.64 eV and 1.65 eV, 28% and 16% larger than the band gap of ZB InAs and InP, respectively. These values are considerably higher than expected by theoretical calculations [91, 92] that foresee that in III-V wurtzite semiconductors the band-gap should be larger than that of their ZB correspondent by 5-10%. In GaN, a III-V compound, this difference is experimentally known to be about 6.5% in bulk materials [93,94]

X-ray diffraction (XRD) of InAs/InAsP core-shell structures [89] shows that shells containing a small P content grow pseudomorphically on the sidewalls of the InAs core, while plastic relaxation occurs in the shells containing high P concentrations.

InAsP insertions in InP NWs were also grown by MBE on InP(111)B substrates [95]. In that work, InP/InAs$_{1-x}$P$_x$/InP (0.5≤x≤0.65) axial heterostructures were investigated, and a few of them were covered by an InP shell that forms during the growth of the last InP segment if the growth temperature is increased from 390 °C. to 420 °C. Both InP and InAsP parts have wurtzite crystalline structure. While InP presents a small amount of stacking faults, these defects are numerous in the InAsP insertions. The NWs show RT luminescence of both InAsP and InP parts with a clear red shift of the InAsP PL as the amount of As in the alloy is increased. Carrier confinement in thin InAsP insertions that act as quantum dots has been verified by micro-PL measurements at 10 K.

InAsP NWs have also been grown without the use of any metal NP to induce the growth [90, 96]. In the first work [90] P contents up to x=0.2 have been obtained by depositing a thin layer (8 nm) of SiO$_x$ on the InP(111)B substrates. PL and XRD measurements gave similar values of the alloy composition. In the second paper [96], Mattila and coworkers have investigated the self-catalyzed growth of InAsP nanowires on Si(111) substrates using *in situ* deposited In metal droplets as seeds for nanowire growth by MOCVD. Starting from the growth of InP nanowires the As content in the wires has been varied by increasing the amount of tertiarybutylarsine in the group-V gas mixture. The As incorporation is pointed out by the red shift of the luminescence energy. A thin InP shell forms during the sample cooling down in a phosphorous environment that passivates the InAsP NW surface thus giving rise to a strong PL signal.

Light emitting diodes based on InAsP/InP heterostructure NWs were fabricated by MOCVD using the VLS process [97]. To induce electroluminescence, InAsP quantum dot sections were embedded between n-InP and p-InP sections. The electron-☐hole recombination is restricted to the quantum dot-sized InAsP section. These heterostructure NWs show interesting structural inhomogeneities: rotational stacking defects are seen in the InP segments, with a higher density of stacking defects in n-InP; the p-InP has zinc blende crystal structure, whereas the InAsP segment is wurtzite; finally, the n-type InP shows a mixture of zinc blende and wurtzite structure. Very sharp low-temperature micro-PL is observed for very thin (12 nm) InAsP layers in contrast with broader emissions from larger segments. The sharp emission is interpreted as a sign of carrier confinement in the axial direction. The heterostructure with the 12 nm-wide InAsP segment shows electroluminescence at 10 K.

A detailed optical study was performed by the same group on samples all having short InAsP sections in order to have quantum-dot like behaviors [98, 99]. PL emissions with line widths of a few tens of μeV have been observed in samples grown by the different techniques. Biexciton recombination was also observed. An example of the PL spectra obtained in these quantum dot structures and showing both exciton (X) and biexciton (XX) recombination is shown in Fig. **10**. The different assignments are made after the power dependence of their intensity.

PL measurements in magnetic field show a diamagnetic shift compatible with a diameter of ~20 nm, in reasonable agreement with the quantum dot size given by growth settings.

The low solubility of the group V-species in Au has allowed Glas and coworkers to perform an important experiment using InAsP NWs [100]. Small variations of the As/P content in the alloy (x=0.34 ±0.03) have been studied by means of high-angle annular dark field (HAADF) intensity oscillations. A nonlinear dependence of the growth rate on time has been verified, due to the increasing collection at the Au-NP of ad-atoms impinging on the NW sidewalls. Most importantly, the instantaneous growth rate, given by the experiment, show a strongly sub-Poissonian statistics, which indicates that the nucleation events are not independent of each other and are anticorrelated in time.

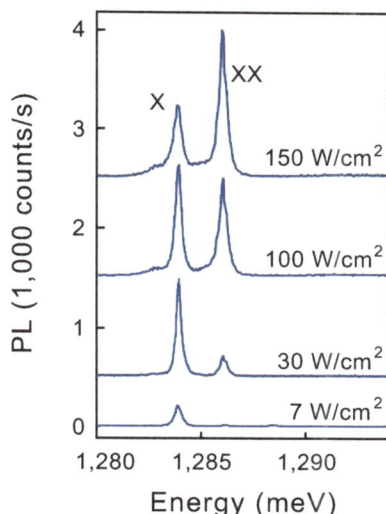

Figure 10: Photoluminescence spectra from InAsP NWs for increasing excitation power, taken at 4.2 K. The two emission peaks correspond to the exciton (X) and biexciton (XX) transitions. Reprinted in part with permission from Ref. [98]. Copyright 2009 American Chemical Society..

GaAsSb

The band gap of GaAs$_x$Sb$_{1-x}$ can be tuned between 0.9 and 1.8 μm, that is in the telecommunication region. It is worth mentioning that GaSb presents a type II band alignment with binary arsenide's.

Zinc blende GaAs$_{0.75}$Sb$_{0.25}$ NWs were grown by Au-assisted MBE [101] on GaAs(111)B substrates. A very interesting feature of those NWs is that they have a zinc blende structure instead of the most common, among III-V NWs, wurtzite structure that is usually found to present numerous stacking faults. Beneath the GaAsSb NWs, short GaAs NWs were grown. The growth temperature was kept the same (540 °C) for both parts of the NWs.

As a follow-up of the previous work, single zinc blende GaAsSb inserts were embedded in wurtzite GaAs NWs by adding a further GaAs segment, with no growth interruption at the interfaces [102]. Because of the spread of the NW diameters in a same growth, the authors were able to investigate the length of the single GaAsSb inserts as a function of the NW diameter, grown under the very same conditions. The experimental points shown in Fig. **11** are reported along with a simulation based on the growth model proposed by Dubrovskii and Sibirev [103]. It has to be noticed, however, that a quite broad spread of the Sb content (0.06<x<0.25) in the inserts was found.

The particular structural characteristic of the GaAsSb inserts and the differences between the defect density in the GaAs grown beneath or on the GaAsSb insert have been explained with the chemical potential difference (supersaturation), that gradually varies with time when the Sb source is open, whereas the material-dependent critical supersaturation value is instantaneously changed between its corresponding values in GaAsSb and in GaAs. The micro-PL spectrum from a single NW shows a 4 meV wide peak at 1.254 eV. Because of the type II alignment of the GaAs/GaAsSb interface, the PL emission is attributed to exciton recombination between confined holes in the GaAsSb insert and electrons in the GaAs NW segments at the GaAs/GaAsSb interfaces.

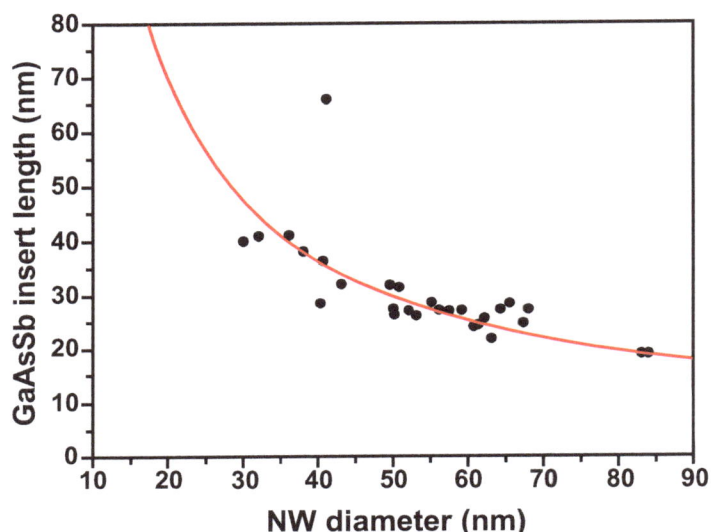

Figure 11: The length of single GaAsSb inserts in GaAs as a function of the NW diameter. The solid line is a simulation according to the VLS growth model proposed by Dubrovskii and Sibirev [103]. Reprinted with permission from Ref. [102]. Copyright 2008 American Chemical Society.

The growth of four GaAsSb inserts in GaAs NWs has then allowed the same group [104] to investigate interface abruptness, growth rate and alloy composition as a function of the insert position in the NW. The growth rate of both GaAsSb and GaAs is found to increase with the NW length, which is with growth time. This feature agrees with what was recently observed by Glas and coworkers for InAsP [99] but the authors suggest that it could be due to the geometry of the MBE machine. The chemical interfaces between GaAs and GaAsSb sections are sharp, as expected in the case of group V exchange. A large variation of the Sb content is instead observed, this value increasing when moving the insert position in the NW toward the tip, a feature that remains unexplained.

Self-catalyzed GaAs/GaAsSb axial and core-shell NW heterostructures were grown by gas source MBE on Si(111) substrates [105]. A Ga droplet was observed at the NW tip suggesting that Ga NPs are the seeds responsible for the 3D NW growth. The presence of a small amount of Sb in the GaAs segment beneath the GaAsSb is indicative of the formation of radial GaAsSb growth. NWs with a higher number of GaAs and GaAsSb sequences have been also grown with an AlGaAs shell to passivate the whole NW surface. However no PL from these NWs was reported.

As well as in the Au-catalyzed growth [101, 102], also the self-catalyzed GaAsSb insertions have a twin-free zinc blende structure, while the GaAs parts show a wurtzite phase with stacking faults, even if sandwiched between two zinc blende GaAsSb insertions. It is worth noticing that the transition from GaAsSb to GaAs shows some stacking faults, whereas the transition from GaAs to GaAsSb is abrupt.

GaAsN

Epitaxial GaAsN is an alloy that has been widely investigated in the last decade with the scope of fabricating efficient GaAs-based emitters operating in the second and third window of the silica-fiber based optical communication system. However, up to date only one paper has been published on GaAsN NWs [106]. These NWs were grown by magnetron-sputtering deposition on GaAs(111)B and on Si(111) substrates *via* the VLS process. The wires grow along the <111> direction with a strongly tapered shape. The length of the wires, which remains below 800 nm, has a dependence on the wire diameter at the top, which is similar to that shown in Fig. **10** for GaAsSb. The N incorporation in the NWs is pointed out by PL measurements, that show how under otherwise equal growth conditions, a lower growth temperature allows the incorporation of a larger amount of nitrogen (x=0.027 at 470 °C; x=0.004 at T=560 °C), in agreement with what is also generally found in the growth of epitaxial layers of this alloy.

CONCLUSIONS

Although the development of ternary III-V NWs is still in the first stages, a noticeable amount of work has been already published. However, the large number of possible alloys and growth techniques, features reflected in the often fragmented information on these materials, make it possible to draw only a very few general conclusions.

The growth of morphologically and compositionally homogeneous III-III-V alloy NWs presents challenging aspects: the competition behaviors due to the different species incorporation in the catalyst nanoparticles, and relative release, and the different ad-atom diffusion lengths make extremely critical the control of the growth parameters. The self-catalyzed growth of these NWs is only slightly favored from the absence of the first competing mechanism in its process. From what is written in this chapter, indeed, the competition mechanism due to different diffusion lengths appears dominant. We have also seen in the case of AlGaAs that different results have been obtained if the NWs are grown by MBE or MOCVD. The Al composition gradient has a different direction in the two cases because of the different interplay of ad-atom diffusion and catalyst-related incorporation. In MBE the diffusion mechanisms appear to be the dominant competition mechanism, while for MOCVD, NP-related mechanisms play a more important role.

From the experiments as a function of the growth temperature in InGaAs and InGaP, the proper choice of a relatively high growth temperature seems to be the way of compensating the longer diffusion length of In with the higher thermal stability of the Ga-related bonds. Such a compensation results in a better alloy homogeneity and the method could be extended to other III-III-V alloys. A different optimal growth temperature depending on the alloy composition has been also reported for AlGaN.

The lower solution of group V species in Au and, more importantly, the fact that most of the growth technique used to grow III-V NWs work in conditions of group-V overpressure, allow an easier control of the growth of III-V-V ternary NWs. As consequences, larger alloy homogeneity and sharper interfaces are generally obtained in III-V-V alloys rather than in III-III-V materials.

Another intriguing aspect pointed out by the present review is that it is possible to observe important differences in the lattice structure of different sections of heterostructure NWs (see the cases of InGaAs/InAlAs, InP/InAsP, and GaAsSb/GaAs). In some cases, wurtzite or zinc blende may be indeed found in different material embedded in the same wire, in other cases different defect densities have been observed. Even doping seems to play a role in the determination of the lattice structure, as in the case of InAsP/InP NWs.

Competing mechanisms in species incorporation and in crystal structure formation, the latter still being far to be understood, make the growth of homogeneous ternary NWs a challenging task. But just for this reason, the complex aspects involved in the growth of this type of nanowires will stimulate researchers to find the proper growth conditions for optimal growth and, at the same time, they will provide a large amount of information that will greatly help the development of an exhaustive growth model for all nanowires.

REFERENCES

[1]   Duan, X.; Lieber, C.M. General synthesis of compound semiconductor nanowires. *Adv. Mater.* **2000**, *12(4)*, 298-302.

[2]   Wu, Z.H.; Sun, M.; Mei, X.Y.; Ruda, H.E. Growth and photoluminescence characteristics of AlGaAs nanowires. *Appl. Phys. Lett.* **2004**, *85(4)*, 657-659.

[3]   Chen, C.; Shehata, S.; Fradin, C.; Couteau, C.; Weihs, G.; LaPierre, R. R. Self-directed growth of AlGaAs core-shell nanowires for visible light applications. *Nano Lett.* **2007**, *7(9)*, 2584-2589.

[4]   Shitara, T.; Neave, J. H.; Joyce, B. A. Reflection high-energy electron diffraction intensity oscillations and anisotropy on vicinal AlAs(001) during molecular-beam epitaxy. *Appl. Phys. Lett.* **1993**, *62(14)*, 1658.

[5]   Plante, M. C.; LaPierre, R. R. Growth mechanisms of GaAs nanowires by gas source molecular beam epitaxy. *J. Cryst. Growth* **2006**, *286(2)*, 394-399.

[6]   Chen, C.; Braidy, N.; Couteau, C.; Fradin, C.; Weihs, G.; La Pierre, R. Multiple quantum well AlGaAs nanowires. *Nano Lett.* **2008**, *8(2)*, 495-499.

[7]   Lim, S.K.; Tambe, M.J., Brewster, M.M.; Gradečak, S. Controlled growth of ternary alloy nanowires using metalorganic chemical vapor deposition. *Nano Lett.* **2008**, *8(5)*, 1386-1392.

[8]   Hoang, T.B.; Titova, L.V.; Yarrison-Rice J.M.; Jackson, H.E.; Govorov, A.O.; Kim, Y.; Joyce, H.J.; Tan, H.H.; Jagadish, C.; Smith, L.M. Resonant Excitation and imaging of nonequilibrium exciton spins in single core–shell GaAs–AlGaAs nanowires. *Nano Lett.* **2007**, *7(3)*, 588-595.

[9]   Tambe, M.J.; Lim, S.K.; Smith, M.J.; Allard, L.F.; Gradečak, S. Realization of defect-free epitaxial core-shell GaAs/AlGaAs nanowire heterostructures. *Appl. Phys. Lett.* **2008**, *93*, 151917.

[10]   Ouattara, L.; Mikkelsen, A.; Sköld N.; Eriksson, J.; Knaapen, T.; Ćavar, E.; Seifert, W.; Samuelson, L.; Lundgren, E. GaAs/AlGaAs Nanowire heterostructures studied by scanning tunneling microscopy. *Nano Lett.* **2007**, *7(9)*, 2859-2864.

[11]   Titova, L. V.; Hoang, Thang B.; Jackson, H. E.; Smith; L. M.; Yarrison-Rice J. M.; Kim, Y.; Joyce, H. J.; Tan H. H.; Jagadish C. Temperature dependence of photoluminescence from single core-shell GaAs–AlGaAs nanowires. *Appl. Phys. Lett.* **2006**, *89*, 173126

[12]   Perera, S.; Fickenscher, M. A.; Jackson, H. E.; Smith, L. M.; Yarrison-Rice, J. M.; Joyce, H.J.; Gao, Q.; Tan, H.H.; Jagadish, C.; Zhang, X.; Zou J. Nearly intrinsic exciton lifetimes in single twin-free GaAs/AlGaAs core-shell nanowire heterostructures. *Appl. Phys. Lett.* **2008**, *93*, 053110.

[13]   Zhou, H.L.; Hoang, T.B.; Dheeraj, D.L.; van Helvoort, A.T.J.; Liu, L.; Harmand, J.-C.; Fimland, B.O.; Weman H.; Wurtzite GaAs/AlGaAs core–shell nanowires grown by molecular beam epitaxy. *Nanotechnology* **2009**, *20*, 415701.

[14]   Moewe, M.; Chuang, L.C.; Crankshaw, S.; Chase, C.; Chang-Hasnain, C. Atomically sharp catalyst-free wurtzite GaAs/AlGaAs nanoneedles grown on silicon. *Appl. Phys. Lett.* **2008**, *93*, 023116.

[15]   Prete, P.; Marzo, F.; Paiano, P.; Lovergine, N.; Salviati, G.; Lazzarini, L.; Sekiguchi, T. Luminescence of GaAs/AlGaAs core–shell nanowires grown by MOVPE using tertiarybutylarsine. *J. Cryst. Growth* **2008**, *310*, S114-S118.

[16]   Tateno, K.; Gotoh, H.; Watanabe, Y. GaAs/AlGaAs nanowires capped with AlGaAs layers on GaAs(311)B substrates. *Appl. Phys. Lett.* **2004**, *85(10)*, 1808-1810.

[17]   Motohisa, J.; Takeda, J.; Inari, M.; Noborisaka, J.; Fukui, T. Growth of GaAs/AlGaAs hexagonal pillars on GaAs(111)B surfaces by selective-area MOVPE. *Physica E*, **2004**, *23(3–4)*, 298–304.

[18]   Noborisaka, J.; Motohisa, J.; Hara, S.; Fukui, T. Fabrication and characterization of freestanding GaAs/AlGaAs core-shell nanowires and AlGaAs nanotubes by using selective-area metalorganic vapor phase epitaxy. *Appl. Phys. Lett.* **2005**, *87*, 093109.

[19]   Tomioka, K.; Kobayashi, Y.; Motohisa, J.; Hara, S.; Fukui, T. Selective-area growth of vertically aligned GaAs and GaAs/AlGaAs core–shell nanowires on Si(111) substrate. *Nanotechnology* **2009**, *20*, 145302.

[20]   Tomioka, K.; Motohisa, J.; Hara, S.; Hiruma, K.; Fukui, T. GaAs/AlGaAs core multishell nanowire-based light-emitting diodes on Si. *Nano Lett.* **2010**, ASAP, DOI: 10.1021/nl9041774

[21]   Sladek, K.; Klinger, V.; Wensorra, J.; Akabori, M.; Hardtdegen, H.; Grützmacher, D.; MOVPE of n-doped GaAs and modulation doped GaAs/AlGaAs nanowires *J. Crys. Growth* **2010**, *312*, 635-640.

[22]   Jabeen, F.; Rubini, S.; Martelli, F. Growth of III– V semiconductor nanowires by molecular beam epitaxy. *Microelectron. J.* **2009**, *40*, 442-445.

[23]   Verschuren, C. A.; Leys, M. R.; Vonk, H.; Wolter, J. H. Interfacet surface diffusion in selective area epitaxy of III–V semiconductors. *Appl. Phys. Lett.* **1999**, 74 2197-2199.

[24]   Kim, Y.; Joyce, H.J.; Gao, Q.; Hoe Tan, H.; Jagadish, R.; Paladugu, M.; Zou, J.; Suvorova, A.A. Influence of nanowire density on the shape and optical properties of ternary InGaAs nanowires. *Nano Lett.* **2006**, *6(4)*, 599-604.

[25]   Regolin, I.; Khorenko, V.; Prost, W.; Tegude, F.J; Sudfeld, D.; Kästner, J.; Dumpich, G. Composition control in metal-organic vapor-phase epitaxy grown InGaAs nanowhiskers. *J. Appl. Phys.* **2006**, *100*, 074321. Sudfeld, D.; Regolin, I.; Kästner, J.; Dumpich, G.; Khorenko, V.; Prost, W.; Tegude, F.-J. Single InGaAs nanowhiskers characterized by analytical transmission electron microscopy. *Phase Transitions* **2006**, *79*, 727-737.

[26]   Akabori, M.; Takeda, J.; Motohisa, J.; Fukui, T. InGaAs nano-pillar array formation on partially masked InP(111)B by selective area metal–organic vapour phase epitaxial growth for two-dimensional photonic crystal application. *Nanotechnology* **2003**, 14, 1071-1074. Motohisa, J.; Noborisaka, J.; Takeda, J.; Inari, M.; Fukui, T. Catalyst-free selective-area MOVPE of semiconductor nanowires on (111)B oriented substrates. *J. Cryst. Growth.* **2004**, *272*, 180-185

[27]   Sato, T.; Motohisa, J.; Noborisaka, J.; Hara, S.; Fukui, T. Growth of InGaAs nanowires by selective-area metalorganic vapor phase epitaxy. *J. Cryst. Growth* **2008**, *310*, 2359-2364.

[28]   Jabeen, F.; Rubini, S.; Grillo, V.; Felisari, L.; Martelli, F. Room temperature luminescent InGaAs/GaAs core-shell nanowires. *Appl. Phys. Lett.* **2008**, *93*, 083117.

[29]   Tateno, K.; Zhang, K.; Nakano, H. Growth of GaInAs/AlInAs Heterostructure Nanowires for Long-Wavelength Photon Emission. *Nano Lett.* **2008**, *8 (11)*, 3645-3650.

[30]   Regolin, I.; Sudfeld, D.; Lüttjoann, S.; Khorenko, V.; Prost, W.; Kästner, J.; Dumpich, G.; Meier, C.; Lorke, A.; Tegude, F.-J. Growth and characterization of GaAs/InGaAs/GaAs nanowhiskers on (111)GaAs. *J. Cryst. Growth* **2007**, *298*, 607-611.

[31]   Bauer, J.; Gottschalch, V.; Paetzelt, H.; Wagner, G. VLS growth of GaAs/(InGa)As/GaAs axial double-heterostructure nanowires by MOVPE. *J. Cryst. Growth* **2008**, *310*, 5106-5110.

[32]   Heiß, M.; Gustafsson, A.; Conesa-Boj, S.; Peiró, F.; Morante, J.R.; Abstreiter, G.; Arbiol. J.; Samuelson, L.; Fontcuberta i Morral, A. Catalyst-free nanowires with axial $In_xGa_{1-x}As/GaAs$ heterostructures. *Nanotechnology* 2009, *20*, 075603.

[33]   Joyce, H.J.; Kim, Y.; Gao, Q.; Tan, H.H.; Jagadish C. Growth of Ill-V nanowires and nanowire heterostructures by metalorganic chemical vapor deposition. *Proceedings of the 2nd IEEE International Conference on Nano/Micro Engineered and Molecular Systems* **2007**, pp. 567-570.

[34]   Yang, L.; Motohisa, J.; Takeda, J.; Tomioka, K.; Fukui, T. Size-dependent photoluminescence of hexagonal nanopillars with single InGaAs/GaAs quantum wells fabricated by selective-area metal organic vapor phase epitaxy. *Appl. Phys. Lett.* **2006**, *89*, 203110.

[35]   Yang, L.; Motohisa, J.; Takeda, J.; Tomioka, K.; Fukui, T. Selective-area growth of hexagonal nanopillars with single InGaAs/GaAs quantum wells on GaAs(111)B substrate and their temperature-dependent photoluminescence. *Nanotechnology* **2007**, *18*, 105302.

[36]   Noborisaka, J.; Sato, T.; Motohisa, J.; Hara, S.; Tomioka, K.; Fukui, T. Electrical characterizations of InGaAs nanowire-top-gate field-effect transistors by selective-area metal organic vapor phase epitaxy. *Jap. J. Appl. Phys.* **2007**, *46(11)*, 7562-7568.

[37]   Yang, L.; Motohisa, J.; Tomioka, K.; Takeda, J.; Tomioka, K.; Fukui, T.; Geng, M.M.; Jia, L.X.; Zhang, L.; Liu, Y.L. Fabrication and excitation-power-density-dependent micro-photoluminescence of hexagonal nanopillars with a single InGaAs/GaAs quantum well. *Nanotechnology* **2008**, *19*, 275304.

[38]   Yang, L.; Motohisa, J.; Fukui, T.; Jia, L.X.; Zhang, L.; Geng, M.M.; Liu, Y.L. Fabry-Pérot microcavity modes observed in the micro-photoluminescence spectra of the single nanowire with InGaAs/GaAs heterostructure. *Optics Express* **2009**, *17(11)*, 9337-9346.

[39]   Krogstrup, P.; Yamasaki, J.; Sørensen, C.B.; Johnson, E.; Wagner, J.B.; Pennington, R.; Aagesen, M.; Tanaka, N.; Nygård, J. Junctions in axial III-V heterostructure nanowires obtained *via* an interchange of group III elements. *Nano Lett.* **2009**, *9(11)*, 3689-3693.

[40]   Cornet, D.M.; LaPierre, R.R. InGaAs/InP core–shell and axial heterostructure nanowires. *Nanotechnology* **2007**, *18*, 385305.

[41]   Dick, K. A.; Kodambaka, S.; Reuter, M. C.; Deppert, K.; Samuelson, L.; Seifert, W.; Wallenberg, L. R.; Ross, F. M. The morphology of axial and branched nanowire heterostructures. *Nano Lett.* **2007**, *7*, 1817-1822.

[42]   Paladugu, M.; Zou, J.; Guo, Y. N.; Auchterlonie, G. J.; Joyce, H. J.; Gao, Q.; Tan, H. H.; Jagadish, C.; Kim, Y. Novel growth phenomena observed in axial InAs/GaAs nanowire heterostructures. *Small* **2007**, *3*, 7.

[43]   Paladugu, M.; Zou, J.; Guo, Y.; Zhang, X.; Kim, Y.; Joyce, H. J.; Gao, Q.; Tan, H. H.; Jagadish, C. Nature of heterointerfaces in GaAs/InAs and InAs/GaAs axial nanowires heterostructures. *Appl. Phys. Lett.* **2008**, *93*, 101911.

[44]   Sköld, N.; Wagner, J.B.; Karlsson, G.; Hernán, T.; Seifert, W.; Pistol, M.-E.; Samuelson, L. Phase Segregation in AlInP Shells on GaAs Nanowires. *Nano Lett.* **2006**, *6(12)*, 2743-2747.

[45] Markowitz, P.D.; Zach, M.P.; Gibbons, P.C.; Penner, R.M.; Buhro, W.E. Phase separation in $Al_xGa_{1-x}As$ nanowhiskers grown by the solution-liquid-solid mechanism. *J. Am. Chem. Soc.* **2001**, *123*, 4502-4511.

[46] Nakaema, M. K. K.; Godoy M. P. F.; Brasil, M. J. S. P.; Iikawa, F.; Silva, D.; Sacilotti, M.; Decobert, J.; Patriarche, G. Optical and structural investigation of $In_{1-x}Ga_xP$ free-standing microrods. *J. Appl. Phys.* **2005**, *98*, 053506.

[47] Sköld, N.; Karlsson, L.S.; Larsson, M.W.; Pistol, M.E.; Seifert,W.; Trägårdh, J.; Samuelson, L. Growth and optical properties of strained $GaAs-Ga_xIn_{1-x}P$ core-shell nanowires. *Nano Lett.* **2005**, *5(10)*, 1943-1947.

[48] Svensson, C.P.T.; Mårtensson, T.; Trägårdh, J.; Larsson, C.; Rask, M.; Hessman, D.; Samuelson, L.; Ohlsson, J. Monolithic GaAs/InGaP nanowires light emitting diodes on silicon. *Nanotechnology* **2008**, *19*, 305201.

[49] Fakhr, A.; Haddara, Y.M.; LaPierre, R.R. Dependence of InGaP nanowire morphology and structure on molecular beam epitaxy growth conditions. *Nanotechnology* **2010**, *21*, 165601.

[50] Hong, L.; Liu, Z.; Zhang, X.T., Hark, S.K. Self-catalytic growth of single-phase AlGaN alloy nanowires by chemical vapor deposition. *Appl. Phys. Lett.* **2006**, *89*, 193105.

[51] Ristiç, J.; Sánchez-García, M.A.; Calleja, E.; Sanchez-Páramo, M J.; Calleja, J.M.; Jahn, U.; Ploog, K.H. AlGaN nanocolumns grown by molecular beam epitaxy: optical and structural characterization. *Phys. Stat. Sol. (a)* **2002**, *192 (1)*, 60-66.

[52] Ristiç, J.; Calleja, E.; Sánchez-García, M.A.; Ulloa, J.M.; Sanchez-Páramo, M J.; Calleja, J.M.; Jahn, U.; Trampert, A.; Ploog, K.H. Characterization of GaN quantum discs embedded in $Al_xGa_{1-x}N$ nanocolumns grown by molecular beam epitaxy. *Phys. Rev. B* **2003**, *68*, 125305.

[53] Rivera, C.; Jahn, U.; Flissikovski, T.; Pau, J.L.; Muñoz, E.; Grahn, H.T. Strain-confinement mechanism in mesoscopic quantum disks based on piezoelectric materials. *Phys. Rev. B* **2007**, *75*, 045316.

[54] Ristiç, J.; Calleja, E.; Sánchez-García, M.A.; Trampert, A.; Fernández-Garrido, S.; Rivera, C.; Jahn, U.; Ploog, K.H Columnar AlGaN=GaN Nanocavities with AlN=GaN Bragg reflectors grown by molecular beam epitaxy on Si(111). *Phys. Rev. Lett.* **2005**, *94*, 146102.

[55] Rigutti, L.; Tchernycheva, M.; De Luna Bugallo, A.; Jacopin, G.; Julien, F.H.; Furtmayr, F.; Stutzmann, M.; Eickhoff, M.; Songmuang, R., Fortuna, F., Photoluminescence polarization properties of single GaN nanowires containing $Al_xGa_{1-x}N$/GaN quantum discs. *Phys. Rev. B* **2010**, *81*, 045411.

[56] Bertness, K.A.; Roshko, A.; Sanford, N.A.; Barker, J.M.; Davydov, A.V.; Spontaneously grown GaN and AlGaN nanowires. *J. Cryst. Growth* **2006**, *287*, 522–527.

[57] Park, Y.S.; Hwang, B.R.; Lee, J.C.; Im, H.; Cho, H.Y.; Kang, T.W., Na, J.H.; Park, C.M. Self-assembled $Al_xGa_{1-x}N$ nanorods grown on Si(001) substrates by using plasma-assisted molecular beam epitaxy. *Nanotechnology* **2006**, *17*, 4640-4643.

[58] Choi, H.-J; Johnson, J.C.; He, R.; Lee, S.-K.; Kim, F.; Pauzauskie, P.; Golgberger, J.; Saykally, R.J.; Yang, P. Self-organized GaN quantum wire UV lasers. *J. Phys. Chem. B* **2003**, *107*, 8721-8725.

[59] Lari, L.; Murray, R.T.; Gass, M.H.; Bullough, T.J.; Chalker, P.R.; Kioseoglou, J.; Dimitrakopulos, G.P.; Kehagias, Th.; Komninou, Ph.; Karakostas, Th.; Chèze, C.; Geelhaar, L.; Riechert, H. Defect characterization and analysis of III-V nanowires grown by Ni-promoted MBE. *Phys. Stat. Sol. (a)* **2008**, *205 (11)*, 2589–2592.

[60] Li, Y.; Xiang, J.; Qian, F.; Gradečak, S.; Wu, Y.; Yan, H.; Blom, D.A.; Lieber, C.M. Dopant-free GaN/AlN/AlGaN radial nanowire heterostructures as high electron mobility transistors. *Nano Lett.* **2006**, *6(7)*, 1468-1473.

[61] Lim, S.K.; Brewster, M.; Qian, F.; Lieber, C.M.; Gradečak, S. Direct correlation between structural and optical properties of III-V nitride nanowire heterostructures with nanoscale resolution. *Nano Lett.* **2009**, *9(11)*, 3940-3944.

[62] Su, J.; Gherasimova, M.; Cui, G.; Tsukamoto, H.; Han, J., Onuma, T.; Kurimoto, M.; Chichibu, S.F.; Broadbridge, C., He, Y.; Nurmikko, A.V. Growth of AlGaN nanowires by metalorganic chemical vapor deposition. *Appl. Phys. Lett.* **2005**, *87*, 183108.

[63] Davydov, V. Yu.; Klochikhin, A. A.; Seisyan, R. P.; Emtsev, V. V.; Ivanov, S. V.; Bechstedt F.; Furthmüller, J.; Harima, H.; Mudryi, A. V.; Aderhold, J.; Semchinova, O.; Graul J. Absorption and emission of hexagonal InN. Evidence of narrow fundamental band gap. *Phys. Stat. Sol.* **2002**, *229*, R1-R3.

[64] Qian, F.; Li, Y.; Gradečak, S.; Wang, D.; Barrelet, C.J.; Lieber, C.M. Gallium nitride-based nanowire radial heterostructures for nanophotonics. *Nano Lett.* **2004**, *4(10)*, 1975-1979.

[65] Qian, F.; Gradečak, S.; Li, Y.; Wen, C.-Y.; Lieber, C.M. Core/Multishell nanowire heterostructures as multicolor, high-efficiency light-emitting diodes. *Nano Lett.* **2005**, *5(11)*, 2287-2291.

[66] Lim, S.K.; Brewster, M.; Qian, F.; Lieber, C.M.; Gradečak, S. Direct correlation between structural and optical properties of III-V nitride nanowire heterostructures with nanoscale resolution. *Nano Lett.* **2009**, *9(11)*, 3940-3944.

[67] Qian, F.; Li, Y.; Gradečak, S.; Park, H.-G.; Dong, Y.; Ding, Y.; Wang, Z.L.; Lieber, C.M. Multi-quantum-well nanowires heterostructures for wavelength-controlled lasers. *Nature Mater.* **2008**, *7*, 701-706.

[68]   Northrup, J. E.; Neugebauer, J. Strong affinity of hydrogen for the GaN (0001) surface: Implications for molecular beam epitaxy and metalorganic chemical vapour deposition. *Appl. Phys. Lett.* **2004**, *85*, 3429–3431.

[69]   Chang, Y.-L.; Wang, J.L.; Li, F.; Mi, Z. High efficiency green, yellow, and amber emission from InGaN/GaN dot-in-a-wire heterostructures on Si(111). *Appl. Phys. Lett.* **2010**, *96*, 013106.

[70]   Armitage, R.; Tsubaki, K.; Multicolour luminescence from InGaN quantum wells grown over GaN nanowires arrays by molecular-beam epitaxy. *Nanotechnology* **2010**, *21*, 195202.

[71]   Bardoux, R.; Kaneta, A.; Funato, M.; Kawakami, Y.; Kikuchi, A.; Kishino, K. Positive binding energy of a biexciton confined in a localization center formed in a single $In_xGa_{1-x}N$/GaN quantum disk. *Phys. Rev. B* **2009**, *79*, 155307.

[72]   Kim, H.-M.; Lee, H.; Kim, S.I.; Ryu, S.R.; Kang, T.W.; Chung, K.S. Formation of InGaN nanorods with indium mole fractions by hydride vapor phase epitaxy. *Phys. Stat.. Sol.* **2004**, *241*, 2802-2805.

[73]   Cai, X.M.; Ye, F.; Jing, S.Y.; Zhang, D.P.; Fan, P.; Xie, E.Q. CVD growth of InGaN nanowires. *J. Alloys Compd.* **2009**, *467*, 472-476.

[74]   Kuykendall, T.; Ulrich, P.; Aloni, S.; Yang, P. Complete composition tenability of InGaN nanowires using a combinational approach. *Nature Mat.* **2007**, *6*, 951-956.

[75]   Kamimura, J.; Kouno, T.; Ishizawa, S.; Kikuchi, A.; Kishino, K.; Growth of high-In-content InAlN nanocolumns on Si(111) by RF-plasma-assisted molecular-beam epitaxy. *J. Cryst. Growth* **2007**, *300*, 160-163.

[76]   Kamimura, J.; Kishino, K.; Kikuchi, A. Growth and properties of InAlN nanocolumns emitting in optical communication wavelengths. *Nano Optoelectronics workshop* **2008**, P23.

[77]   Bohnen, T.; Yazdi, G.R.; Yakimova, R.; van Dreumel, G.W.G.; Hageman, P.R.; Vlieg, E.; Algra, R.E.; Verheijen, M.A.; Edgar, J.H. ScAlN nanowires: A cathodoluminescence study. *J. Crys. Growth* **2009**, *311*, 3147-3151.

[78]   Nelson, R.J.; Holonyak, N. Jr.; Groves, W.O. Free-exciton transitions in the optical absorption spectra of $GaAs_{1-x}P_x$. *Phys. Rev. B* **1976**, *13*, 5415-5419.

[79]   Mårtensson, T.; Svensson, C.P.T.; Wacaser, B.; Larsson, M.W.; Seifert, W.; Deppert, K.; Gustafsson, A.; Wallenberg, L.R.; Samuelson, L. Epitaxial III-V nanowires on silicon. *Nano Lett.* **2004**, *4(10)*, 1987-1990.

[80]   Svensson, C.P.T.; Seifert, W.; Larsson, M.W.; Wallenberg, L.R.; Stangl, J.; Bauer, G.; Samuelson, L. Epitaxially grown $GaP/GaAs_{1-x}P_x$/GaP double heterostructure nanowires for optical applications. *Nanotechnology* **2005**, *16*, 936-939.

[81]   Chen, C.; Plante, M.C.; Fradin, C.; LaPierre,R.R.; Layer-by-layer and step-flow growth mechanisms in GaAsP/GaP nanowire heterostructures. *J. Mater. Res.* **2006**, *21(11)*, 2801-2809.

[82]   Mohseni, P.K.; Maunders, C.; Botton, G.A.; LaPierre, R.R. GaP/GaAsP/GaP core/multishell nanowire heterostructures on (111) silicon. *Nanotechnology* **2007**, *18*, 445304.

[83]   Mohseni, P.K.; LaPierre, R.R. A growth interruption technique for stacking fault-free nanowire superlattices. *Nanotechnology* **2009**, *20*, 025610.

[84]   Mohseni, P. K.; Rodrigues, A.D.; Galzerani, J.C.; Pusep, Y.A.; LaPierre, R.R.; Structural and optical analysis of GaAsP/GaP core-shell nanowires. *J. Appl. Phys.* **2009**, *106*, 124306.

[85]   Hua B.; Motohisa, J.; Kobayashi, Y.; Hara, S.; Fukui,T. Single GaAs/GaAsP coaxial core-shell nanowire lasers. *Nano Lett.* **2009**, *9(1)*, 112-116.

[86]   See, e.g.: (a) Olego, D.; Cardona, M. Photoluminescence in heavily doped GaAs. I. Temperature and hole-concentration dependence. *Phys. Rev. B* **1980**, *22*, 886-893; Holtz, P.O.; Ferreira, A.C.; (b) Sarnelius, B.E.; Buyanov, A.; Monemar, B.; Mauritz, O.; Ekenberg, U.; Sundaram, M.; Campman, K.; Merz, J.L.; Gossard, A.C. Many-body effects in highly acceptor-doped $GaAs/Al_xGa_{1-x}As$ quantum wells. *Phys. Rev. B* **1998**, *58*, 4624-4628.

[87]   Persson, A.I.; Björk, M.T.; Jeppesen, S.; Wagner, J.B.; Wallenberg, L.R.; Samuelson, L. $InAs_{1-x}P_x$ nanowires for device engineering. *Nano Lett.* **2006**, *6(3)*, 403-407.

[88]   Trägårdh, J.; Persson, A.I.; Wagner, J.B., Hessman, D.; Samuelson, L. Measurements of the band gap of wurtzite $InAs_{1-x}P_x$ nanowires using photocurrent spectroscopy. *J. Appl. Phys.* **2007**, *101*, 123701.

[89]   Keplinger, M.; Mårtensson, T.; Stangl, J.; Wintersberger, E.; Mandl, B.; Kriegner, D.; Holy′, V.; Bauer, G.; Deppert, K.; Samuelson, L. Structural investigations of core-shell nanowires using grazing incidence x-ray diffraction. *Nano Lett.* **2009**, *9(5)*, 1877–1882.

[90]   Mandl, B.; Stangl, J.; Mårtensson, T.; Brehm, M.; Fromherz, T.; Bauer, G.; Samuelson, L.; Seifert, W. Metal free growth and characterization of $InAs_{1-x}P_x$ nanowires. *Proceedings of the 28th International Conference on the Physics of Semiconductors*, Jantsch, W., Schäffler, F. eds. American Institute of Physics, **2007**, p. 97

[91]   Murayama, M.; Nakayama, T. Chemical trend of band offsets at wurtzite/zinc-blende heterocrystalline semiconductor interfaces. *Phys. Rev. B* **1994**, *49*, 4710-.

[92]   Zanolli, Z.; Fuchs,F.; Furthmüller, J.; von Barth, U.; F. Bechstedt, F. Model GW band structure of InAs and GaAs in the wurtzite phase. *Phys. Rev. B* **2007** *75*, 245121.

[93] Maruska, H.P.; Tietjen, J.J. The preparation and properties of vapor-deposited silgle-crystalline GaN. *Appl.Phys.Lett.* **1969**, *15*, 327-329.

[94] Lei, T.; Moustakas, T.D.; Graham, R.J.; He, Y.; Berkowitz, S.J. Epitaxial growth and characterization of zinc-blende gallium nitride on (001) silicon. *J. Appl. Phys.* **1992**, *71*, 4933-4943.

[95] Tchernycheva, M.; Cirlin, G.E.; Patriarche, G.; Travers, L.; Zwiller, V.; Perinetti, U.; Harmand, J.-C. Growth and characterization of InP nanowires with InAsP insertions. *Nano Lett.* **2007**, *7*, 1500-1504.

[96] Mattila, M.; Hakkarainen, T.; Lipsanen, H.; Jiang, H.; Kauppinen, E.I. Catalyst-free growth of In(As)P nanowires on silicon. *Appl. Phys. Lett.* **2006**, *89*, 063119.

[97] Minot, E.D.; Kelkensberg, F.; van Kouwen, M.; van Dam, J.A.; Kouwenhoven, L.P.; Zwiller, V.; Borgström, M.T.; Wunnicke, O.; Verheijen, M.A.; Bakkers, E.P.A.M. Single quantum dot nanowire LEDs. *Nano Lett.* **2007**, *7(2)*, 367-371.

[98] van Weert, M.H.M.; Akopian, N.; Perinetti, U.; van Kouwen, M.P.; Algra, R.E.; Verheijen, M.A.; Bakkers, E.P.A.M.; Kouwenhoven, L.P.; Zwiller V. Selective excitation and detection of spin states in a single nanowire quantum dot. *Nano Lett.* **2009**, *9*, 1989-1993.

[99] Zwiller, V.; Akopian, N.; van Weert, M.; van Kouwen, M.; Perinetti, U.; Kouwenhoven, L.; Algra, R.; Gómez Rivas, J.; Bakkers, E.; Patriarche, G.; Liu, L.; Harmand, J.-C.; Kobayashi, Y.; Motohisa, J. Optics with single nanowires. *C.R. Physique* **2008**, *9*, 804-815.

[100] Glas, F.; Harmand, J.-C.; Patriarche, G.; Nucleation antibunching in catalyst-assisted nanowire growth. *Phys. Rev. Lett.* **2010**, *104*, 135501.

[101] Dheeraj, D.L.; Patriarche, G.; Largeau, L.; Zhou, H.; van Helvoort, A.T.J.; Glas, F.; Harmand, J.-C.; Fimland, B.-O.; Weman, H. Zinc blende GaAsSb nanowires grown by molecular beam epitaxy. *Nanotechnology* **2008**, *19*, 275605.

[102] Dheeraj, D.L.; Patriarche, G.; Zhou, H.; Hoang, T.B.; Moses, A.F.; Grønsberg, S.; van Helvoort, A.T.J.; Fimland, B.-O.; Weman, H. Growth and characterization of wurtzite GaAs nanowires with defect-free zinc blende GaAsSb inserts. *Nano Lett.* **2008**, *8*, 4459-4463.

[103] Dubrovskii, V. G.; Sibirev, N. V. General form of the dependences of nanowire growth rate on the nanowire radius. *J. Cryst. Growth* **2007**, *304*, 504–513.

[104] Dheeraj, D.L.; Patriarche, G.; Zhou, H.; Harmand, J.-C.; Weman, H.; Fimland, B.-O. Growth and structural characterization of GaAs/GaAsSb axial heterostructured nanowires. *J. Cryst. Growth* **2009**, *311*, 1847-1850.

[105] Plissard, S.; Dick, K.A.; Wallart, X.; Caroff, P. Gold-free GaAs/GaAsSb heterostructure nanowires grown on silicon. *Appl. Phys. Lett.* **2010**, *96*, 121901.

[106] Soshnikov, I.P.; Cirlin, G.E.; Nadtochii, A.M.; Dubrovskii, V.G.; Bukin, M.A.; Petrov, V.A., Busov, V.M.; Troshkov, S.I. Properties of GaAsN nanowires grown by magnetron_sputtering deposition. *Semiconductors* **2009**, *43(7)*, 906–910. Original russian text: *Fizika i Tekhnika Poluprovodnikov* **2009**, *43(7)*, 938–942.

CHAPTER 7

III-V Semiconductor Nanowire Transistors

Franz-Josef Tegude* and Werner Prost

Center for Semiconductor Technology and Optoelectronics, University Duisburg-Essen, Lotharstr. 55, D-47057 Duisburg, Germany

Abstract: Semiconductor nanowires may be used as the channel of a field-effect transistor device. In contrast to field-effect transistors made of epitaxial layers, the nanowire approach provides large material diversity and enables a fully surrounding gate contact. In this way, no carriers can escape from the channel and a higher transconductance is routinely observed. In contrast to carbon nanotubes the charge polarity can be selected and a metallic phase that may inhibit the channel depletion is avoided. In this chapter device concepts will be presented based on semiconductor band structure and heterostructures. Recent advances in device technology and self-assembly will be discussed. The corresponding DC performance of the different nanowire approaches is reviewed. Special emphasis is given on the measurement and the performance of nanowire field-effect transistors at high frequencies.

Keywords: Nanowire, vapor-liquid-solid, growth, field-effect transistor, InAs, high frequency electronics, coplanar waveguide.

INTRODUCTION

The transistor is probably the most important invention of the 20th century. This three-terminal device provides high-speed switching and high frequency signal amplification at very lower power consumption. Based on its unique performance in terms of input/output isolation and gain it is today's ultimate answer to the needs of both digital and analog circuits. Therefore, the transistor and especially the field-effect transistor has paved the way to the modern information society. Since its invention countless works were carried out to improve the device performance. Almost all new findings in semiconductor physics and technology were first tested against the transistor, basically following two strategies: (i) reducing the features size according to Moore's rule, and (ii) to address specific needs such as speed, gain, power or efficiency, often using III/V semiconductor heterostructures. After almost 60 years of development the transistor has reached a unique degree of maturity and the standard strategies of improvement are approaching its end. In case of the field-effect transistor the reduction of the minimum feature size, the gate length, can only be performed with detailed impacts on scaling of all other device parameters accordingly, i. e., the device channel depth has to be less than about one third of the gate length in order to avoid a dramatic loss in input/output isolation and gain which is today's most stringent limitation of the minimum feature size in CMOS circuits. On the other hand, the heterostructure field-effect transistor consists of a sophisticated sequence of layers fulfilling in each layer a specific device requirement. This stack of layers is all epitaxially grown basically with the same lattice constant determined by the substrate in use. In the 1980's the growth of GaAs and InP based materials on silicon substrates was studied intensively, mainly implementing relaxed buffer layers [1, 2]. However, the obtained material quality was not sufficient for most applications. Thus, a further material optimization is limited by the stringent demand of a certain lattice constant, but also by related cost issues like expensive substrate material and processing. On the other hand, a larger material variety would not only improve device performance, but also enable a higher degree of heterogeneous integration, especially in optoelectronics and sensors.

In the recent decade semiconductor nanowires were intensively investigated as a field-effect transistor channel in order to overcome the current limitations in scaling and substrate dependence. The scaling issues were aggressively addressed by mainly silicon nanowires [3] and by the carbon nanotube transistor [4]. In this review we will focus on III/V nanowires and their contribution to realize a low-voltage high speed nanowire transistor and circuits on any substrate.

Address correspondence to Franz-Josef Tegude: Center for Semiconductor Technology and Optoelectronics, University Duisburg-Essen, Lotharstr. 55, D-47057 Duisburg, Germany. Tel. +49 203 379 3391; E-mail: franz.tegude@uni-due.de

The paper is organized as follows. In the second section we describe how the III/V semiconductor nanowire can effectively be used as the channel of a field-effect transistor. The current device technologies including a brief summary of currently used nanowire growth techniques for field-effect transistor applications will be given in section 3. The following section presents and discusses the performance of III/V nanowire transistors.

NANOWIRE FET DESIGN

Besides the nearly independent choice of the channel material, the most important advantage for the FET design is the nanowire geometry which allows an optimum control by an omega shaped or a gate-all-around configuration (Fig. 1). The vertical all-around-gate version (Fig. 1b) requires ultra-high resolution lithography and a highly sophisticated technology and has demonstrated impressive performance [5, 6]. This approach based on nanowires as grown on the growth substrate is an excellent candidate for integration due to its perfect position control. The transistors in Fig. 1a can be produced without ultra-high resolution lithography using self-alignment and self-assembly technologies and will be discussed thoroughly in sections 3 and 4.

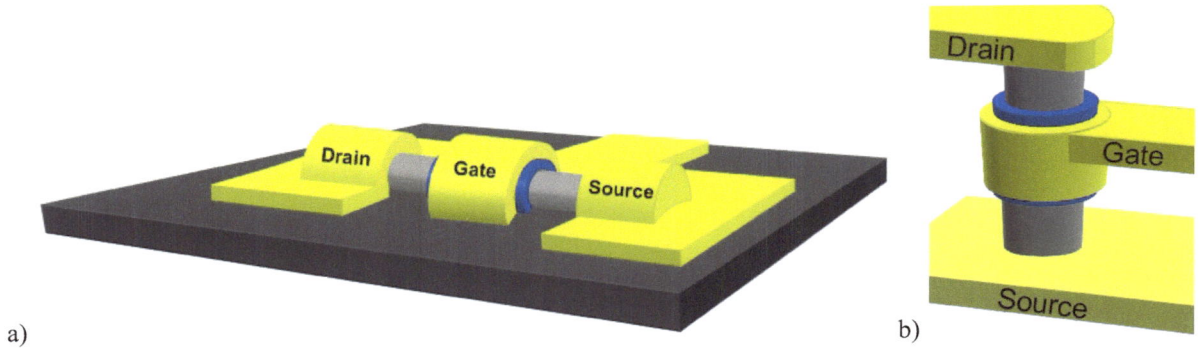

Figure 1: Concepts of nanowire transistors (a) Omega shaped gate MISFET, (b) vertical all-around-gate MISFET.

For the conventional planar MOSFET with gate contact from the top only, the gate control is characterized by a parameter termed natural length $\lambda = \sqrt{(\varepsilon_{Si}/\varepsilon_{Ox})t_{Si}t_{Ox}}$ [7] with $\varepsilon_{Si}, \varepsilon_{Ox}$ the dielectric constants of the channel material and the gate oxide, respectively, and t_{Si}, t_{Ox} the related thicknesses. This length should be five to ten times larger than the FET gate length to avoid excessive short channel effects like threshold voltage shift and the drain induced barrier lowering (DIBL), which significantly can degrade the FET sub threshold behavior. For gate-all-around nanowire FET this natural length transforms to [8]:

$$\lambda = \sqrt{\frac{2\varepsilon_{Si}t_{Si}^2 \ln\left(1+\frac{2t_{Ox}}{t_{Si}}\right)+\varepsilon_{Ox}t_{Si}^2}{16\varepsilon_{Ox}}} \tag{1a}$$

$$\lambda = \sqrt{\frac{2\varepsilon_{InAs}t_{InAs}^2 \ln\left(1+\frac{2t_{Ox}}{t_{InAs}}\right)+\varepsilon_{Ox}t_{InAs}^2}{16\varepsilon_{Ox}}} \tag{1b}$$

Lind *et al.* have transferred eq. (1a) from Si MOSFET to InAs nanowire MISFET devices [9] in order to demonstrate that this improved gate control allows for shorter gate length at relaxed oxide thickness. This excellent gate control can be employed to produce extremely steep potential drops and correlated band bending in semiconductors. Fig. **2** shows the schematic cross section of Tunnel-FET which is represented by a pin-structure with a gated i-region [16]. The band structure shows that for a gate voltage representing the off state the barriers at the n^+i as well as the ip^+ junction are to high for carrier transport over these barriers (Fig. **2b**). But for a gate voltage representing the on state, the n^+i barrier height is removed and the ip^+ barrier can be tunneled by valance band electrons from the p+ region into the conduction band of the i-region, due to the steep potential drop. Because the on current is a tunneling current, tunnel-FETs can provide a very steep transfer characteristic. This may outperform the 60 mV/dec sub threshold slope that is representative for drift-diffusion current transport in normal FET devices at room temperature. The operation principle can be supported by using material combinations with staggered band offsets for the ip^+ junctions favoring the use of III-V-semiconductor materials combinations.

Instead of insulator gate isolation, Schottky contacts (MESFET) [10, 11] and heterojunctions (HFET, HEMT) can be used like in conventional planar device structures [12, 13]. In the nanowire case these form core-shell structures with the nanowire channel, maintaining the improved omega or "all-around" gate control. Important to note is that also these nanowire core-shell structures are not that seriously restricted by the lattice match requirements, allowing for materials combination best suited for HFET function, like high conduction band offsets and Schottky barriers. In this context the surface and interface properties of the channel material have to be considered. Besides properties like density of states, mobility, carrier type and concentration, the surface properties, mainly the Fermi level position, are important: it determines the type, accumulation or depletion, of the device. Further, it determines the quality of the source and drain ohmic contacts, which are especially important for a down scaled device with small contact areas. For low power and simultaneously high speed FETs, materials with low effective mass and low density of states, like InAs and InSb, may be especially well suited. For nanowire channel diameters of about 10 nm and an all-around gate these already operate in the so called quantum capacitance limit, where this quantum capacitance C_Q is smaller than the geometric capacitance C_G. In this case the power delay as well as the energy delay products are reduced, resulting in excellent low power, high speed performance [14, 15].

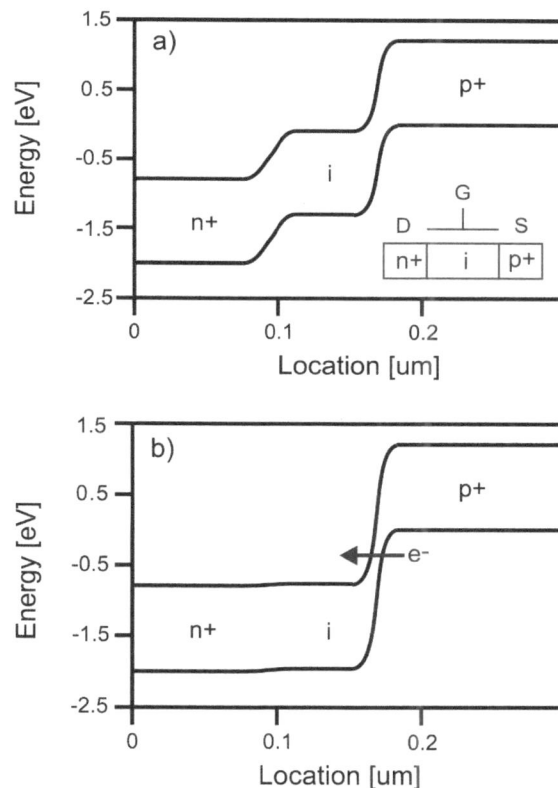

Figure 2: Band diagram of a Tunnel-FET, **(a)** Off state, **(b)** on state.

DEVICE TECHNOLOGY

The drain current of any type of field-effect transistor is controlled by a potential applied to the gate-channel capacitance. The strategy towards ultimate device performance is threefold:

(1)   reducing the gate length and scaling other device parameters properly,

(2)   restricting the capacitive gate control to mobile carriers in the channel, and

(3)   selecting a channel material offering excellent mobile carrier transport properties.

The ultra short gate length (strategy 1) is a general need for high integration and high speed device development. The nanowire approach gives easy access to a nanometer scale channel depth, basically the nanowire diameter. Moreover, the drain-induced barrier lowering is reduced by a surrounding gate that relaxes scaling needs in terms of oxide and nanowire thickness as compared to the gate length.

To restrict the capacitive charge control to mobile carriers in the channel (strategy 2) is the major motivation for the nanowire approach. The nanowire channel is co-axially surrounded by the gate and any escape into a substrate as in the case of epitaxial grown layered devices is inhibited. On the other hand, strategy (2) also gives rise to the main challenges of III/V nanowire design. The gate metallization has to be guided to the intrinsic gate-channel capacitance with a very low parasitic capacitance having in mind that the intrinsic gate capacitance is about 1fF or less. This challenge is even more severe for any parasitic gate-to-drain capacitance which especially degrades the rf-performance. In addition, even in the intrinsic gate-to-channel area, any low mobile or immobile charge stemming from surface- or interface-states of the nanowire channel or the gate dielectric results in a parasitic capacitance which degrades the device performance.

III/V semiconductors offer a huge number of materials with excellent transport properties (strategy 3). Moreover, according to the specific application the optimum material may be selected; *i.e.*, the low band gap InAs with very high mobility is very well suited for high-speed but low voltage applications while other materials may address other applications. However, this material has to fulfill the requirements of the other strategies too, which reduces the number of promising candidates substantially.

As an example the applicability as a channel material is discussed here for GaAs and InAs in some detail. The GaAs surface and its interface to a gate dielectric exhibits a high density of parasitic surface and interface states. The high surface potential results in a deep space charge region which may fully deplete the nanowire channel. Due to the limited n-type doping capability of GaAs in general and GaAs nanowires in particular only thick GaAs nanowires exhibit conductivity in the non-gated area. Therefore, despite the huge number of reports on GaAs nanowires [17 and references therein] there are very few reports on FET [18]. In contrast to GaAs the Fermi level at the InAs surface is pinned within the conduction band. This enhances the nanowire conductivity down to very low diameters and simplifies ohmic contact formation substantially. In addition, InAs offers a shallow n-type background doping which is high enough for FET operation and does not require intentional doping. Therefore, InAs is today the preferential III/V semiconductor nanowire channel material. A few reports are provided for alternatives like GaN [11, 19] and InGaAs [10]. Beyond III/V materials Si [3] is the major nanowire channel material and among the II/VI semiconductors many reports are available for ZnO nanowire FET [20-22].

Growth

Here, we focus on InAs nanowire growth for high performance nanowire FET. Basically, all types of modern III/V growth apparatus like molecular beam epitaxy (MBE) [23], chemical beam epitaxy (CBE) [24-26], and metal organic vapor phase epitaxy (MOVPE) [6, 27-28] have been adopted for nanowire growth. In addition, Hang *et al.* 2007 [29] reported successful InAs FET fabrication from polydecene solution-liquid-solid mechanism. Except selective area epitaxy, which requires holes in a dielectric layer like SiO_2 [30], all other methods require a seed. This is mainly an Au nanoparticle formed by high temperature annealing of a thin Au layer [31], aerosol deposition [32], electron beam lithography and lift-off [31], or from a colloidal solution [27]. In Si CMOS technology Au is attributed to reduce minority carrier life time and hence numerous alternatives are tested like Ni [33] or Bi [29] or even metal-free seeds like SiO_2 [6]. The majority of reports is based on the vapor-liquid-solid growth mechanism [33] in an MOVPE apparatus [34]. To navigate towards an optimum combination of apparatus, seed, and growth mode is a difficult task that is affected by a number of parameters beyond the InAs nanowire quality; *i.e.*, aerosol techniques or colloidal deposition are not applicable for seed formation if nanowire position control is aimed at during initial growth. The high temperature annealing of a thin homogenous Au film is not allowed if nanowires with identical diameters shall be formed. On the other hand, ultra-high resolution electron beam lithography needed for position control of monodisperse seeds will not be the choice for mass production of nanowire transistors.

Moreover, there is no easy access to precise quality assessment of InAs nanowires yet [28]. Even for bulk InAs there is a huge scatter of data due the impact of the substrate and the high intrinsic carrier concentration affecting the van-der-Pauw-Hall measurement. In case of nanowires, there is no access to Hall mobility data. Basically, nanowire transistors or test structures are used for mobility evaluation based on a capacitive charge control. This method is hampered from various aspects. The extracted mobility is rather a measure of the obtained device performance than of the carrier transport within the wire. All kinds of parasitics affect the nanowire transistor performance and hence the deduced mobility. In addition, a precise physical transistor model is lacking which is mandatory for a physically meaningful extraction of the low-field mobility from measured device parameters that compares to Hall data. Nevertheless, InAs nanowires offer a mobility in the range of 3,000 cm²/(Vs) up to 13,000 cm²/(Vs) with a carrier concentration in low to mid 10^{17} cm^{-3} [22, 27-28,].

In the following the vapor-liquid-solid growth mode in a low-pressure MOVPE is described in more detail as a common example of InAs nanowires growth [27]. InAs substrates with (001) orientation have been used. Prior to growth, commercial colloidal Au nanoparticles are deposited on the InAs substrate as seed particles. The Au nanoparticles are monodisperse with a shallow size distribution. A wide range of particle sizes is available. In our experiments nanoparticles of 30 nm to 100 nm diameters were used. The InAs surface is hydrophobic which substantially reduces their adhesion. Therefore no wafer spinning but a hot plate was used in order to dry the surface. Unfortunately, the hot plate process results in a more inhomogeneous distribution of nanoparticles compared to the spinning. After loading into the low-pressure MOVPE system, the samples were annealed at 620 °C in order to form an Au–In composite nanoparticle. The annealing was carried-out under tertiarybuthylarsine (TBAs) flow to prevent InAs substrate surface decomposition. After annealing, the temperature was ramped down to a growth temperature of 480 °C $\leq T_g \leq$ 400 °C. Trimethylindium was used as group-III precursor and TBAs for group-V at a constant V/III ratio of six. In some experiments ditertiarybutylsilane was added to the gas-phase for n-type doping, although compared to the n-type background impurity no direct evidence for intentional doping could be identified. The nanowire diameter is equal to the monodisperse nanoparticle size that may be selected in a wide range. Sometimes, seed particles may partially merge with each other during the annealing step giving rise for higher diameter seeds and nanowires.

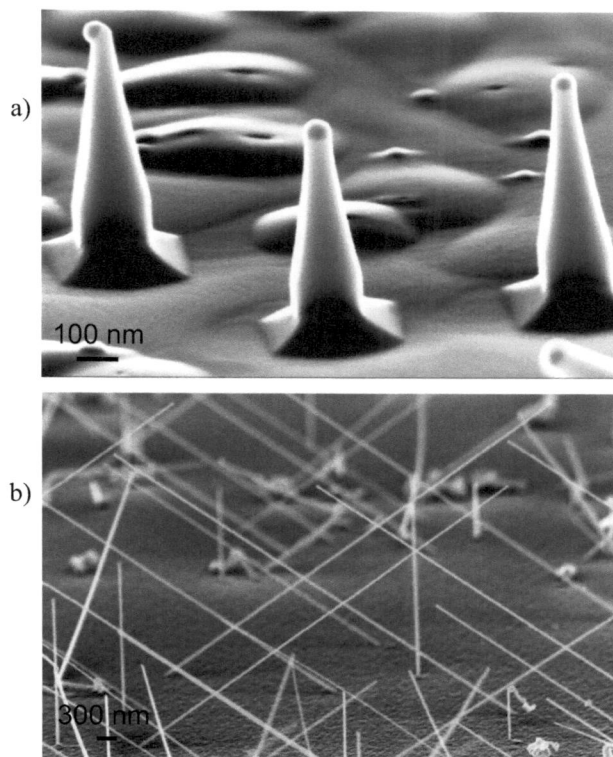

Figure 3: Scanning electron micrograph of InAs nanowires grown on (001) substrate at different growth temperature **(a)** 480 °C, **(b)** 400 °C.

In Fig. **3** a micrograph of typical InAs nanowires is depicted. At a growth temperature of 480 °C, the decomposition of TMIn is still quite effective and results in an additional conventional epitaxial layer growth on the side walls (Fig. **3a**). If the growth temperature is reduced to 400 °C the tapering effect is suppressed. In contrast to VLS growth on most other substrates a perpendicular growth is observed on InAs (001) substrates in addition to the preferential growth in (111) direction. The reason for this anomaly is still under discussion [35, 36]. The maximum growth rate of 1.3 µm/min is achieved at T_G = 420 °C. At the typical growth temperature for devices (T_G = 400 °C) the rate is reduced to 0.8 µm/min which is still about 40 times higher than in conventional layer growth without any degradation of the crystal. The crystalline quality of the InAs nanowires is investigated by transmission scanning electron microscopy [37].

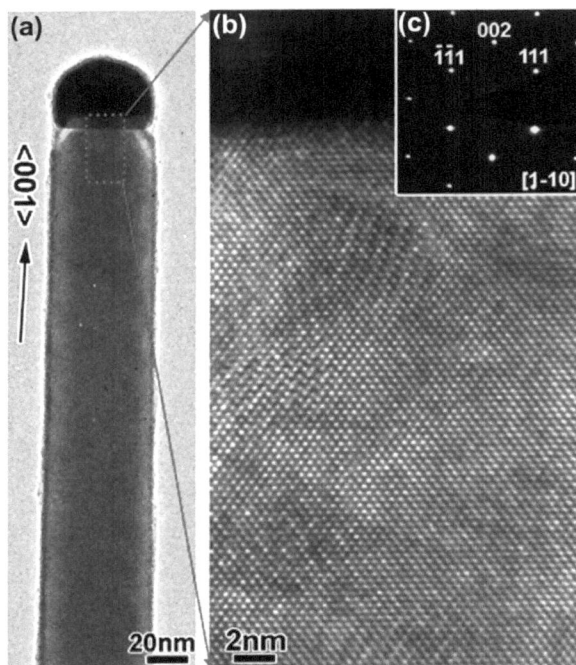

Figure 4: InAs nanowire grown on a InAs (001) substrate: **(a)** low magnification bright-field TEM micrograph of a perpendicular grown nanowire, **(b)** high resolution TEM image at the InAs/Au interface, **(c)** electron diffraction pattern [37].

A bright-field image of an individual InAs nanowire and its corresponding electron diffraction pattern and HRTEM micrograph are shown in Fig. **4a-c**, representing tens of these observed ⟨001⟩-oriented InAs nanowires. The identified [1-10] zone axis and these sharp diffraction spots indicate the cubic zinc-blende (ZB) structure of the NWs and their high crystalline nature, respectively. By checking the image contrast, diffraction patterns, and HRTEM micrographs of different parts of the entire nanowire, it is confirmed that the nanowire is stacking-faults-free and {111}-twinning-free. The highly defect-free state of the InAs NWs is of particular importance for exploiting fabrication of nanosized electronic (optic-) devices. More detailed investigations on InAs nanowire growth can be found elsewhere [28, 38-40].

Gate Dielectric and Device processing

Except some MESFET [11, 18] the nanowire devices are designed as a metal-insulator-semiconductor field-effect transistor (MISFET). InAs has a low band gap with a carrier accumulation at the surface such that an insulating thin film as a gate dielectric is indispensable. The deposition of the insulating film has to be carried out at low temperatures in order to avoid a damage of the critical InAs surface. The appropriate choice and deposition technique has a high impact on the nanowire MISFET performance. There are a number of different dielectrics developed for high-k application in CMOS or III/V-MIS technology adopted for nanowire MISFET such as SiN_x, [5, 27, 41], SiO_2 [10, 13, 28, 32], HfO_x [31], AlO_x sometimes in combination with HfO_x [41], and MgO [42]. In a back gate arrangement often a conductive n-silicon substrate is simply covered with SiO_2 prior to nanowire deposition. In this case even thermal oxide can be used because there is no need to obey a thermal budget and perfect oxides are formed that, however, have a relatively low dielectric constant. The deposition of high k-materials on sophisticated nanowires like InAs or in the presence of Au-contacts requires deposition techniques offering relatively low substrate temperatures such as room-temperature electron-cyclotron resonance plasma enhanced chemical vapour deposition (ECR-CVD) [41], atomic layer deposition (ALD) [31, 41], or even molecular beam epitaxy [42].

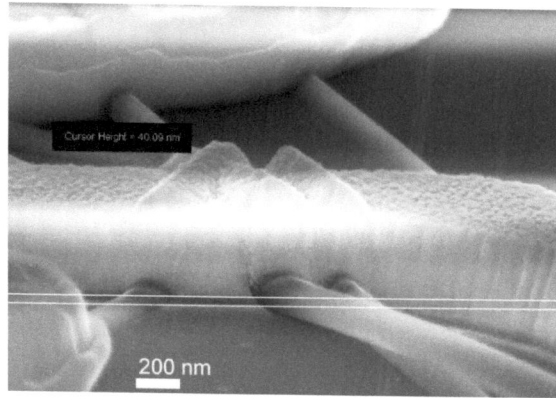

Figure 5: SEM micrograph of the gate region of a NW MISFET fabricated by FASA. The shadow at the metal/nanowire edge may indicate the thickness of the Al_2O_3 gate dielectric.

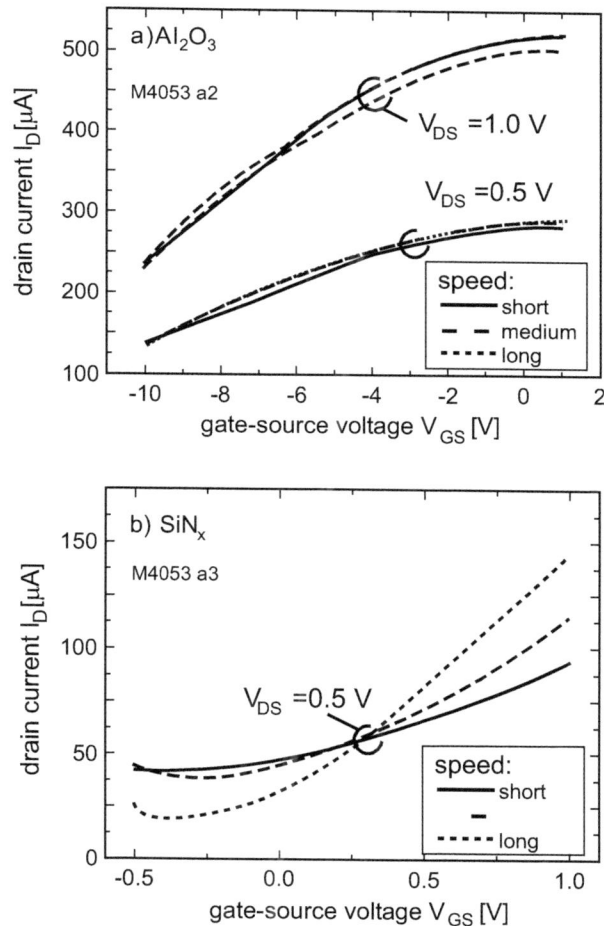

Figure 6: Transfer characteristics of NW MISFET with different gate dielectric consisting of up to 10 parallel wires: **(a)** Al_2O_3, **(b)** SiN_x.

It is well known that the III/V interface to dielectric materials suffer from a high density of low-mobile surface states which result in a degraded sub threshold swing and in a hysteresis of the I-V characteristics. The SiN_x gate dielectric enables the highest gain [27] but it suffers from severe hysteresis effect in the I-V characteristics. Here we report on a comparison of SiN_x and Al_2O_3 gate dielectrics. The Al_2O_3 deposition has been carried out at the University of Nagoya by atomic layer deposition [43, 44]. InAs nanowires were grown using the VLS mode in an MOVPE apparatus using Au colloidal nanoparticles of about 50 nm diameter. The wires were shaved from the growth substrate and transferred to a silicon substrate with a 4 μm thick SiO_2 cap layer using the field-assisted selective deposition process. 30 nm Al_2O_3 was deposited at 250 °C by atomic layer deposition [43]. The samples were covered with 30 nm SiN_x by ECR-CVD at 300 K for comparison. After the deposition of the dielectric a 0.5 μm and a 1 μm long gate was defined by electron beam lithography. In Fig. **5** the gate region of a fabricated nanowire MISFET with Al_2O_3 gate dielectric is shown. In this case 4 wires are connecting the source and drain contact.

The purpose of the study is to compare the transient behaviour of nanowire MISFET with Al_2O_3 versus SiN_x gate dielectric. As an initial study, the measurement mode of the HP 4145B parameter analyzer has been varied during the take up of the transfer characteristics (cf. Fig. **6**). There are three measurement modes (short, medium, and long) available indicating a different waiting time at each measurement point. In Fig. **6b** the well known problem of SiN_x gate dielectric is shown: the transfer characteristic severely depends on the measurement speed which is related to a low charging speed of traps in the gate dielectric or at its interface. The sample with Al_2O_3 dielectric exhibits almost no variation regardless of the drain bias selected.

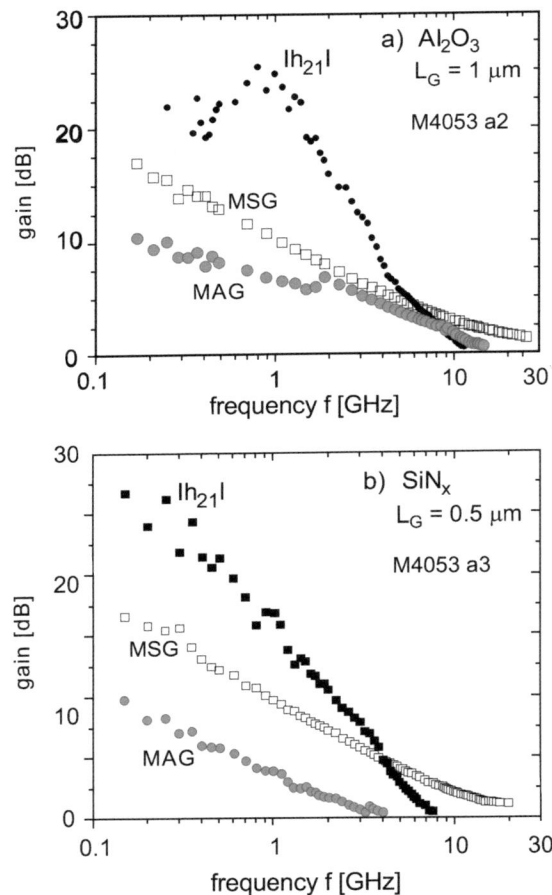

Figure 7: Maximum stable gain (MSG) and current gain (h_{21}), and maximum available gain (MAG) of the NW MISFET with **(a)** Al_2O_3 and **(b)** SiN_x gate dielectric.

RF measurements were performed on a HP 8510 network analyzer using an on wafer test set for frequencies up to 50 GHz. From measured scattering parameters the maximum stable gain (MSG) and current gain (h_{21}) are extracted. The

NW MISFET with about 1 μm gate length and an Al_2O_3 gate dielectric exhibit a MSG of 10 dB at 1 GHz despite the degraded threshold voltage while the cut-off frequency of current gain reaches f_T = 12 GHz (cf. Fig. **7a,b**). These data are not the best obtained but they are clearly comparable to data from NW MISFET with SiN_x gate dielectric.

DEVICE PERFORMANCE

DC Measurements and Performance

First the dc-characteristics and performance of InAs nanowire MISFETs fabricated as described in the previous section will be presented, concentrating on single nanowire FETs [27]. InAs is excellently suited for n-channel FETs because of its high electron mobility and Fermi level pinning within the conduction band, which results in low knee voltages and, due to the missing Schottky barrier, in very low ohmic contact resistances. All devices are deposited on an insulating substrate, e.g. a GaAs or a Si substrate covered with 300 nm thick silicon nitride (SiN_X). The SiN_x insulating layer reduces the leakage current to the low pA-region that could be neglected in further I-V-measurements. Electrical contacts to the deposited nanowires were patterned by electron beam lithography. Ohmic contacts were formed by depositing the metal system Ti/Au. After the deposition, the contacts were annealed at 300 °C for 30 sec. The contacted nanowires were covered by SiN_x gate dielectric with thicknesses investigated being 90 nm, 60 nm, 30 nm, and 20 nm. Finally, a Ti/Au gate metal was evaporated forming an omega-shaped gate partly wrapped around the InAs wire. The scanning electron micrograph of a realized FET is shown in Fig. **8**.

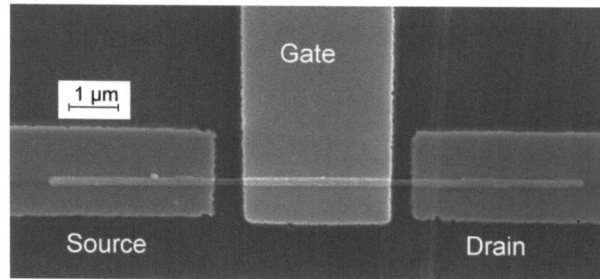

Figure 8: SEM image of an omega-shaped gate n-InAs nanowire FET.

Using the same batch of grown n-InAs nanowires, a number of FET were fabricated for each SiN_x dielectric layer thickness. The gate length of realized transistors varied from 1 μm to 5 μm. The DC characteristics of the single n-InAs NW channel FET devices were measured on-wafer at room temperature with the parameter analyzer HP 4145B. Fig. **9** shows a typical output characteristics $I_D - V_{DS}$ with a SiN_X dielectric layer thicknesses h = 30 nm.

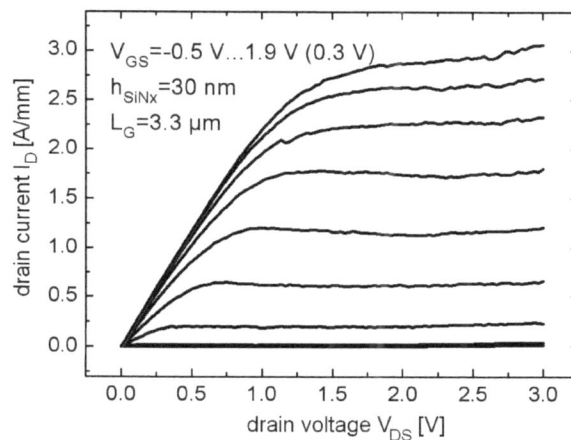

Figure 9: I-V characteristics of a single n-InAs nanowire field-effect transistor with a wire diameter of 50 nm and 30 nm SiN_x.

The devices show a good pinch-off behavior. The threshold voltage is V_T = -0.25 V and shifts to positive values with decreasing silicon nitride thickness. At high gate bias (V_{GS} = 1.9 V) a maximum drain current of 130 μA is measured corresponding to a current density in the wire of 6.6×10^6 A/cm^2. The measured maximum transconductance of the FET is $g_m = 97.5$ µS (cf. Fig. **9**).

For FET devices the normalized drain current $I_D^* = I_D / w_g$ and normalized transconductance $g_m^* = g_m / w_g$ are important figures of merit, where w_g is the channel width. In the case of a cylindrical channel the normalization may be obtained by replacing the channel width by the channel diameter which is fixed here to $d = 50\ nm$: $I_D^* = I_D / d$ and $g_m^* = g_m / d$. The single n-InAs NW channel FET provide a high output current $I_D^* = 3$ A/mm and a very high transconductance of $g_m^* \geq 2$ S/mm if the gate dielectric thickness is less than $h = 30$ nm and the gate length is $L_G = 2$ µm or less given (cf. Fig. **10**).

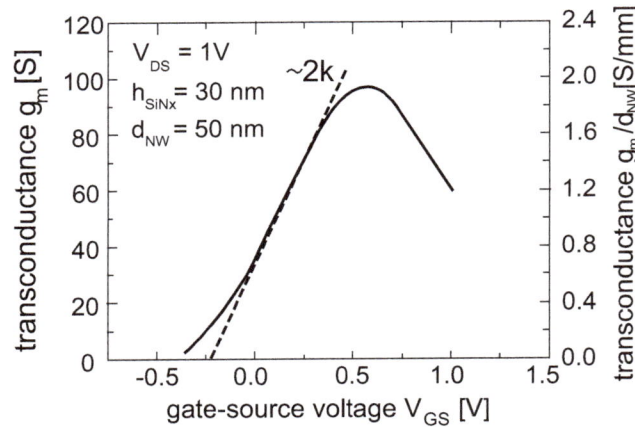

Figure 10: Transconductance vs. gate voltage with 2k scaling factor.

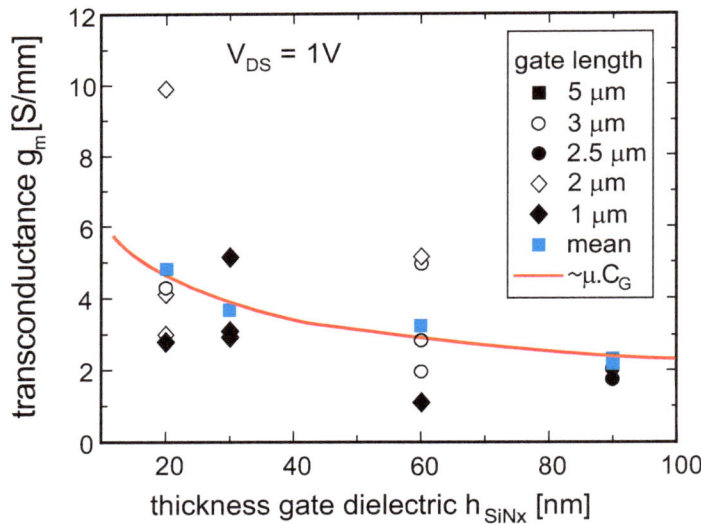

Figure 11: Normalized transconductance of single n-InAs nanowire field-effect transistor versus the SiN$_x$ gate insulator thickness h$_{SiN}$.

The measured maximum transconductance of about 2 S/mm [27] is among the highest ever reported for any semiconductor nanowire FET. The high performance obtained is attributed to the excellent transport properties of the InAs NW channel. Long-channel MOSFET equations may be used to model the device characteristics and to extract the carrier mobility of the channel which is not easily measured by other means. The transconductance g_m of a long channel MOSFET in saturation is given as $g_m = 2k(V_{GS} - V_T)$ with the scaling factor *2k*, the applied gate bias V$_{GS}$, and the threshold voltage V$_T$. Experimentally, the scaling factor *2k* may be taken from the slope of the transfer characteristic g_m-V_{GS} according to Fig. **10**. Following the long-channel MOSFET equations, the scaling factor is given as $2k = \mu \cdot C_G \cdot L_G^{-2}$ where C_G is the gate capacitance, μ the low field mobility, and L_G is the gate length, respectively. The capacitance of a cylindrical shaped gate-channel configuration is determined as

$C_G = 2\pi \cdot \varepsilon_0 \varepsilon_r \cdot L_G \cdot \left(\ln \left(\alpha + 2h/d \right) \right)^{-1}$ where ε_0 is the vacuum permittivity and $\varepsilon_r = 7.5$ is the dielectric constant of

the SiN_x gate isolation. In case of a coaxial gate all-around structure the correction factor is $\alpha = 1$. Here, the SiN_x dielectric and the gate metal is only partly surrounding the nanowire and the effective gate capacitance is reduced. An evaluation using an electrostatic field simulation results to $\alpha = 1.55$. So the transconductance of the InAs NW channel FET can be modeled in dependence of the geometrical device parameters and the results can be fitted to experimental data using the low-field mobility as the only fitting parameter. Fig. **11** shows both experimental and modeled data of the transconductance vs. the dielectric thickness for different gate lengths.

The modeled data assume a gate length of $L_G = 2$ μm. The best agreement to the average measured data is obtained with a low field mobility of $\mu = 13,000$ cm²/Vs. Fig. **11** also shows that there is a substantial scattering of experimental data attributed to variations in the fabrication process. In addition, the InAs nanowires grown both in (100) and (111) direction are randomly used for device fabrication and there might be a contribution from crystal orientation dependent transport properties.

RF Measurements and Performance

In spite of the fascinating dc characteristics of InAs nanowire field-effect transistors, the high speed potential is not yet demonstrated. The major challenges towards a reliable RF characterization of single nanowire transistors, regardless of the material system, are the dominant parasitic capacitances and the small signal power. Therefore, high frequency measurements of NW-FETs are barely reported [19, 45-47]. A typical two gate finger device investigated is shown in Fig. **12**. For on-wafer RF measurements a coplanar waveguide pattern with a characteristic impedance of 50 Ω was used according to the requirements of scattering parameter measurements. A coplanar tip of 50 μm pitch was selected to reduce contact pad size and therefore signal-to-ground capacitance. In order to point out the capacitances arising from the coplanar waveguide pattern, field theory calculations of an "open" structure without the device contacts (see Fig. **13a**) were carried out by ADS MOMENTUM. The gate-to-drain coupling capacitance C_{io} reaches 1.3 fF while the input and output capacitances C_{pg}, C_{pd} are 8.9 fF and 8.6 fF, respectively. The contact pattern give a lower limit for the parasitic load, still present in case of further optimized layout of the inner device contacts. Compared to the typical intrinsic gate-source capacitance of about a few hundred ato Farad estimated by electrostatic field simulation, the parasitic capacitance is at least one order of magnitude higher.

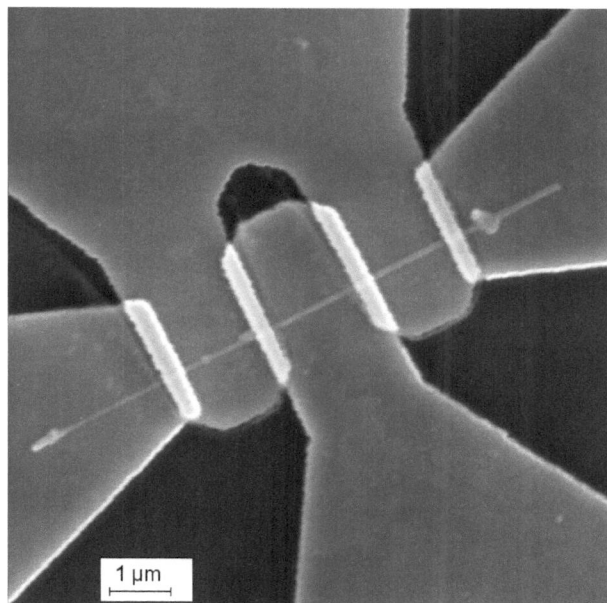

Figure 12: SEM micrograph of an InAs self-aligned two-finger gate NW-FET with a gate length of 1.4 μm.

The open element of Fig. **13a** is not sufficient for a precise de-embedding of the intrinsic InAs NW-FET. The self-aligned two finger gate overlapping drain and source electrodes (see Fig. **12**) contributes to the parasitic capacitances and therefore is incorporated in the open structure used for de-embedding as shown in Fig. **13b**. The calibration procedure applied to this "generic open" element in the frequency range from 45 MHz to 40 GHz revealed the given small signal equivalent circuit. Due to the overlap of the self-aligned gate, the gate-drain coupling increases while the pad capacitances varies form 9 fF up to17 fF. In addition, the contact pattern may be improved by shielding effects provided by the source pads [47]. The coupling capacitance is effectively suppressed down to C_{io} = 1.7 fF (cf. Fig. **13c**).

The self-aligned technique avoids high series resistance due to non-gated regions and therefore increases the low output current, which is one of the other challenging properties of NW-FETs in terms of RF characterization. A further improvement in signal power was achieved with the two-finger gate configuration presented above. With use of the generic open and short structures, the parasitics could be de-embedded from the DUT-measurements resulting in corrected S-parameters exhibiting typical field-effect transistor behavior. The maximum stable gain (MSG) of the investigated NW-FET is shown in Fig. **14**. A maximum stable gain higher than 30 dB at low frequency and a maximum oscillation frequency of 15 GHz are obtained. However, these data are very preliminary due to the high parasitic capacitances of the contact pattern.

The extraction of current gain h_{21} from measured scattering parameters is a difficult technical task. h_{21} may be calculated as follows:

$$h_{21} = \frac{-S_{21}}{\left(1-S_{11}\right)\left(1+S_{22}\right)+S_{12}S_{21}} \cdot \tag{2}$$

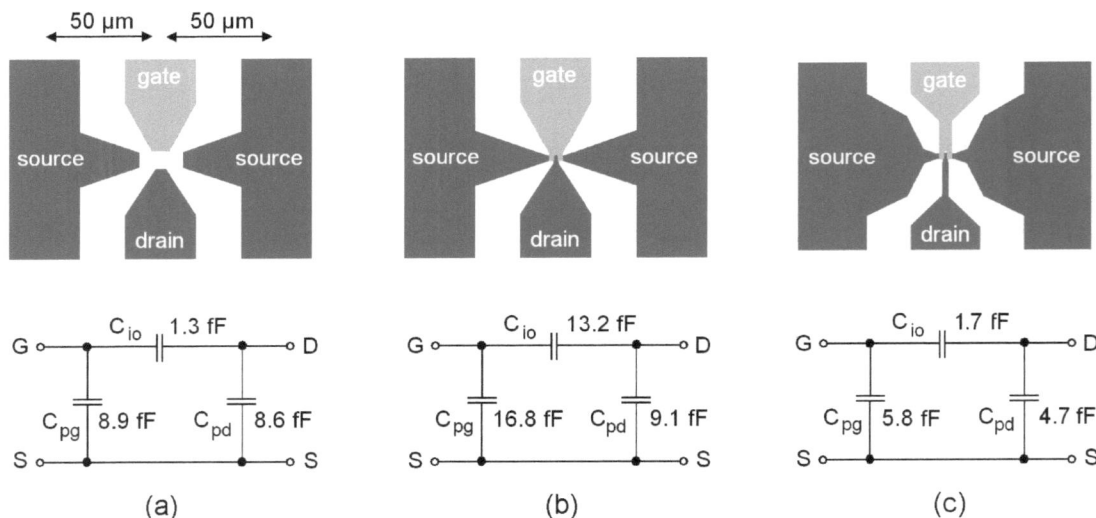

Figure 13: Layout of Coplanar contact pattern for 50 μm pitch layout with parasitic capacitances: **(a)** "Open" calibration element without and **(b)** with the device contacts in the inner part.

Due to the highly capacitive pad environment, S_{11} is close to unity. Therefore, even marginal fluctuation may result in numerical instabilities or even the zero crossing of the term ($1-S_{11}$) in eq. 2. In order to overcome this limitation, S_{11} has to be determined very precisely.

The small signal equivalent circuit of Fig. **15** was used to extract the transconductance and the intrinsic gate-source capacitance as a means of verifying the performed de-embedding. An intrinsic gate-source capacitance per unit gate length C_{gs}/L_g of about 340 aF/μm was determined for 30 nm SiN_x gate dielectric in good agreement with electrostatic field simulation. Therefore, and due to the high maximum oscillation frequency, we conclude that no significant capacitance arises due to the charge trapping effects sometimes observed in the DC measurements.

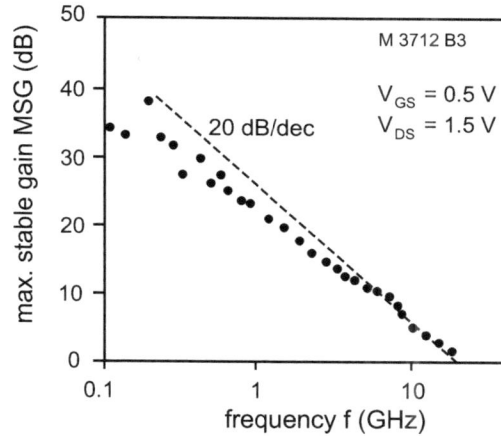

Figure 14: Maximum stable gain MSG of a NW-FET with 1.4µm gate length versus frequency.

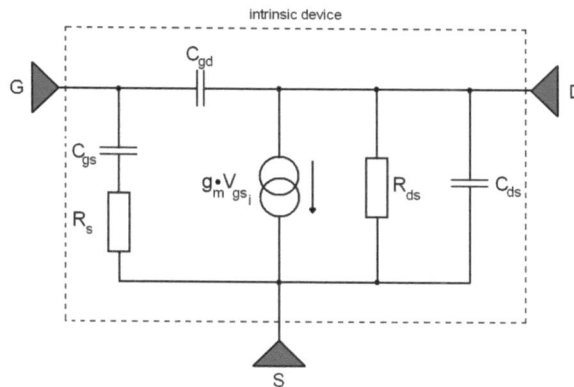

Figure 15: Standard small signal equivalent circuit used for the RF characterisation of the intrinsic InAs NW-FETs

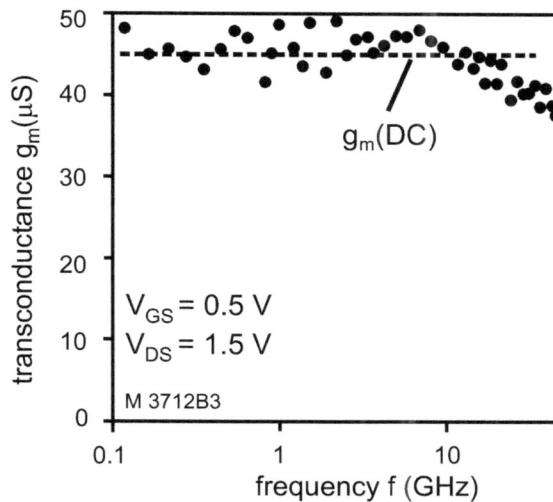

Figure 16: Transconductance g_m versus frequency: a good agreement with the DC transconductance of 45 µS is observed in a wide frequency range.

By an optimized transistor design, e.g. reducing the input capacitance, the absolute value of S_{11} may be reduced. Further, an advanced nanowire device model is indispensable towards a complete and reliable RF study. Nevertheless, the extracted transconductance g_m versus frequency shown in Fig. **16** looks reliable: the transconductance is about 45 μS, as derived from DC measurements, in a broad frequency range up to 2 GHz. The decrease beyond is believed to be due to an insufficient accuracy of the small signal equivalent circuit at higher frequencies and the increasing influence of the input-to-output coupling capacitance C_{gd}. As a figure of merit, the unity current gain cut-off frequency f_T is calculated by

$$f_T = \frac{g_m}{2\pi \cdot C_{gs}}.$$ (3)

Using (2) and the values derived from the RF characterization, the unity current gain cut-off frequency of the transistor presented above is calculated to f_T = 7.5 GHz.

The transistor characterized above does not reflect the performance limits of InAs NW-FETs due to the gate length of 1.4 μm. The cut-off frequency is expected to increase with decreasing gate length as a result of lower gate capacitance and higher transconductance, but in this case parasitics are increasingly dominating. Investigations in these directions are discussed in detailed elsewhere [46].

SUMMARY AND OUTLOOK

The technology and performance of III/V nanowire transistors has been reviewed with emphasis on InAs nanowire MISFET. InAs nanowires for FET application can be grown by vapor-liquid-solid growth mode at ultra high growth rate of 50 μm/h. Despite the high growth rate they exhibit perfect crystalline quality resulting in excellent transport properties. The low-field mobility reaches more than 10,000 cm²/Vs. Based on these data excellent InAs nanowire MISFET have been fabricated using various technologies. An ultra high drain current density of I_D/d = 3 A/mm and transconductance g_m/d = 2 S/mm have been demonstrated. The stability and reliability of the devices was found to heavily depend on the technology of gate dielectric fabrication. While SiN_x has offered highest gain, alternative high-k materials deposited by ALD result in a lower hysteresis of the I-V characteristics. InAs nanowires may be transferred to an insulating substrate in order to fabricate rf devices. A coplanar contact pattern allows for on-wafer scattering parameter measurements. However, even optimized coplanar contacts still results in a parasitic feedback capacitance that is almost twice the intrinsic gate-source capacitance of about 1 fF. This limits the measured cut-off frequencies to the 10 GHz regime in this study. A more precise access to the intrinsic device data will be available if complete nanowire circuits are fabricated and monolithically integrated with output buffers feeding the 50 Ohm measurement environment.

ACKNOWLEDGEMENTS

This work has been supported by the Deutsche Forschungsgemeinschaft in the frame of (i) the Sonderforschungsbereich SFB 445 and (ii) the Japanese/German cluster on Nanoelectronics, project "Nanowire/CMOS Heterogeneous Integration for Next-Generation Communication Systems". W. P. was on leave at Sophia University, Tokyo, during manuscript preparation and is in debt to the hospitality and fruitful discussions of his host Takao Waho. The authors are indebted to Kai Blekker, Benjamin Münstermann, Christoph Gutsche, Ingo Regolin, and Thai Do for their excellent experimental contributions to this paper

REFERENCES

[1]   Poate, J. M.; Fan, J. C. C., Eds, *Heteroepitaxy on Silicon*, 67 MRS Symposia Proc.; Materials Research Society. Pittsburgh, **1986**.

[2]   Kai, M.; Urata, R.; Miller, D.A.B.; Harria, J.S. Low-temperature growth of GaAs on Si used for ultrafast photoconductive switches. *IEEE J. Quantum Electronics* **2004**, *40*, S. 800-4.

[3]   Lu, W.; Xie, P.; Lieber, C.M. Nanowire Transistor Performance Limits and Applications. *IEEE Trans. Electron Dev.* **2008**, *55*(11), 2859-2876.

[4]   Thompson, S.E.; Chau, R.S.; Ghani, T.; Mistry, K.; Tyagi, S.; Bohr, M. T. In Search of "Forever," Continued Transistor Scaling One New Material at a Time. *IEEE Trans. Electron Dev.* **2005**, *18*(1), 26-36.

[5]   Thelander, C.; Rehnstedt, C.; Froberg, L.E.; Lind, E.; Martensson, T.; Caroff, P.; Lowgren, T.; Ohlsson, B.J.; Samuelson, L.; Wernersson, L.E. Development of a Vertical Wrap-Gated InAs FET. *IEEE Trans. Electron Dev.* **2008**, *55*(11). 3030-3036.

[7]   Yan, R. H.; Ourmazd, A.; Lee, K.F. Scaling the Si MOSFET: from bulk to SOI to bulk. *IEEE Trans. Electron Dev.* **1992**, *39*(7), 1704-1710.

[8]   Shin, M. Full-quantum simulation of hole transport and band-to-band tunneling in nanowires using the k center dot p method. *J. Appl. Phys.* **2009**, *106*(5), Art. No. 054505.

[9]   Lind, E.; Persson, M.P.; Niquet, Y.M.; Wernersson, L.E. Band Structure Effects on the Scaling Properties of [111] InAs Nanowire MOSFETs. *IEEE Trans. Electron Dev.* **2009**, *56*(2), 201-205.

[10] Noborisaka, J.; Sato, T.; Motohisa, J.; Hara, S.; Tomioka, K.; Fukui, T. Electrical characterizations of InGaAs nanowire-top-gate field-effect transistors by selective-area metal organic vapor phase epitaxy. *Jpn. J. Appl. Phys.*, **2007**, *46*(11), 7562–7568.

[11] Blanchard, P. T.; Bertness, K. A.; Harvey, T. E.; Mansfield, L. M.; Sanders, A. W.; Sanford, N. A. MESFETs made from individual GaN nanowires. *IEEE Trans. Nanotechnol.* **2008**, *7*(6), 760–765.

[12] Froberg, L.E.; Rehnstedt, C.; Thelander, C.; Lind, E.; Wernersson, L.E.; Samuelson, L. Heterostructure barriers in wrap gated nanowire FETs. *IEEE Electron Dev. Lett.* **2008**, *29*(9), 981-983.

[13] Jiang, X.; Xiong. Q.; Nam, S.; Qian, F.; Li, Y.; Lieber, C. M. InAs/InP Radial Nanowire Heterostructures as High Electron Mobility Devices. *Nano Lett.* **2007**, *7*(10), 3214-3218.

[14] Khayer, M.A.; Lake, R.K. The quantum and classical capacitance limits of InSb and InAs nanoire FETs. *IEEE Trans. Electron Dev.* **2009**, *56*(10), 2215-2223.

[15] Knoch, J.; Riess, W.; Appenzeller, J. Outperforming the conventional scaling rules in the quantum capacitance limit. *IEEE Electron Dev. Lett.* **2008**, *29*(4), 372-374.

[16] Appenzeller, J.; Knoch, J.; Björk, M. T.; Riel, H.; Schmid, H.; Riess, W. Toward Nanowire Electronics. *IEEE Trans. Electron Dev.* **2008**, *55*(11), 2827-2845.

[17] Soci, C.; Bao, X.Y.; Aplin, D. P. R.; Wang, D. A Systematic Study on the Growth of GaAs Nanowires by Metal−Organic Chemical Vapor Deposition. *Nano Lett.*, **2008**, *8*(12), 4275–4282.

[18] Fortuna, S.A.; Li, X.L.; GaAs MESFET With a High-Mobility Self-Assembled Planar Nanowire Channel. *IEEE Electron Dev. Lett.*, **2009**, *30*(6), 593.

[19] Vandenbrouck, S.; Madjour, K.; Théron, D.; Dong, Y.; Li, Y.; Lieber, C. M.; Gaquiere, C. 12 GHz FMAX GaN/AlN/AlGaN Nanowire MISFET. *IEEE Electron Dev. Lett.* **2009**, *30*(4), 322.

[20] Choe, M.; Jo, G.; Maeng, J.; Hong, W.K.; Jo, M.; Wang, G.; Park, W.; Lee, H.; Hwang, H.; Lee, T. Electrical properties of ZnO nanowire field effect transistors with varying high-k Al_2O_3 dielectric thickness. *J Appl. Phys.*, **2010**, *107*(3), 034504.

[21] Fan, Z.Y.; Lu, J.G. Chemical sensing with ZnO nanowire field-effect transistor. *IEEE Trans. Nanotechn.* **2006**, *5*(4), 393 – 396.

[22] Prost, W.; Blekker, K.; Do, Q.T.; Regolin, I.; Tegude, F.-J.; Müller, S.; Stichtenoth, D.; Wegener, K.; Ronning, C. Modeling the carrier mobility in nanowire channel FET. *Mater. Res. Soc. Symp. Proc.* **2007**, *1017*.

[23] Sorensen, B.S.; Aagesen, M.; Sorensen, C.B.; Lindelof, P.E.; Martinez, K.L.; Nygard, J. Ambipolar transistor behavior in p-doped InAs nanowires grown by molecular beam epitaxy. *Appl. Phys. Lett.* **2008**, *92*(1), Art. No. 012119.

[24] Lind, E.; Persson, A.I.; Samuelson, L.; Wernersson, L.E. Improved subthreshold slope in an InAs nanowire heterostructure field-effect transistor. *Nano Lett.*, **2006**, *6*(9), 1842-1846.

[25] Bryllert, T.; Wernersson, L.E.; Lowgren, T.; Samuelson, L. Vertical wrap-gated nanowire transistors. *Nanotechnology* **2006**, *17*(11), S227-S230.

[26] Bryllert, T.; Wernersson, L.E.; Froberg, L.E.; Samuelson, L. Vertical high-mobility wrap-gated InAs nanowire transistor. *IEEE Electron Dev. Lett.* **2006**, *27*(5), 323-325.

[27] Do, Q.T.; Blekker, K.; Regolin, I.; Prost, W.; Tegude, F.J. High transconductance MISFET with a single InAs nanowire channel. *IEEE Electron Dev. Lett.* **2007**, *28*(8), 682-684.

[28] Dayeh, S.A. Electron transport in indium arsenide nanowires. *Sem. Sci. & Techn.* **2010**, *25*, 024004.

[29] Hang, Q.; Wang, F.; Buhro, W. E.; Janes, D. B. Ambipolar conduction in transistors using solution grown InAs nanowires with Cd doping. *Appl. Phys. Lett.* **2007**, *90*, 062108.

[30] Tanaka, T.; Tomioka, K.; Hara, S.; Motohisa, J.; Sano, E.; Fukui, T. Vertical Surrounding Gate Transistors Using Single InAs Nanowires Grown on Si Substrates. *Appl. Phys. Express* **2010**, *3*, 025003.

[31] Sorensen, B.S.; Aagesen, M.; Sorensen, C.B.; Lindelof, P.E.; Martinez, K.L.; Nygard, J. Ambipolar transistor behavior in p-doped InAs nanowires grown by molecular beam epitaxy. *Appl. Phys. Lett.* **2008**, *92*(1), Art. No. 012119.

[32] Lind, E.; Persson, A.I.; Samuelson, L.; Wernersson, L.E. Improved subthreshold slope in an InAs nanowire heterostructure field-effect transistor. *Nano Lett.* **2006**, *6*(9), 1842-1846.

[33] Wagner R. S., Ellis, W. C. Vapor-Liquid-Solid mechanism of single crystal growth. *Appl. Phys. Lett.* **1964**, *4*, 89-90.

[34] Yazawa, M.; Koguchi, M.; Hiruma, K. Heteroepitaxial ultrafine wire-like growth of InAs on GaAs substrates. *Appl. Phys. Lett.* 1991, *58*(10), 1080.

[35] Björk, M. T.; Ohlsson, B.; Sass, T.; Persson, A. I.; Thelander, C.; Magnusson, M. H.; Deppert, K.; Wallenberg, L. A.; Samuelson, L. One-dimensional heterostructures in semiconductor nanowhiskers. *Appl. Phys. Lett.* **2002**, *80*(6), 1058.

[36] Samuelson; L. One-dimensional assembly of quantum dots and Heterostructures. *Solid State Physics/ the Nanometer Consortium*, Lund University.

[37] Li, Z.A. HRTEM measurements of nanowires. Faculty of Physis, University of Duisburg-Essen, **2010**.

[38] Dick, K. Epitaxial Growth and Design of Nanowire and Complex Nanostructures. Doctoral Thesis, Solid State Physics, Lunds University, Lund **2007**.

[39] Dick, K. A.; Deppert, K.; Samuelson, L.; Seifert, W. Optimization of Au-assisted InAs nanowires grown by MOVPE. *J. Cryst. Growth*, **2006**, *297*, 326-333.

[40] Jensen, L. E.; Björk, M. T.; Jeppesen, S.; Persson, A. I.; Ohlsson, B. J.; Samuelson, L. Role of Surface Diffusion in Chemical Beam Epitaxy of InAs Nanowires. *Nano Lett.* **2004**, *4*(10), 1961-194.

[41] Wiersch, A.; Heedt, C.; Schneiders, S.; Tilders, R.; Buchali, F.; Kuebart, W.; Prost, W.; Tegude, F.-J. Room-temperature deposition of SiN$_x$ using ECR-PECVD for III/V semiconductor microelectronics in lift-off technique. *J. Non-Crystalline Solids*, **1995**, *187*, 334.

[42] Do, Q.T.; Blekker, K.; Regolin, I.; Schuster, E.; Peters, R.; Prost, W.; Tegude, F.-J. Magnesium oxide (MgO) as gate dielectric for n-doped single InAs nanowire field-effect transistor. *7th Topical Workshop on Heterostructure Microelectronics*, Japan, August **2007**.

[43] Miyazaki, E.; Goda, Y.; Kishimoto, S.; Mizutani, T. AlGaN/GaN MOSFETs with Al$_2$O$_3$ Gate Oxide Deposited by Atomic Layer Deposition; Comparative Study. *Topical workshop on Heterostructure Microelectronics*, Nagoya, **2009**.

[44] Sugiura, S.; Hayashi, Y.; Kishimoto, S.; Mizutani, T.; Kuroda, M.; Ueda, T.; Tanaka, T. Fabrication of normally-off mode GaN and AlGaN/GaN MOSFETs with HfO$_2$ gate insulator. *Sol. Stat. Electronics* **2010**, *54*(1), 79-83.

[45] Chaste, J.; Lechner, L.; Morfin, P.; Fève, G.; Kontos, T.; Berroir, J. M.; Glattli, D. C.; Happy, H.; Hakonen, P.; Placais, B. Single Carbon Nanotube Transistor at GHz Frequency. *Nano Lett.* **2008**, *8*, 525-528.

[46] Blekker, K.; Münstermann, B.; Matiss, A.; Do, Q.T.; Regolin, I.; Brockerhoff, W.; Prost, W.; Tegude, F.-J. High Frequency Measurements on InAs Nanowire Field-Effect Transistors Using Coplanar Waveguide Contacts. *IEEE Trans. Nanotechnology* **2009**, *DOI* 10.1109/TNANO.2009.2032917.

[47] Blekker, K.; Matiss, A.; Münstermann, B.; Regolin, I.; Li, B.; Do, Q.T.; Prost, W.; Tegude, F.J. Coplanar Contact Pattern for single InAs Nanowire FET. *66th Device Research Conference*, Santa Barbara, CA, **2008**.

CHAPTER 8

III-V Semiconductor Nanowire Light Emitting Diodes and Lasers

Junichi Motohisa[1,*], Katsuhiro Tomioka[2,3,4], Bin Hua[3], Kumar S.K. Varadwaj[2], Shinjiroh Hara[1,3], Kenji Hiruma[1,3] and Takashi Fukui[1,3]

[1]*Graduate School of Information Science and Technology, Hokkaido University, Japan;* [2]*PRESTO-JST, Japan Science and Technology Agency, Japan;* [3]*Research Center for Integrated Quantum Electronics, Hokkaido University, Japan and* [4]*Present Address: Suzhou Institute of Nano-tech and Nano-bionics (SINANO),Chinese Academy of Science (CAS),China*

Abstract: We describe the growth and optical properties of III-V semiconductor nanowires and their application to nanoscale photonic devices such as Fabry-Perot cavity, waveguides, optically-pumped lasers, and light-emitting diodes. The nanowires were grown by selective-area metalorganic vapor phase epitaxy (SA-MOVPE) on the (111) oriented substrates. Nanowires containing heterostructures in their radial direction, that is, core-shell heterostructures, have also been realized by controlling the growth mode during SA-MOVPE. The nanowires were characterized by micro-photoluminescence measurements and those detached from the grown substrate showed resonant peaks associated with Fabry-Perot cavity modes. It was simultaneously shown that core-shell hetereostructured nanowires exhibited stronger photoluminescence than bare nanowires due to reduced surface non-radiative recombination. Furthermore, core-shell nanowires exhibited lasing oscillation originating from the cavity formed by both end facets at pulsed-laser excitation. Meanwhile, electroluminescence from core-shell nanowires was also demonstrated.

Keywords: Metalorganic vapor phase epitaxy, selective-area growth, core-shell heterostructures, fabry-perot cavity, waveguide, laser, light emitting diode.

INTRODUCTION

Semiconductor nanowires are attracting recent interest as a promising candidate as the building blocks for nanoelectronics and nanophotonics [1-5]. This is because their unique features originating from the size and superior properties are suitable for the applications to various kinds of devices. For instance, in photonics, their small lateral dimension is useful to confine light in a very small region that is close to the diffraction limit. At the same time, they allow the light propagation along their elongation axis [6]. A possibility to go beyond the diffraction limit is also attempted and manipulation of light in sub-wavelength dimensions is already demonstrated [7]. From the materials point of view, semiconductors are active materials for light sources and detectors, and their material diversity for nanowires make it possible to cover ultraviolet (UV) [8-10], visible [11], and near infrared (NIR) [12,13] regions. Tunability of emission wavelength can further be explored utilizing semiconductor alloy and heterostructures [14]. In addition, their small footprints can free us from constraints of lattice-matching issues in their formation [15], which is inevitable in the growth of thin films, and further enhances the material freedom. It is also noted that formation of electrical contacts to individual or array of nanowires is straightforward, thus the current injection or extraction is relatively easy. All of these features are important to realize photonic integration for information processing in ultimately small size as well as nanophotonic devices, and open up a possibility to connect microscopic and macroscopic worlds in nano-sensing or bio-sensing. On the other hand, the light confinement can be tuned, for example, by the size of nanowires, thus it is expected to achieve high light extraction efficiencies [16,17] or to reduce reflectivity [18]. These are big advantages for light-emitting devices or solar cells as compared to their counterpart based on thin film technology.

Among many kinds of semiconductors, III-V semiconductors are the most successful materials for photonics applications. This owes to advanced epitaxial growth techniques as well as their material superiority. That is, it is possible to realize high quality thin films with controlled doping, alloys with accurately controlled composition, and

*Address correspondence to Junichi Motohisa: Graduate School of Information Science and Technology, Hokkaido University, Japan. E-mail: motohisa@rciqe.hokudai.ac.jp

heterostructures with atomically flat interfaces. These will also hold for nanowires. Therefore, it is necessary to develop a technology to form well-defined and high-quality III-V semiconductor nanowires for the success of nanowire-based photonics.

The major approach for the realization of nanowires is the catalyst-assisted vapor-liquid-solid (VLS) method [19,20], where sub-micron or nano-sized whiskers are formed through crystallization of materials *via* the catalyst. In this method, a large amount of nanowires possessing extremely small size and reasonably high quality are synthesized in a very short time scale with a rather simple approach. Control of the growth in the axial or radial direction of the nanowires is demonstrated to form complicated heterostructures, such as one-dimensional superlattices and/or core-shell structures [21]. Various kinds of electron and photonic devices have been demonstrated so far. However, it is frequently discussed if the metal catalyst is incorporated into the nanowires, since most of the metal catalysts are the deep impurities in III-V semiconductors. For example, gold is reported to be deep acceptors in GaAs [22-24]. Those deep impurities might be carrier killers or non-recombination centers, and lead to deterioration of nanowire qualities. Therefore, some attempts to grow nanowires without metal catalysts or self-catalyzed growth have been reported [25-27].

We have been working on the growth of III-V semiconductor nanowires by using an alternative method, that is, selective-area metal organic vapor phase epitaxy (SA-MOVPE) [17, 28-30]. As we will elaborate in the next section, SA-MOVPE is a method to carry out epitaxial growth on patterned or partially masked substrates prepared by lithographic method. For the growth of nanowires, they are formed in the mask opening with a size from sub-micron to a few tens of nanometer. Therefore, SA-MOVPE is considered to be a combination of bottom-up and top-down technology for fabrication of nanostructures. Thanks to this combination, it is possible to grow an array of nanowires in predetermined positions without any aid of a catalyst. Furthermore, this method fully utilizes the nature of epitaxial growth, and it is expected to exhibit superior crystalline quality as well as good control of the growth with atomic precision to form abrupt doping profiles and heterojunctions.

In this chapter, we describe the growth of III-V semiconductor nanowires by SA-MOVPE and their applications to photonic devices. In Section 2, SA-MOVPE growth of freestanding nanowires vertical to the substrate is described. The nanowires have hexagonal cross section with diameter d of 20 to 400 nm, which can be controlled by the design of the masked substrate and growth conditions. Core-shell heterostructured nanowires have also been grown *via* control of growth mode. We also show the core-shell heterostructures are extremely important and useful for photonic devices using III-V semiconductors. In Section 3, we show that individual lay-down nanowires exhibit Fabry-Perot cavity modes and waveguiding properties. These result in the achievement of lasing oscillation in core-shell nanowires. Section 4 describes the application of nanowires for light-emitting diodes (LEDs). Section 5 is a summary of this chapter.

SA-MOVPE GROWTH OF NANOWIRES

Process for SA-MOVPE

In selective-area growth, crystal growth is carried out on partially masked substrates. For the mask, amorphous films, such as SiO_2 or SiN, are used. Since the surface energy is higher on amorphous films, the growth species, which is supplied from vapor phase (in the case of vapor phase epitaxy, VPE) preferentially attach and react on the semiconductor surfaces. Thus, the growth proceeds epitaxially on the opening regions of the mask where semiconductor surfaces are exposed. One of the important features is that the growth rate strongly depends on the orientation of the crystallographic facets and growth conditions. This enables us to realize various kinds of semiconductor nanostructures by controlling the orientation of the substrate, size and direction of the mask opening, and the growth conditions [31-35]. In the case of growth of III-V semiconductor nanowires, we use (111)-oriented substrates and choose the growth conditions so that the growth rate in the (111)-direction is much higher than the other directions [33,34]. It is noted that the (111) oriented surface has six {-110} surfaces perpendicular to (111) in zincblende crystals, and their growth rate is generally the slowest. Therefore, the grown nanowires are vertical to the substrates and exhibit hexagonal cross section surrounded by {-110} sidewalls facets.

The process for the SA-MOVPE growth of nanowires is shown in Fig. **1** and performed as follows. First, SiO_2 was formed by plasma sputtering or plasma-enhanced chemical vapor deposition on (111)-oriented substrates. Thickness of the SiO$_2$ was 20 to 30 nm. Then, periodic circular opening were defined in the SiO$_2$ mask by electron-beam (EB) lithography and wet-chemical etching techniques. Circular openings were arranged in a triangular lattice (see Fig. **1(b)**), with the pitch a ranging from 0.4 to 3 μm. The diameter, d, of nanowires is directly related to the opening diameter d_0, and a smaller d_0 results in a longer nanowire height h. Minimum d or d_0 we have achieved is about 20 nm [36] (see Fig. **2(b)**), but for applications to lasers or LEDs, d_0 around 100~300nm were used. The opening patterns were defined within 100 × 100 μm square regions, and patterns with different d_0 and a were repeated with periodicity of 200 μm. SiO$_2$ was removed from the regions between each pattern. Finally, MOVPE growth was carried out. We employed the growth in a horizontal, low-pressure system working at 0.1 atm. For GaAs nanowires, trimethylgallium (TMGa) and arsine (AsH$_3$) were used as source materials. We chose rather high growth temperature, T_G, (typically, $T_G = 750°C$) and low AsH$_3$ partial pressures, [AsH$_3$], (typically, [AsH$_3$] = 2.5×10^{-4} atm) in order to have faster growth rate and to grow in the (111)B direction [17,28,29]. For materials containing Al, In, and P, trimethylaluminum (TMAl), trimethylindium (TMIn), and tertiarybutylphosphine (TBP) were used for source materials, respectively. The growth was done in two different MOVPE systems. Although detailed conditions for optimum nanowire growth may be dependent on the growth system as well as the material system, we have achieved good reproducibility and uniformity in both growth systems as long as the aforementioned criteria for growth conditions is satisfied [30].

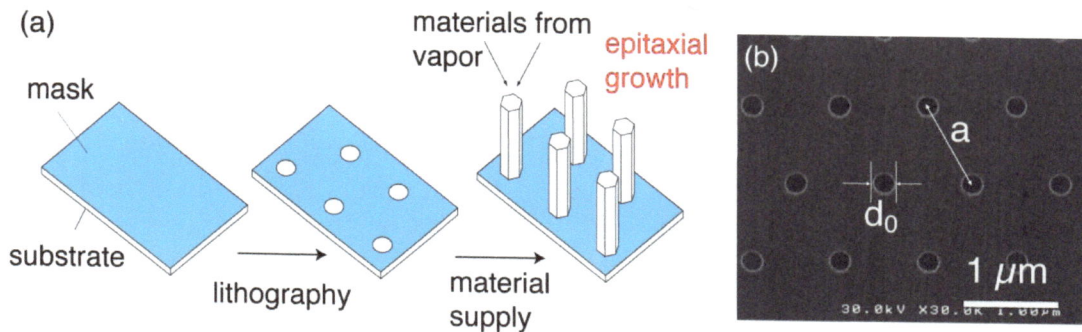

Figure 1: (a) Schematic illustration of process for SA-MOVPE growth for nanowires. **(b)** Secondary electron microscopy (SEM) image of a typical mask pattern for nanowires. Opening diameter d_0 of the mask was 100 nm and pitch a was 1 μm.

Fig. **2(a)** shows a typical secondary electron microscopy (SEM) image of GaAs nanowires grown on a GaAs (111) B substrate. The pattern period a was 1 μm, and the diameter d_0 was 200 nm. One can see a uniform array of GaAs nanowires with diameter d=200nm having hexagonal cross section. The diameter d of the nanowires becomes smaller as the initial opening diameter d_0 is reduced. Fig. **2(b)** also shows an SEM image of InAs nanowires grown on InAs (111)B, whose diameter was about 20 nm. As shown in these examples, SA-MOVPE technique enables us to realize arrays of freestanding III-V semiconductor nanowires with extreme uniformity and size controllability.

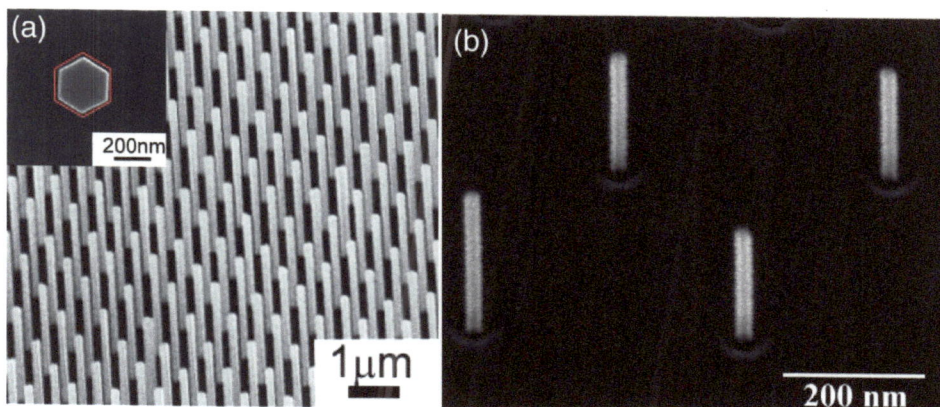

Figure 2: (a) SEM image of the GaAs nanowire array with diameter 200nm grown on GaAs (111)B substrate. Inset shows a top view of a nanowire, showing its cross section is hexagonal. **(b)** SEM image of InAs nanowire with diameter 20 nm, grown on InAs (111)B substrates.

Core-Shell Heterostructured Nanowires and Their Characterization

One of the important features of semiconductor nanowires is in the formation of two types of heterostructures [21]. The first type is called vertical heterostructures, in which heterostructures are introduced in their axial direction. The other is radial or core-shell heterostructures. Formation of these relies on the fact that the growth condition can control the direction of the growth in either the axial or the radial direction of nanowires. Furthermore, their combination is also possible to realize complex structures. In this section, we focus on core-shell heterostructure nanowires since they are important to reduce non-radiative recombination at the surface and to increase emission efficiency. We describe a GaAs/GaAsP core-shell heterostructure [13] as an example, although we have shown various kind of material combination including core-multishell structures [37,38].

For GaAs/GaAsP core-shell heterostructures, we grew GaAs and GaAsP in succession. The same procedure and conditions as we described in the previous section was employed for core GaAs nanowires, while lower temperature and higher partial pressures for precursors of group V atoms were used to enhance lateral growth. In the case of the GaAsP shell, growth temperature was 650 °C, and partial pressures were 1.0×10^{-6}, 2.2×10^{-4} and 2.5×10^{-4} atm for TMG, TBP and AsH$_3$, respectively. Figs. **3(a)** and **(b)** shows SEM images of the nanowires before and after the growth of GaAsP. The growth time of GaAs was 60min and GaAsP, 5 min. A uniform array of vertical standing nanowires is prepared with smooth top surface and sidewall. Top view images of a nanowire show a clear hexagonal cross section. The length of the core-shell wire is almost equal to that of the GaAs one, but the diameter of the former is approximately 100 nm larger than that of the latter, suggesting a 50 nm thick GaAsP shell around the GaAs core wire.

To examine the wire structure and composition, transmission electron microscopy (TEM) and energy dispersive x-ray spectroscopy (EDXS) were then applied. A bright-field TEM image and EDXS linescans obtained from a single nanowire axial plane exposed by focused ion beam (FIB) etching are shown in Fig. **3(c)** and **3(d)**. Although the heterointerface of GaAs and GaAsP was not clearly distinguished in the bright-field and high-resolution TEM images, EDXS linescan clearly shows the formation of core-shell structures. Green, blue and red lines correspond to EDXS counts of elemental Ga, As and P, respectively. The length and diameter of this nanowire are 2.8 µm and 390 nm, respectively. In Fig. **3(d)**, results are shown for an EDXS linescan along the axis of the nanowire (along the solid line). Here the nanowire right end corresponds to its top end during growth. Another linescan performed across the diameter of the nanowire is shown in Fig **3e**. The scan position is labeled with a dotted line arrow in Fig. **3(d)**. The axial EDXS scan indicates that the nanowire core is entirely composed of GaAs along the center axis and P content on its top surface can be negligible. However the radial linescan clearly shows presence of P element around the GaAs core. These results prove that the nanowire structure should be a GaAs core surrounded by a GaAsP shell. The P concentration in GaAsP shell is around 10 atomic %, *i.e.* GaAs$_{0.8}$P$_{0.2}$, and the thickness of outer GaAsP shell is estimated to be 50 nm, which is consistent with that SEM observation before and after the growth of GaAsP.

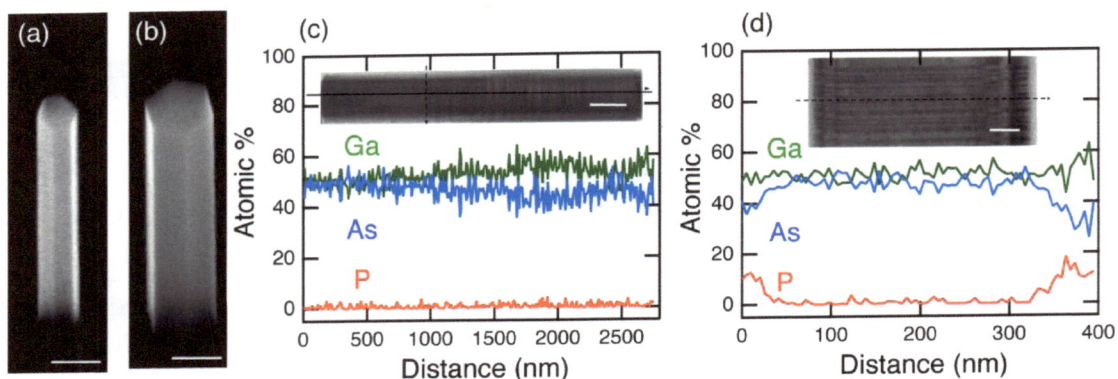

Figure 3: SEM images of GaAs **(a)** and GaAs/GaAsP **(b)** nanowires. Scale bars are 200 nm. **(c)** Bright-field transmission electron microscopy (TEM) image and axial EDXS linescans of a single nanowire showing variations in Ga (green), As (blue) and P (red) compositions. Scale bar, 300 nm. **(d)** TEM image and radial EDXS scan corresponding to the dotted line in (c) indicate that a GaAs core is surrounded by a GaAsP shell. Scale bar, 50 nm. Reprinted in part with permission from Nano Letters 2009, 9, 112-116. Copyright 2009, American Chemical Society.

We carried out micro-photoluminescence (μ-PL) characterization on single GaAs and GaAs/GaAsP core-shell nanowires at 4K. Typical results are shown in Fig. **4**. Here, nanowires were mechanically cut down and dispersed onto 2-μm-thick-SiO$_2$-covered Si substrates, and He-Ne laser was used for excitation (see next section for experimental detail). We can see a series of peaks in both nanowires. These are due to the Fabry-Perot cavity resonance in nanowires and will be discussed later in detail. We can see the PL emission intensity was larger by more than 100 times in core-shell nanowires as compared to bare GaAs nanowires. This can be explained by passivation of GaAs surfaces and reduction of nonradiative surface recombination by the GaAsP shell. In general, III-V semiconductors possess a high-density of surface states, and they are nonradiative recombination centers, which deteriorate emission efficiencies. This is indeed problematic in nanowires since the surface to volume ratio is much larger than their bulk counterpart. It is shown here, however, that it can be overcome by using core-shell heterostructures by preventing the coupling of photoexcited carriers in the core with the surface. A similar effect is observed in GaAs/AlGaAs core-shell heterostructures, but AlGaAs surface was not very stable presumably due to oxidation, and reduction of PL intensity was observed after repeated intense photoexcitation. Stable PL intensity was confirmed with present GaAsP shell or GaAs/AlGaAs/GaAs core-multishell structures and resulted in lasing oscillations, which will also be described later.

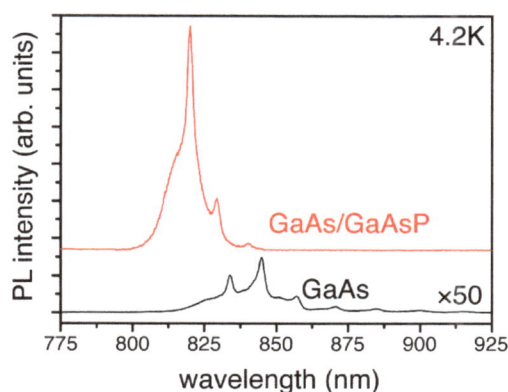

Figure 4: PL spectra of single GaAs and GaAs/GaAsP nanowires was recorded at 4.2 K. Compared to bare GaAs nanowire, GaAs/GaAsP core-shell nanowire showed stronger PL intensity by a factor of over two orders of magnitude. This result was attributed to the passivation of GaAs core due to GaAsP shell layer. Reprinted in part with permission from Nano Letters 2009, 9, 112-116 (supporting info). Copyright 2009, American Chemical Society.

NANOWIRE CAVITY AND NANOLASERS

Experimental Setup

To investigate a possibility for nanowire photonic devices, we investigated optical properties of single nanowires, which were mechanically cut down and dispersed onto a SiO$_2$-coated Si substrate. The substrate had Ti/Au markers which enabled us to locate individual wires and to perform optical measurements and SEM on the same nanowires. Experimental setup for μ-PL measurements is shown in Fig. **5**. It was carried out using a 632.8 nm continuous-wave He-Ne laser for excitation or pulsed Ti:sapphire laser with a pulse width of about 200 fsec. The excitation beam was focused to a spot ~ 2 μm in diameter with a ×50 microscope objective on the sample placed in a cryostat. The excitation power was adjusted using different combinations of neutral density filters. The emission through the same microscope objective was collected into a liquid-N$_2$ cooled charge coupled device (CCD) for spectral analysis or a CCD camera for imaging. In both cases, laser light was filtered using appropriate color glass filters. SEM images of a typical nanowire used for PL measurements in this study are shown in Fig. **5**. The lay-down nanowire on Si substrate is well defined and extremely uniform.

Nanowire Fabry-Perot Cavity

Fig **6(a)** illustrates 4.2 K PL spectra of a lay-down GaAs nanowire (L = 5.0 μm, d = 300 nm). The excitation power density was changed stepwise from 1050 to 0.33 W/cm^2. Emission of some GaAs nanowires was relatively weak (see inset), but most of the nanowires exhibited relatively stronger emission in the near-infrared region. In addition, they showed a series of periodic peaks in PL spectra ranging from 830 to 940 nm [39]. The excitation power

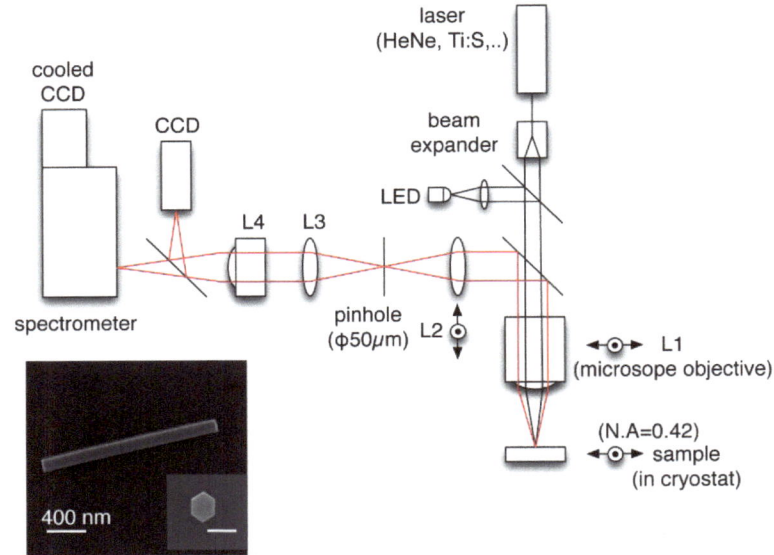

Figure 5: Experimental setup for pattern for μ-PL measurement. Left bottom image shows a typical SEM image of the nanowire lay-down on the substrates. Reprinted in part with permission from Applied Physics Letters 2007, 91, 131112. Copyright 2007, American Institute of Physics.

densities have no influence on the peak wavelengths, and the temperature dependence of the peak position was much smaller than was expected from the band-gap energy of GaAs. Hence, it is concluded, these distinct peaks correspond to the resonant peak of different longitudinal modes of the Fabry-Perot cavity in GaAs nanowires. That is, the ends of the nanowire are smooth enough to function as two reflecting mirrors, and the nanowire can define a Fabry-Perot optical cavity with modes $m = 2nL/\lambda$, where m is an integer, L is the length of the cavity and the nanowire length, λ is the wavelength, and n is the refractive index. Temperature dependence of the peak position in the μ-PL spectra was also studied and found that they showed redshift as the temperature (2-2.6 nm from 4.2 to 100 K). This redshift was much smaller than the redshift that is expected from the temperature dependence of band gap energy of GaAs (9.1 nm from 4.2 to 100K), and thus attributed to the temperature dependence of refractive index. This also shows that the observed peaks originated from Fabry-Perot cavity resonance in nanowires.

If one takes into account the material dispersion, the mode spacing $\Delta\lambda$ for a Fabri-Perot cavity with length L is given by:

$$\Delta\lambda = \lambda 2/[2L(n-\lambda(dn/d\lambda))], \tag{1}$$

where n is the refractive index and $dn/d\lambda$ is the first-order dispersion of the index. Group refractive index is defined as $n-\lambda(dn/d\lambda)$. This expression provides a good description of the observed mode spacing and the refractive index when the measured nanowire length is equated with L. Fig. **6(b)** shows the 4.2 K PL spectra of those GaAs nanowires with different lengths ranging from 2.7 μm to 5 μm. As expected, we obtain an inverse relationship between mode spacing and nanowire length. Based on the experimental results of mode spacing and nanowire length, the estimation of wavelength-dependent group refractive indices of GaAs nanowires are plotted in Fig. **6(c)**. For comparison, the wavelength dependence of the theoretical index is also given, which was calculated from the Sellmeier equation:

$$n(\lambda) = [A+B\lambda^2/(\lambda^2-C^2)]^{1/2}, \tag{2}$$

where A, B, C are empirical coefficients. For GaAs at room temperature, A = 8.950, B = 2.054, C^2 = 0.390, and λ is the light wavelength in the vacuum in μm [40]. Here we neglect the temperature dependence on refractive index since its influence is negligible as mentioned above. Most of the experimental data seem to agree with the theoretical tendency of the wavelength-dependent index except that the experimental values appear 0.2-1.1 higher than the theoretical ones. These discrepancies can be explained by the dispersion of nanowire waveguides [41].

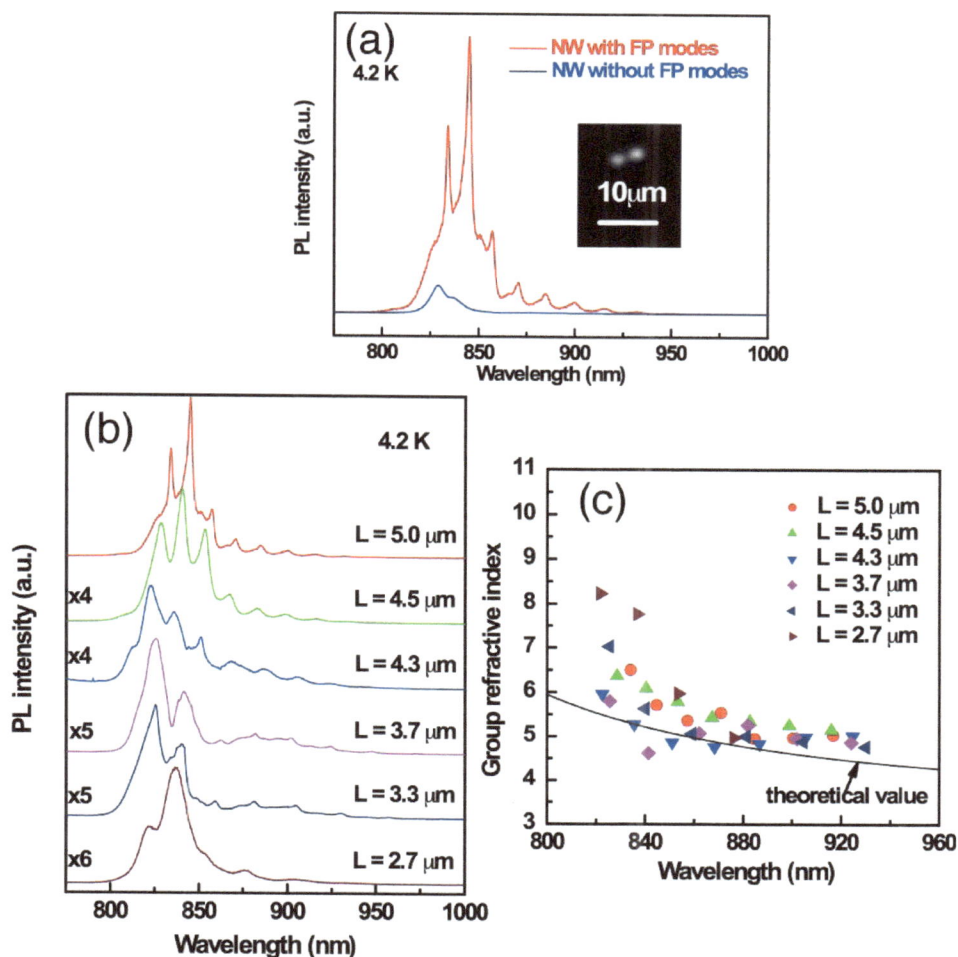

Figure 6:. (a) PL spectra of lay-down GaAs nanowires with and without Fabry-Perot mode. The nanowires exhibiting Fabry-Perot mode shows stronger emission by about 50 times. Inset shows a spectrally integrated image of a nanowire with Fabry-Perot resonance. **(b)** PL spectra of nanowires with different length. **(c)** Calculated group refractive indices of different length nanowires according to Fabry-Perot cavity modes. The theoretical index of GaAs is also given for comparison. Reprinted with permission from phys. stat. sol. (c) 5, No. 9, 2722-2725 (2008) Copyright 2008, Wiley-VCH Verlag GmbH & Co. KGaA.

In Fig. **6(b)**, the width of the mode peak significantly decreases as the nanowire length increases. This indicates that the enhancement of the nanowire length can improve quality factor Q due to reduction of mirror loss. In the case of a 5.0 μm-long nanowire, the Q at the wavelength of 845 nm is estimated to be around 220 at 4.2 K.

It is found that the nanowires with a diameter less than 200 nm are difficult to form Fabry-Perot cavities. In the case of GaAs nanowires, Fabry-Perot cavity modes died away at room temperature. However, they were observed up to room temperature in InP nanowires [42]. This is because the surface state density is smaller in InP than in GaAs, and emission efficiency is higher. Furthermore, as described in the previous section, strong PL enhancement and pronounced cavity mode was observed in core-shell nanowires. Therefore, it is thought that significant loss in the nanowire cavity is from the surface non-radiative recombination.

Nanowire Waveguide

As described in the previous section, formation of nanowire cavity implies light propagation inside nanowires. To investigate and explore it in detail, we studied a reversed T-branch consisting of two InP nanowires with perpendicular position, which is shown in Fig. **7(a)** [42]. The T-branch was formed randomly during the dispersion of InP nanowires on Si substrate. The two wires have the diameter of around 500 nm and 750 nm, respectively. In the experiment, the position of the He-Ne laser excitation is scanned from b to f on a T-branch as indicated in Fig. 7(a), and corresponding emission images are shown Figs. **b-f**, respectively. In Fig. **7(b)** and **(c)**, we can see the luminescence image from the two ends of InP nanowire with the excitation laser focused on b and c, respectively. The localization of bright spontaneous emission at the two ends of the wire suggests the strong waveguiding effect as well as the existence of Fabry-Perot cavity modes, which were confirmed by PL spectra in a separate measurement. For spot c, we can see the emission was also from the nanowire body. We also note that the emission from the nanowire in another branch was not observed. When we focused the excitation laser spot at d, both of the nanowire was excited and we can see the emission from both edges of the two nanowires, showing the propagation of light through nanowires. Furthermore, when the focus was moved to the upper nanowire, photoluminescence from InP nanowires was observed at the edge of the top nanowire. At the same time, emission was observed from both edges of the bottom nanowire as shown in Fig. **(e)** and **(f)**. This means the light propagated from the top end to the bottom end of upper wire then coupled to the lower wire, and then the light was emitted from the two ends of the lower wire. The fact that the light from the top nanowire couple to the bottom nanowire, but it does not from the bottom to top, indicates that the light is strongly confined in the nanowire, but could be coupled evanescently at the nanowire sidewall.

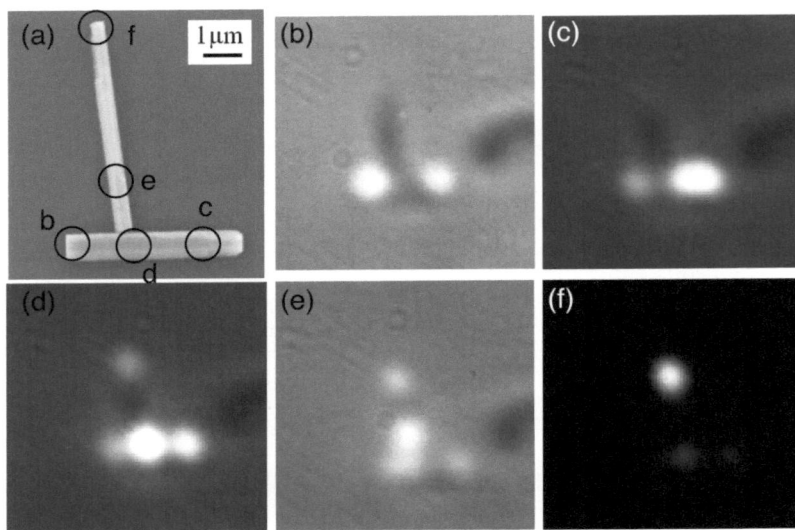

Figure 7: (a) SEM image of an inverse T-branch consisting of two nanowires. **(b)-(f)** Spectrally integrated emission images obtained for excitation spot focused on b-f in Fig. 7(a). Reprinted in part with permission from Nano Lett 2007, 7, 3598-3602. Copyright 2007, American Chemical Society.

Lasing Oscillations

Upon pulsed excitation, the core-shell nanowires exhibited sharp single peaks, and the peak became larger and narrower with excitation intensity [13]. Fig. **8(a)** shows typical results of PL from a GaAs/GaAsP core-shell nanowire of 330 nm in diameter and 5.5 μm in length. At low excitation power densities the PL spectra display a broad emission band at about 820 nm. However, above a certain threshold (8.4 kW/cm^2 in this nanowire), a sharp and narrow peak centered at ~ 816 nm appears and the peak intensity increases rapidly with excitation energy. This emission peak is exactly consistent with GaAs band gap energy (~ 1.52 eV at 4.2 K), indicating the onset of stimulated emission from the GaAs core of the nanowire. Stimulated emission or lasing is evidenced by the appearance of the simultaneous line narrowing and superlinear increase of intensity at pump densities above the threshold, as shown in Fig. **8(b)**. Furthermore, temperature dependent PL spectra of nanowires were recorded from 4.2 to 150 K, at which sharp peaks were observed. Similar to the case of a Fabry-Perot cavity, the redshift was observed but the amount was much smaller than that expected from the temperature dependence of the GaAs band gap. These results agree with the previous observation of temperature dependent cavity modes in single GaAs nanowires. Considering the expression of $m = 2nL/\lambda$, the peak position is strictly determined by refractive index n, for a certain nanowire length L.

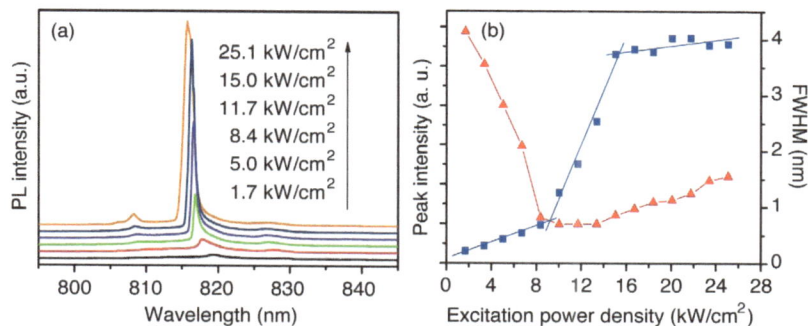

Figure 8: (a) PL spectra from a single GaAs/GaAsP core-shell nanowire with a length of 5.5 μm, as a function of increasing excitation power density at 4.2 K. **(b)** Plots of emission peak intensity (■) at the center of 816 nm and full width at half maximum (FWHM) (▲) of PL spectra versus excitation power density. Reprinted in part with permission from Nano Letters 2009, 9, 112-116. Copyright 2009, American Chemical Society.

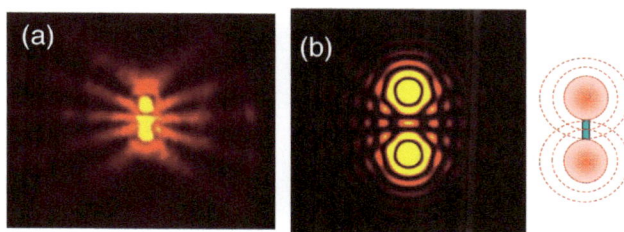

Figure 9: (a) Far-field optical images of lasing emission from single GaAs/GaAsP core-shell nanowires. The length of nanowire is 5.5 μm. The incident laser spot is focused perpendicularly at the center of the nanowires. **(b)** Simulated interference patterns, assuming a spherical non-directional emission from the wire end facets. Reprinted in part with permission from Nano Letters 2009, 9, 112-116. Copyright 2009, American Chemical Society.

Specific interference patterns were observed in lasing nanowires. A typical result is shown in Fig. **9(a)**. The length of the nanowire was 5.5 μm. This interference is explained by the diffraction and interference of emission from both end facets of the nanowire, and by non-directional spherical emission from the nanowire edges [43]. For more quantitative understanding, we simulated diffraction and interference patterns following an approach described in Ref. 35 and presented in Fig. **9(b)**. Here, NA of the microscope objective and the difference of the phase at two edges are taken into account. One can see that the interference patterns around the nanowires are nicely reproduced. Furthermore, these interference patterns are expected to depend on the wire length, which was also reproduced by the simulation [13]. The longer nanowire obviously exhibits more complex interference patterns, compared to the shorter one. This also confirms the lasing in nanowires, in which two coherent point light sources are present at the edges of nanowires.

Finally, we show a polarization dependent lasing emission in GaAs/AlGaAs/GaAs core-multishell nanowires. In Fig. **10(a)**, we plot PL spectra polarized perpendicular ($I_\perp$) or parallel ($I_\parallel$) to nanowire elongation axis above lasing threshold excitation. Total diameter of the nanowire was 380nm. Giant polarization anisotropy was observed. The peak corresponding to lasing is unchanged in intensity, width and position with perpendicular polarized emission. The polarized component parallel to the nanowire contains a broad spontaneous emission band with intensity two orders of magnitude lower than perpendicular intensity. The polarization ratio ρ, which is defined by $(I_\perp-I_\parallel)/(I_\perp+I_\parallel)$, was 0.99 at the lasing peak. The plot of intensity at different emission polarization angles θ shows periodic ($\sin 2\theta$) dependence, as shown in Fig. **10(b)**. The polarized far-field emission image (upper inset of Fig. **10(a)**) further shows unperturbed spherical lasing with interference patterns in perpendicular polarization, but a smooth emission similar to the case of cw excitation (inset of Fig. **6(a)**) for parallel polarization (lower inset of Fig. **10(a)**). This polarization anisotropy is in contrast to the emission (and absorption) polarized parallel to the nanowire with much smaller diameter [44]. This is because the polarization in spontaneous emission is mainly determined by the dielectric mismatch with the nanowires and background, whereas lasing is supported by the Fabry-Perot cavity mode, which implies that the wavevector is parallel to the nanowire and thus the electric field is perpendicular to the nanowires.

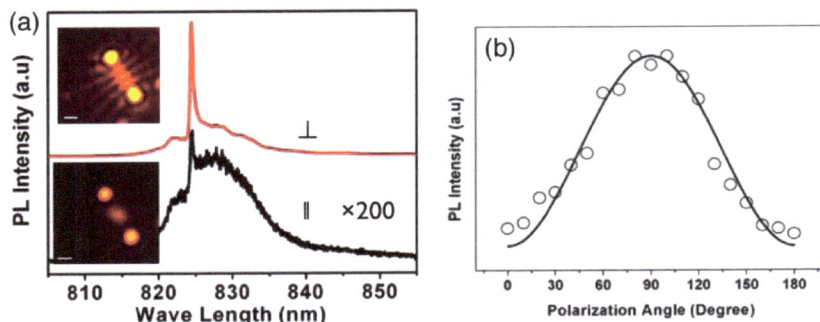

Figure 10: (a) Polarized emission spectra of lasing nanowires. Insets show spectrally integrated images for each polarization. For perpendicular polarization, an interference pattern similar to Fig. 9(b) was observed, but it was not for polarization parallel to the nanowire. Scale bar: 2μm. **(b)** Dependence of lasing peak intensity with respect to the polarization angle θ. Solid line is a fitting for experimental data using sin 2θ.

NANOWIRE LIGHT EMITTING DIODE

Fig **11(a)** illustrates the structure of nanowire-based LEDs. For this structure, GaAs/AlGaAs-based core-mutishell (CMS) nanowires were grown as follows. Firstly, n-GaAs nanowires were grown on n^+-GaAs substrates at 750 °C. Then, core-multishell structures were formed by successive lateral growth of *n*-AlGaAs, *p*-GaAs, *p*-AlGaAs, and *p*-GaAs. To enhance the lateral growth, the growth of the shell was carried out at lower temperature (~700°C). Silane (SiH$_4$) and diethylzinc (DEZn) were used as *n*-type and *p*-type dopants. We expect electroluminescence from the

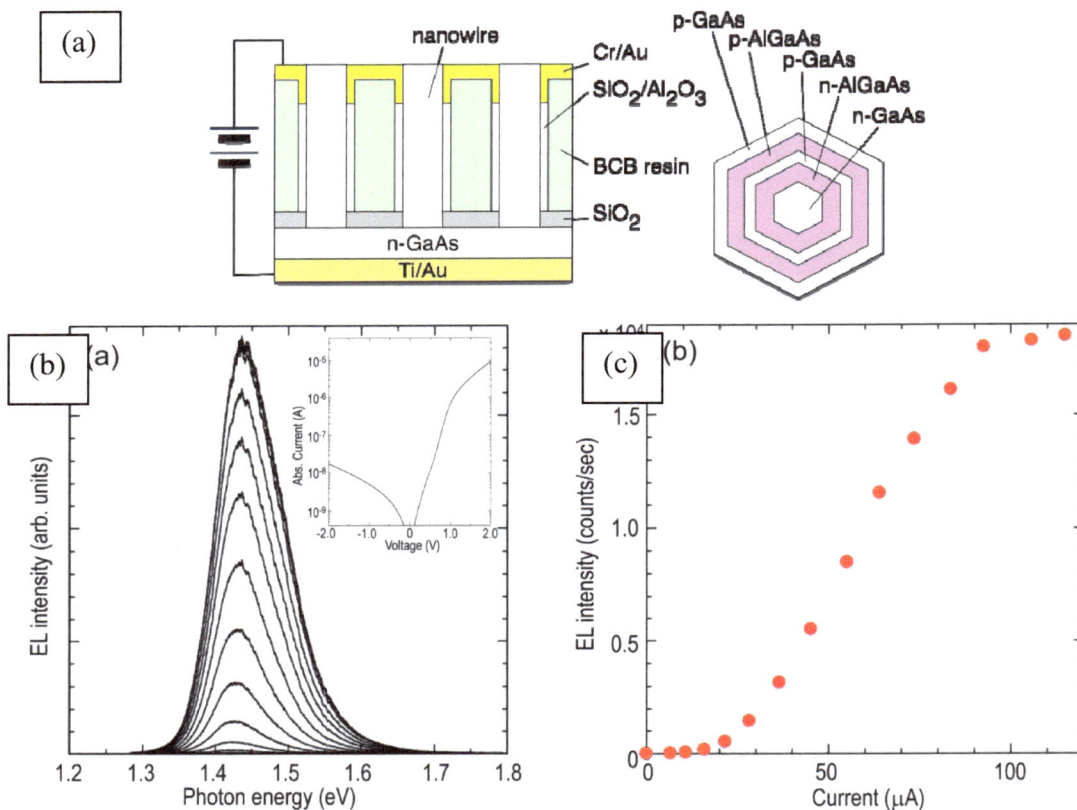

Figure 11: (a) Schematic illustration of cross sectional view of nanowire-based LEDs and core-multishell nanowires. **(b)** Electroluminescence spectra and **(c)** current-light output characteristics of nanowire LEDs. Inset of (b) shows current voltage characteristics represented in semi-log lot.

first *p*-GaAs layer in the shell with estimated thickness of 20-25 nm. Doping density was estimated to be in the order of 10^{18} cm^{-3} for both n- and p-type, although accurate measurement of the doping density is still a challenge for nanowires. Then, LEDs were fabricated by the following step. After the deposition of SiO$_2$ (50 nm) /Al$_2$O$_3$ (20 nm) on the nanowire surface for protection, the sample was buried with poly-resin and successively etched with reactive ion etching (RIE) to expose the top parts of nanowires for contacts. Then, SiO$_2$/Al$_2$O$_3$ was partially etched with BHF solution and 70 nm-wide slit was formed around the CMS-nanowires. The depth of the slit was 1.5 μm. Next, Cr/Au metal was deposited onto the top parts of the RIE-etched nanowires. The thickness of Cr and Au were 10 nm and Au(130 nm), respectively. Finally, Ti(20 nm)/Au(100 nm) were deposited for the backside electrode. Approximately 2×10^5 nanowires are connected in parallel and the total junction area is estimated to be 1.3×10^{-3} cm^2.

Fig **11(b)** shows an electroluminescence spectra of CMS-nanowire LEDs. We can clearly see the peak centered at 1.42 eV. Current-voltage characteristics showed nice rectifying behavior of the *pn* junction, as shown in the inset. Fig. **11(c)** shows current-light intensity characteristics. For low injection level, the light output is small. This is presumably because of non-radiative recombination, which should be further optimized in sample quality or surface passivation. Above 30 μA, light output linearly increases with injection current. Saturation above is due to carrier overflow.

SUMMARY AND OUTLOOK

We have demonstrated the catalyst-free formation of III-V semiconductor nanowires and their application to photonic devices. SA-MOVPE is a versatile method for the formation of high-quality nanowires and, in particular, core-shell heterostructured nanowires, which is important for photonic device applications to enhance emission efficiencies. Although not described here, our nanowires can be grown directly on Si substrate and LEDs on Si substrate have been demonstrated [45]. This opens another direction to nanowire-based photonics since they can be used for optical interconnects and integrated with Si-CMOS technology. In addition, solar cells using core-shell type InP nanowires are already demonstrated using SA-MOVPE [46].

With all these achievements, we still have a lot of issues for nanowire-based photonic integration in an ultimate scale. First of all, lasing threshold should be minimized, and eventually, lasing with current injection at room temperature should be achieved. A mirror loss is one of the major factors that determine the lasing threshold, and the passivation with core-shell structure seems not to be sufficient at present. Thus, appropriate anti-reflection coating at nanowire ends and further optimization of core-shell structure including proper surface treatment is required to minimize loss in the nanowire cavity. Making contacts to nanowires is straightforward on lay-down nanowires but minimization of the process-induced damages and loss associated with metal coating are required for lasing with current injection. As in our study, most of lasing oscillation in nanowires have been demonstrated with those randomly scattered onto substrates. This is not very suitable for integration unless we resort to a smart position control method. A candidate technique for alignment and integration will be to use microfluidics [47], electric field [48], or contact printing [49], which are already demonstrated. Conversely, it is important to utilize as-grown nanowires. This is demonstrated in LEDs and formation of contacts is already achieved as described in the previous section. A challenge toward lasing, however, is to realize mirrors with high-reflectivity at both ends of nanowires,

The lasing wavelength is determined by the gain spectra in bulk semiconductors and the cavity mode. To control the wavelength, introduction of vertical or core-shell heterostructures is necessary, but the issue there will be how to achieve optimum overlap with the electron-hole system and the electromagnetic field inside nanowires. At the same time, effort to extend operational wavelength will be continued using nanowires of InP-based, GaN-based, or other semiconductors including II-VI materials, part of which is already demonstrated [8-12,14]. Lastly, our nanowire lasers are not in the "nano" scale, since it's operating wavelength is comparable to the nanowire diameter, and did not show lasing for further decrease of diameter. Combination with metal cavity or some plasmonic structures [50,51] helps us to further decrease the dimension of photon confinement to explore nanowire lasers operating in a truly subwavelength regime.

ACKNOWLEDGEMENTS

The authors thank Prof. V. Zwiller for stimulating discussions, and Dr. M. van Kouwen, Dr. Y. Ding, Dr. L. Yang, and M. Kobayashi, for valuable experimental contribution to the present work. This work was financially supported in part by a Grant-in-aid for Scientific Research supported by the Japan Society for the Promotion of Science.

REFERENCES

[1]   Xia, Y.; Yang, P.; Sun, Y.; Wu,Y.; Mayers, B.; Gates, B.; Yin,Y.; Kim, F.; Yan, H. One-dimensional nanostructures: synthesis, characterization, and applications. *Adv. Mater.* **2003**, *15*, 353-389.

[2]   Riess, W.; Ohlsson, B.; Gösele, U.; Samuelson, L. Nanowire-based one-dimensional electronics. *Mater.Today* **2006**, *9*, 28-35.

[3]   Li, Y.; Qian, F.; Xiang, J.; Lieber, C. M. Nanowire electronic and optoelectronic devices. *Mater. Today* **2006**, *9*, 18-27.

[4]   Pauzauskie, P. J.; Yang, P. Nanowire photonics. *Mater. Today* **2006**, *9*, 36-45.

[5]   Yan, R.; Gargas, D.; Yang, P. Nanowire photonics. *Nature Photonics* **2009**, *3*, 569-576.

[6]   Law, M.; Sirbuly, D.; Johnson, J.; Goldberger, J.; Saykally, R. J.; Yang, P. Nanoribbon waveguides for subwavelength photonics integration. *Science* **2004**, *305*, 1269-1273.

[7]   Sirbuly, D. J.; Law, M.; Yan, H.; Yang, P. Semiconductor nanowires for subwavelength photonics integration. *J. Phys. Chem. B* **2005**, *109*, 15190-15213.

[8]   Huang, M. H.; Mao, S.; Feick, H.; Yan, H.; Wu, Y.; Kind, H. ;Weber, E.; Russo, R.; Yang, P. Room-temperature ultraviolet nanowire nanolasers. *Science* **2001**, *292*, 1897-1899.

[9]   Johnson, J. C.; Choi, H. J.; Knutsen, K.P.; Schaller, R. D.;Yang,P.; Saykally, R. J. Single gallium nitride nanowire lasers. *Nature Mater.* **2002**, *1*, 106-110.

[10] Zimmler, M. A.; Bao, J.; Capasso, F.; Müller, S.; Ronning, C. Laser action in nanowires: Observation of the transition from amplified spontaneous emission to laser oscillation. *Appl. Phys. Lett.* **2008**, *93*, 051101.

[11] Duan, X.; Huang, Y.; Agarwal, R.; Lieber, C. Single-nanowire electrically driven lasers. *Nature* **2003**, *421*, 241-245.

[12] Chin, A. H.; Vaddiraju, S.; Maslov, A. V.; Ning, C. Z.; Sunkara, M. K.; Meyyappan, M. Near-infrared semiconductor subwavelength-wire lasers. *Appl. Phys. Lett.* **2006**, *88*, 163115.

[13] Hua, B.; Motohisa, J.;Kobayashi,Y.; Hara, S.; Fukui,T. Single GaAs/GaAsP Coaxial Core-Shell Nanowire Lasers. *Nano Lett.* **2009**, *9*, 112-116.

[14] Qian, F.; Gradecak, S.; Li, Y.; Wen, C.-Y.; Lieber, C. M. Core/Multishell Nanowire Heterostructures as Multicolor, High-Efficiency Light-Emitting Diodes. *Nano Lett.* **2005**, *5*, 2287-2291.

[15] Glas, F. Critical dimensions for the plastic relaxation of strained axial heterostructures in free-standing nanowires. *Phys. Rev. B* **2006**, *74*, 121302(R).

[16] Fan, S.; Villeneuve,P.; Joannopoulos, J.; Schubert, E. High Extraction Efficiency of Spontaneous Emission from Slabs of Photonic Crystals. *Phys. Rev. Lett.* **1997**, *78*, 3294- 3297.

[17] Motohisa, J.; Takeda, J.; Inari, M.; Noborisaka, J.; Fukui, T. Growth of GaAs/AlGaAs hexagonal pillars on GaAs(111)B surfacesby selective-area MOVPE. *Physica E* **2004**, *23*, 298-304.

[18] Hu, L.; Chen, G. Analysis of optical absorption *in silicon* nanowire arrays for photovoltaic applications. *Nano Lett.* **2007**, *7*, 3249-3252.

[19] Wagner, R. S.; Ellis, W. C. Vapor-liquid-solid mechanism of single crystalgrowth. *Appl. Phys. Lett.* **1964**, *4*, 89-90.

[20] Hiruma, K.; Yazawa, M.; Katsuyama, T.; Ogawa, K.; Haraguchi, K.; Koguchi, M.; Kakibayashi, H. Growth and optical properties of nanometer-scale GaAs and InAs whiskers. *J. Appl. Phys.* **1995**, *77*, 447-462.

[21] Lauhon, L. J.; Gudiksen, M. S.; Lieber, C. M. Semiconductor nanowire heterostructures. *Philos. T. R. Soc. A* **2004**, *362*, 1247-1260.

[22] Sze, S. M. *Physics of Semiconductor Devices*, Wiley, NewYork, 1981.

[23] Hiesinger, P. Hall Effect Levels in Ag- and Au-Doped p-Type GaAs. *Phys. Stat. Sol. (a)* **1976**, *33*, K39-K41.

[24] Yan, Z. X.; Milnes, A. G. Deep Level Transient Spectroscopy of Silver and Gold Levels in LEC Grown Gallium Arsenide. *J. Electrochem. Soc.* **1982**, *129*, 1353-1356.

[25] Novotny, C.J.; Yu, P. K. L. Vertically aligned, catalyst-free InP nanowires grownby metalorganic chemical vapor deposition. *Appl. Phys. Lett.* **2005**, *87*, 203111.

[26] Mårtensson, T.; Wagner, J. B.; Hilner, E.; Mikkelsen, A.; Thelander, C.; Stangl, J.; Ohlsson, B. J.; Gustafsson, A.; Lundgren, E.; Samuelson, L. Epitaxial growth of indium arsenide nanowires on silicon using nucleation templates formed by self-assembled organic coatings. *Adv. Mater.* **2007**, *19*, 1801-1806.

[27] Yoshizawa, M.; Kikuchi, A.; Mori, M.; Fujita, N.; Kishino, K. Growth of Self-Organized GaN Nanostructures on $Al_2O_3(0001)$ by RF-Radical Source Molecular Beam Epitaxy. *Jpn. J. Appl. Phys.* **1997**, *36*, L459-L462.

[28] Motohisa, J.; Noborisaka, J.; Takeda, J.; Inari, M.; Fukui, T. Catalyst-free selective-area MOVPE of semiconductor nanowires on (111) B oriented substrates. *J. Cryst. Growth* **2004**, *272*, 180-185.

[29] Noborisaka, J.; Motohisa, J.; Fukui, T. Catalyst-free growth of GaAs nanowires by selective-area metalorganic vapor-phase epitaxy. *Appl. Phys. Lett.* **2005**, *86*, 213102.

[30]   Motohisa, J.; Fukui, T. Catalyst-free selective area MOVPE of semiconductor nanowires. *Proceed. SPIE* **2006**, *6370*, 63700B.

[31]   Fukui, T.; Ando, S. New GaAs quantum wires on {111}B facets by selective MOCVD. *Electron. Lett.* **1989**, *25*, 410-412.

[32]   Fukui, T.; Ando, S.; Tokura, Y.; Toriyama,T. GaAs tetrahedral quantum dot structures fabricated using selective area metalorganic chemical vapor deposition. *Appl. Phys. Lett.* **1991**, *58*, 2018-2020.

[33]   Ando, S.; Kobayashi, N.; Ando, H. Selective area metalorganic chemical vapor deposition growth for hexagonal-facet lasers. *J. Cryst. Growth* **1994**, *145*, 302-307.

[34]   Hamano, T.; Hirayama, H.; Aoyagi, Y. New technique for fabrication of two-dimensional photonic bandgap crystals by selective epitaxy. *Jpn. J. Appl. Phys. Part 2 Lett.* **1997**, *36*, L286-L288.

[35]   Nakajima, F.; Ogasawara, Y.; Motohisa, J.; Fukui, T. GaAs dot-wire coupled structures grown by selective area metalorganic vapor phase epitaxy and their application to single electron devices. *J. Appl. Phys.* **2001**, *90*, 2606-2611.

[36]   Tomioka, K.; Kobayashi, Y.; Motohisa, J.; Hara, S.; Fukui, T. Selective-area growth of vertically aligned GaAs and GaAs/AlGaAs core-shell nanowires on Si (111) substrate. *Nanotechnology* **2009**, *20*, 145302.

[37]   Noborisaka, J.; Motohisa, J.; Hara, S.; Fukui, T. Fabrication and characterization of free-standing GaAs/AlGaAs core-shell nanowires and AlGaAs nanotubes by using selective-area metalorganic vapor phase epitaxy. *Appl. Phys. Lett.* **2005**, *87*, 093109.

[38]   Mohan,P.; Motohisa, J.; Fukui,T.Fabrication of InP/InAs/InP core-multishell heterostruc-ture nanowires by selective area metalorganic vapor phase epitaxy. *Appl. Phys. Lett.* **2006**, *88*, 133105.

[39]   Hua, B.; Motohisa, J.; Ding, Y.; Hara, S.; Fukui, T. Characterization of Fabry-Pe´rot microcav-ity modesin GaAs nanowires fabricated by selective-area metal organic vapor phase epitaxy. *Appl. Phys. Lett.* **2007**, *91*, 131112.

[40]   Marple, D. T. F. *J. Appl. Phys.* **1964**, *35*, 1241.

[41]   Yang, L.; Motohisa, J.; Fukui, T.; Jia, L. X.; Zhang, L.; Geng, M. M.; Chen, P.; Liu,Y. L. Fabry-Pe´rot microcavity modes observed in the micro-photoluminescence spectra of the single nanowire with InGaAs/GaAs heterostructure. *Optics Express* 2009, *17*, 9337-9346.

[42]   Ding,Y.; Motohisa, J.; Hua, B.; Hara, S.; Fukui, T. Observation of microcavity modes and waveguides in InP nanowires fabricated by selective-area metalorganicvapor-phase epitaxy. *Nano Lett.* **2007**, *7*, 3598-3602.

[43]   Van Vugt, L.K.; Rühle,S.; Vanmaekelbergh, D. Phase-Correlated Nondirectional Laser Emission from the End Facets of a ZnO Nanowire. *Nano Lett.* **2006**, *6*, 2707-2711.

[44]   Wang, J.; Gudiksen, M. S.; Duan, X.; Cui, Y.; Lieber, C. M. Highly polarized photoluminescence and photodetection from single indium phosphide nanowires. *Science* **2001**, *293*, 1455.

[45]   Tomioka, K.; Motohisa, J.; Hara, S.; Hiruma, K.; Fukui, T. GaAs/AlGaAs Core Multishell Nanowire-Based Light-Emitting Diodes on Si. *Nano Lett.* **2010**, *10*, 1639-1644.

[46]   Goto, H.; Nosaki, K.; Tomioka, K.; Hara, S.; Hiruma, K.; Motohisa, J.; Fukui,T. Growth of Core-Shell InP Nanowires for Photovoltaic Application by Selective-Area Metal Organic Vapor Phase Epitaxy. *Appl. Phys. Express* **2009**, *2*, 035004.

[47]   Huang, Y.; Duan, X.; Wei, Q.; Lieber, C. M. Directed assembly of one-dimensional nanostructures into functional networks. *Science* **2001**, *291*, 630-633.

[48]   Smith,P.; Nordquist, C.; Jackson, T.; Mayer, T. Electric-field assisted assembly and alignment of metallic nanowires. *Appl. Phys. Lett.* **2000**, *77*, 1399-1401.

[49]   Javey, A.; Nam, S. W.; Friedman, R. S.; Yan, H.; Lieber, C. M. Layer-by-layer assembly of nanowires for three-dimensional, multifunctional electronics. *Nano Lett.* **2007**, *7*, 773-777.

[50]   Hill, M. T.; Oei, Y. S.; Smalbrugge, B.; Zhu, Y.; de Vries, T.; van Veldhoven, P. J.; van Otten, F.W.M.; Eijkemans,T.J.; Turkiewicz,J.P.; deWaardt, H. Lasingin metallic-coated nanocavities. *Nat Photonics* **2007**, *1*, 589-597.

[51]   Oulton, R. F.; Sorger, V. J.; Zentgraf, T.; Ma, R. M.; Gladden, C.; Dai, L.; Bartal, G.; Zhang, X. Plasmon lasers at deep subwavelength scale. *Nature* **2009**, *461*, 629-632.

CHAPTER 9

III-V Semiconductor Nanowire Solar Cells

S. Crankshaw and A. Fontcuberta i Morral*

Laboratoire des Matériaux Semiconducteurs, Ecole Polytechnique Fédérale de Lausanne, 1015 Lausanne, Switzerland

Abstract: This chapter reviews the implementation of semiconductor nanowires as the active optical elements of nanowire-based photovoltaics. Some essential principles for understanding the semiconductor p-n junction photoresponse are briefly covered, followed by a more detailed presentation of the arguments for turning to a nanowire geometry rather than the traditional planar case. These include the decoupling of the light absorption and carrier collection directions for a radial junction geometry, the relaxation of lattice-mismatching constraints in choice of material combinations for axial heterojunctions, and absorption enhancement *via* light scattering in nanowire arrays. The emphasis here is on semiconductor nanowires grown epitaxially from a substrate, rather than solution-based methods. Because of the application-oriented nature of the material discussed, though, this scientific literature review is complemented by some remarks on the relevant patent literature for this field, also including alternative synthesis methods, as a gauge of the readiness for moving from scientific research to technology development. Finally, an outlook on additional modifications or integrations into the semiconductor nanowire solar cell framework is presented, such as the incorporation of plasmonic cell elements.

Keywords: Nanowires, III-V semiconductors, solar cell, photovoltaics, third generation, heterostructures, light trapping.

INTRODUCTION

III-V semiconductor nanowires are attractive building blocks for future-generation optoelectronic devices such as solar cells. Already the quintessential optical material for the semiconductor laser industry, the III-V's such as GaAs and InP constitute a well-studied material platform in three and two dimensions but offer new design possibilities in the nanowire geometry that is the topic of this chapter. Photovoltaic devices convert (solar) radiation into direct current electricity, and have gained renewed interest in the last few years as one means of obtaining reliable renewable sources of energy to lower global greenhouse gas emissions. For a widespread deployment of photovoltaics to become a reality, the efficiency:cost ratio must be competitive in the marketplace, and hence the development of so-called "third generation" solar cells. These represent alternatives to traditional c-Si bulk (first generation) and thin film (second-generation) devices. The purpose is to significantly improve the efficiency:cost ratio by exploiting new physical principles and solutions, arising largely from nanotechnology [1,2]. It is in this context of finding the optimal combination of optical materials and device design that semiconductor nanowires become candidates as components of third generation solar cells [3].

Particular attention has been given to core-shell radial p-i-n structures, in which the direction of light absorption is orthogonal to the carrier collection [4]. Such structures can enable highly efficient carrier collection, as the minority carrier diffusion length can in all cases be kept shorter than the optical absorption length. As a consequence, efficient radial p-i-n nanowire structures can be obtained with reduced material quality, in turn leading to reduced fabrication costs. There are additional advantages in using nanowires for photovoltaic applications such as the improvement in the light collection and the inherent minimization of the amount of material used [5]. Moreover, due to the new radial degree of freedom for strain relaxation associated with the small footprint of a nanowire on its growth substrate, lattice-mismatched material combinations become possible which cannot be produced as thin films. This enables both the formation of strain-free multijunction solar cells and the utilization of non-epitaxial substrates, which can also lower costs [6,7].

*****Address correspondence to A. Fontcuberta i Morral:** Ecole Polytechnique Fédérale de Lausanne, Institut des Matériaux, Laboratoire des Matériaux Semiconducteurs, MXC 330 Station 12, 1015 Lausanne, Switzerland. Email: anna.fontcuberta-morral@epfl.ch

Jianye Li, Deli Wang and Ray R. LaPierre (Eds)

This chapter is devoted to solid-state III-V nanowire-based solar cells. After reviewing some general principles necessary for understanding the photoresponse, we present in some detail the conceptual basis arguments for nanowire-based solar cell devices. We will follow with a review of what has been achieved to date and the state-of-the-art in terms of fabrication. The patent landscape of nanowire-based photovoltaics will also be reviewed before presenting the perspectives of this area of research in conclusion.

BASIC CONCEPTS

Solid-state solar cells are photodiodes optimized for the large-area conversion of solar radiation into electrical energy. Consider a solar cell made from a semiconductor with a bandgap E_g. Under the solar illumination, the material will only absorb photons of energy $h\nu \geq E_g$, where h is Planck's constant and ν the frequency of the light. The current-voltage (I-V) characteristics under dark and illuminated conditions are, respectively:

$$I = I_s \left[\exp\left(\frac{eV}{kT}\right) - 1 \right]$$

(1)

$$I = I_s \left[\exp\left(\frac{eV}{kT}\right) - 1 \right] - I_L$$

(2)

where I_L corresponds to the current generated by the light absorption. Under illumination, the diode characteristic is shifted due to the production of a photocurrent. Important parameters characterizing the cell are the short-circuit current, I_{SC}; the open circuit voltage, V_{OC}; the fill factor, FF; and the efficiency, η. The energy conversion efficiency is obtained by dividing the maximum electrical power produced by the cell by the incident optical power. The most typical illumination conditions are a spectral distribution close to an air-mass of 1.5 (AM 1.5), a total irradiance of 1 kW/m^2 and a cell temperature of 25°C. The FF is the ratio between the efficiency and the product of the open-circuit voltage and the short-circuit current:

$$FF = \frac{\eta}{V_{OC} \cdot I_{SC}} \, x\, 1kW / m2$$

(3)

The open-circuit voltage is given by:

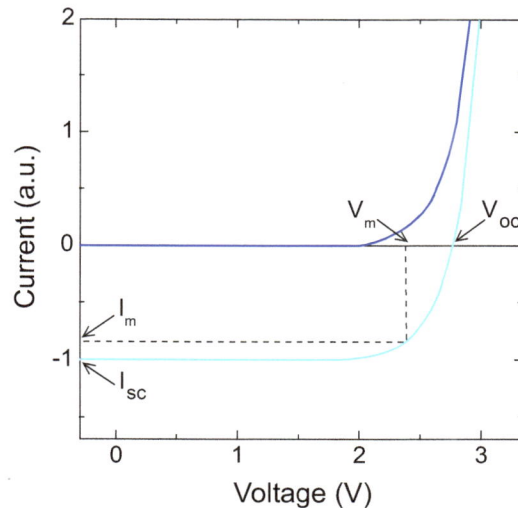

Figure 1: Generic example of an I-V curve. Upon exposure to light, the characteristic diode I-V curve is shifted down by an amount corresponding to the photocurrent. I_{SC} and V_{OC} represent the short-circuit current and open-circuit voltage, respectively, while I_m and V_m represent the operating points for maximum power generation.

$$V_{oc} = \frac{kT}{e} \ln\left(\frac{I_L}{I_s} + 1\right)$$

(4)

The FF is a measurement of the "squareness" of the I-V curve in the operation regime and gives an idea of the degree of improvement that can still be realized on the device. It is directly affected by the values of the series and shunt resistance of the solar cell. For example, a higher fill factor can be obtained by increasing the shunt resistance and decreasing the series resistance. For clarity, these quantities are depicted in Fig. **1**, below.

THE CASE FOR NANOWIRE-BASED SOLAR CELLS

In this section, we review the main conceptual arguments in favor of the use of nanowires for third generation solar cell applications. These are:

     i)   The possibility of geometrically decoupling light absorption and carrier separation by the use of radial p-n junctions.

    ii)   The possibility of new combinations of lattice-mismatched materials without the generation of defects, having bandgaps optimized for solar absorption.

   iii)   The improved scattering characteristics (light trapping) of dense nanowire or nanocone arrays.

Furthermore, other practical arguments in favor of nanowire based photovoltaics include:

     i)   The possibility of using cheap materials as substrates.

    ii)   The possibility of transferring the devices onto any substrate, including plastic.

   iii)   Minimization of material use, due to the low material fraction of nanowire arrays.

Decoupling Light Absorption and Carrier Separation: the Radial p-n Junction

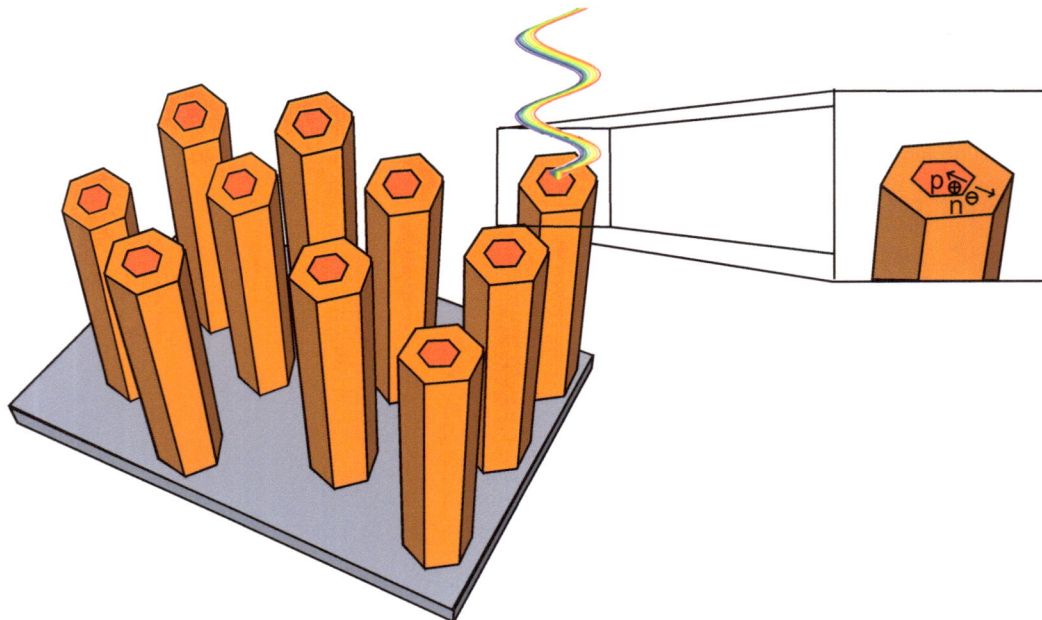

Figure 2: Nanowires presenting a core-shell p-n junction, arranged in a high density, an ideal configuration for a single-bandgap nanowire solar cell.

Although the first patent applications on radial p-n junction nanowires date from 2002 [8,9], it was only in 2005 that Kayes *et al.* first demonstrated the real advantages of radial-junction nanowires for photovoltaics as a means of decoupling the direction of light absorption and carrier separation [4]. A schematic drawing of the nanowire configuration of such a design is shown in Fig. **2**. A high density of nanowires is obtained perpendicularly to a substrate, with each of the nanowires exhibiting a core-shell p-n junction that enables the collection of photo-generated carriers in the radial direction. As depicted in Fig. **2**, the core and the shell must be contacted separately by top and bottom electrodes; more details on the typical fabrication methods are given in the following section.

The solar radiation path is parallel to the nanowire axis for light normally-incident to the substrate, with the length of the nanowires adjustable in principle so as to maximize the optical absorption. Crucially, such a configuration enables the decoupling of absorption and carrier diffusion lengths, which is not possible for planar cells. Indeed, in classical bulk solar cells, the optimal design has always balanced a compromise between light absorption and carrier separation: on the one hand, increasing solar cell thickness leads to a higher absorption; while on the other the thickness of the active layer should not be larger than the minority carrier diffusion length [10,11]. By turning to vertically-oriented nanowires with a radial p-n junction design, it is possible to use semiconductors in which the collection length of photo-generated minority carriers is much shorter than the optical penetration length. Such a condition results in the relaxation of the material electronic properties, which should in turn result in an improvement of the efficiency:cost ratio of the device.

Without further device designs, the core-shell p-n junction approach is limited to a single axial junction. Indeed, the electrical circuit of a multijunction radial p-n solar cell stacked along the nanowire axis would correspond to a parallel circuit connection, with the resulting device extracting current from the two cells but operating both of them at the same voltage. In this case the photoconversion efficiency is not increased with respect to a single radial junction device no matter what the material combination. The efficiency for the single radial junction itself is strongly dependent on the material chosen, and more specifically on the bandgap, E_g. Indeed, for the absorbed photon energies $\hbar\omega$ greater than E_g, part of the absorbed energy is not directly converted in electrical energy due to the thermalization of electrons down to the conduction band edge – *i.e.* hot carriers. The upper bound of the solar cell efficiency can be obtained by applying the detailed balance model, in which only the radiative absorption and emission of photons is considered [12]. In such a framework, the maximum efficiency obtained with only one junction is close to 30%, for material with a bandgap around 1.31 eV [13]. The efficiency curve around the optimal E_g value is quite flat, such that materials with E_g lying between 1.1 and 1.45 eV are good candidates for the highest efficiency single junction solar cells. Optimal materials then include Si, InP, and GaAs.

Silicon has the advantage of being an earth-abundant material, and can leverage extensive fabrication know-how from the logic industry. However, associated with its indirect bandgap is an optical absorption length relevant for the solar spectrum of about 100 μm, in other words requiring physical thicknesses on this order for an efficient photoresponse. On the other hand, the direct bandgap semiconductors InP and GaAs instead have absorption lengths of about 1 μm, and correspondingly require a comparatively low quantity of active material for fabrication. Because of the suitability of their bandgaps, ~1.3-1.4 eV, InP and GaAs have recently attracted significant attention for nanowire-based solar cells. InP is known to exhibit a lower surface recombination rate than GaAs [14,15], and as such one would expect higher photoconversion efficiencies for InP nanowires with respect to GaAs considering the large surface-to-volume ratio, as surface recombination represents carrier losses in the form of a decreased short-circuit current. However, GaAs nanowires can be grown with faceting of the {110} family [16], which also exhibit small surface recombination such that both materials should be suitable for radial p-i-n nanowire devices.

Radial Nanowire Junctions: Experimental Demonstrations

Nanowire solar cell devices have been fabricated and tested both at the single nanowire and array level. While the second provides the platform of testing the final device, the first one allows the study of the pure device neglecting the influence of eventual parasitic effects such as parallel resistances due to unoptimized fabrication techniques. In this section, we review the fabrication techniques and results obtained for nanowire-based solar cells in the single and array configurations. Tables **1** and **2** summarize the efficiencies obtained to-date for nanowire solar cells for different materials, respectively for single nanowire and nanowire arrays measurements.

We start by presenting measurements realized on single nanowires having a core-shell p-n or p-i-n junction structure. The typical contacting scheme is shown in Fig. **3**. The core and shell of the nanowire are contacted independently by a two-step lithography process. In order to reach the core of the nanowire, an etch of the shell is performed in a controlled way, followed by the second lithography procedure. After fabrication, the current-voltage (I-V) characteristics of the device are measured in the dark and under calibrated illumination. One should note here that the measurement of the efficiency in single nanowires is not a straightforward task, owing partly to the difficulty of accurately determining the active area. Typically, the projection of the nanowire active junction is taken, though this can lead to a slight underestimation of the efficiency. A highly absorbing substrate should also be used in order to avoid back reflections of the incident light on the nanowire. Lack of a perfectly absorbing substrate would therefore result in a higher apparent light absorption and therefore in an overestimation of the efficiency.

Pioneering experiments were realized on silicon nanowires [17]. The core was obtained by a vapor-liquid-solid (VLS) process [18] doped p-type. The intrinsic and n-type shells were obtained by switching the growth from VLS to a predominant chemical vapor deposition (CVD) type of process. The structure of the core was single crystalline, while the shell exhibited a nanograin structure, with the grain size between 30 and 80 nm and the total wire diameter ~300 nm. The V_{oc} of the device was quite low (0.26 V), probably due to the existence of numerous interfaces in the active layer of the device, but the 3.4% efficiency remains quite respectable for such a thin nanowire. The authors speculated that the high absorption might be related to the nanocrystalline structure of the shell. Later on, it was shown that the nanowire geometry can lead into enhancement in the absorption [19] due to so-called leaky-mode resonances that confine light within subwavelength high-index nanostructures. This effect may also explain the high efficiency obtained by B. Tian *et al.* Moreover, its mastering may lead to improvements in the nanowire design for photovoltaic applications.

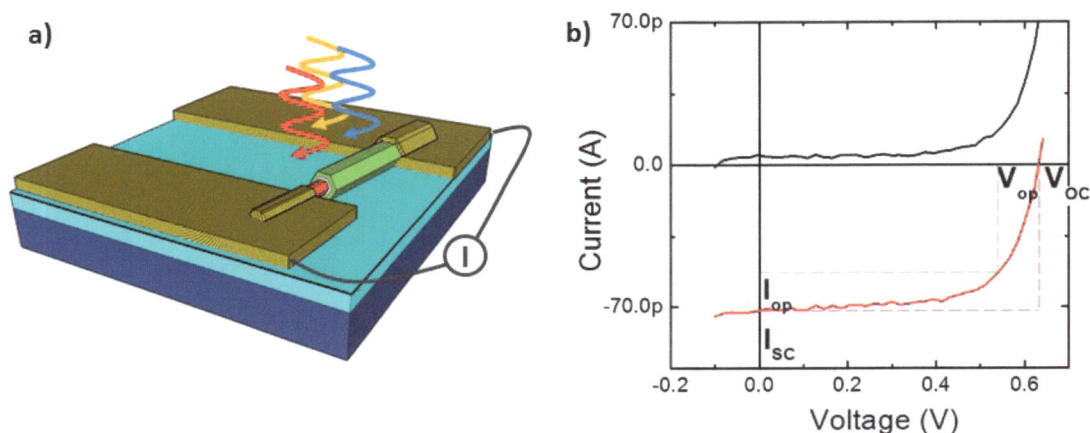

Figure 3: a) Contacting and measurement schematics of single nanowires p-i-n junctions. **b)** I-V curve in the dark and under illumination of a single GaAs nanowire p-i-n junction [20].

Table 1: Summary of results obtained to date in single nanowire *core-shell* (radial) junctions under illumination conditions of 1.5 AM.

Material	Efficiency (%)	FF	V_{OC} (V)	Reference
Si	3.4	0.55	0.26	B. Tian *et al.* (2007) [17]
GaAs	4.5	0.65	0.60	C. Colombo *et al.* (2009) [20]
$In_xGa_{1-x}N$ (0<x<0.27)	≤ 0.19	0.56	1.0-2.0V	Y. Dong *et al.* (2009) [25]

In 2009, the same concept was applied to III-V materials. Core-shell GaAs nanowire structures were obtained by catalyst-free molecular beam epitaxy (MBE) growth [21,20]. The advantage of catalyst-free growth is that switching from axial to radial growth is relatively straightforward [22,23]. The active region of the device was approximately 20 nm thick. An efficiency of 4.5% was reported, along with a V_{OC} and FF of 0.6 V and 0.65, respectively. The MBE growth accounts for extremely sharp and high quality interfaces, which should explain the high value of V_{OC} and minimize the current losses. A similar work was reported by Dong *et al.*, in which a GaN based nanowire pin junction was fabricated with $In_xGa_{1-x}N$ as intrinsic active layer. The synthesis technique was metalorganic CVD

(MOCVD), which is also known to produce high quality core-shell structures [24]. The indium composition of the intrinsic layer was tuned from 0 to 0.27, in order to obtain different bandgaps and thereby modify the absorption of the solar spectrum. One should note here an additional advantage of such structures. The band alignment between the n(p) type GaN and the $In_xGa_{1-x}N$ layers are such that an additional intrinsic electric field is present that favors the drift of the minority carriers. This is a design parameter that deserves to be explored in the future for the optimization of radial p-n (or p-i-n) junction solar cells. Moreover, with this work, the authors demonstrated the ability to rationally tune the absorption and thereby the electrical characteristics of the device, which will certainly be used in the future for nanowire based solar cell engineering.

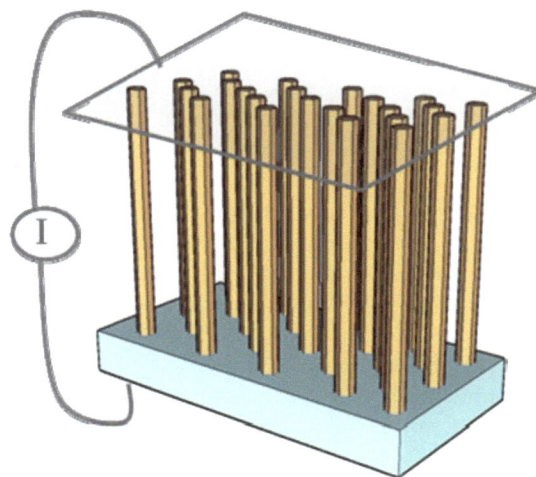

Figure 4: Schematic of the electrical connections for a nanowire array solar cell.

Studies realized on single nanowires provide valuable information on the device and on the suitability of the materials and/or interfaces [10]. The measurements constitute a platform for understanding the properties of the nanowires and help in the design of macroscopic nanowire based solar cells. Nevertheless, the final device will consist of a vertical nanowire array as depicted in Fig. **4**. The contacting procedure is in this case slightly more complicated than for the single nanowire. The general approach consists in fabricating a top and a bottom contact that correspond separately to the core and/or shell of the nanowire, with the top contact being transparent. For mechanical stability of the array, the space between the nanowires is filled with an insulating transparent material, such as PMMA or spin-on-glass [26,27]. To our knowledge, the first device composed of vertical nanowires exhibiting a radial p-n junction dates from 1992 [28]. The authors fabricated a 1 mm^2 array of n-type GaAs nanowires by selective area epitaxy, which were then covered by a p-type layer. The space between the wires was filled with spin-on-glass, and the p- and n-type regions of the nanowires were electrically connected by top and bottom contacts, respectively. The device exhibited a typical diode behavior, with narrow optical emission under forward bias. Unfortunately, the authors did not investigate the properties related to the creation of a photocurrent.

Initial proof-of-principle studies on nanowire arrays for solar applications were realized by Tsakalakos *et al.* [29]. First, p-type nanowires were fabricated by the VLS method on a metal foil, followed by a conformal deposition of a hydrogenated amorphous silicon (a-Si:H) n-type layer by plasma-enhanced CVD (PECVD). Although the efficiency obtained was quite low (0.1%), the technique demonstrated the feasibility of such a device design. A slightly higher efficiency was obtained a short time later by E. C. Garnett *et al.* [30]. The silicon nanowire arrays were first obtained by electroless etching of a silicon substrate, leading to nanoscale columns [31]. A conformal coating was also obtained by depositing an n-type layer of a-Si:H, which was then crystallized by annealing the ensemble at high temperatures under a forming gas. As a result, a slightly higher efficiency was obtained: 0.46%.

The highest efficiency obtained in nanowire arrays has been recently obtained by V. Sivakov *et al.* [32]. The fabrication method was quite original and with a great potential for low cost and large area fabrication of the solar cells. First, a nanowire array was fabricated by electroless etching of a polycrystalline layer that had been obtained by crystallization of an a-Si:H stack deposited on glass. The stack consisted already of a p-n junction, so that the nanowire formation resulted directly in an array of axial p-n junctions. The thickness of the ensemble was about 2 μm. Even if the geometry of the junction was axial instead of radial, an efficiency of 4.4% was obtained.

An alternative method to test the energy conversion properties of nanowire arrays has been demonstrated by Boettcher *et al.* [33]. A nanowire array is submerged in an aqueous electrolyte, where the liquid wets the nanowire surface and thereby creates a junction. As a consequence, the electrical characteristics and conversion properties can be measured without introducing further artifacts/challenges linked to the fabrication of conformal solid state junctions, deposition of transparent electrodes, or metallic grid contacts. Arrays measured under a monochromatic illumination of 60 mW cm^{-2} at 808 nm exhibited an efficiency of 2.6%, a V_{OC} of 0.41 V, and a FF of 0.50. This is to date the highest FF obtained with silicon nanowire arrays. A similar structure was investigated for the potential application on the generation of hydrogen from water splitting. In particular, Si/TiO$_2$ nanowire heterostructures were fabricated and the photo-oxidation of water investigated. It was observed that highly dense nanowire arrays enhanced the photocurrent by a factor 2.5 with respect to equivalent planar surfaces. Water splitting and hydrogen generation upon solar illumination is extremely promising route towards finding reliable renewable energy [34].

Nanowire array solar cells have also been obtained with compound semiconductor materials. Two distinct strategies can be found: i) radial p-n junction arrays or ii) fabrication of a heterojunction between the semiconductor and a silicon substrate that favors the carrier collection. The group of LaPierre obtained the first GaAs nanowire array solar cells, with an efficiency of 0.83% [35,36]. InP and GaAs are close to the optimal bandgap for single-junction solar cell efficiency. InP nanowire based solar cell arrays were fabricated for the first time by the group of Fukui [37]. The nanowire array was obtained by selective area epitaxy, in a similar way as Haraguchi *et al.* The total device area was 2.0x2.6 mm^2. The characteristics of the solar cell were a V_{OC} of 0.43 V, a FF of 0.57 and an overall efficiency of 3.37%. A rather different approach was adopted by Y. B. Tang *et al.* [27] and W. Wei *et al.* [38]. The vertical array solar cell/photodiode device was composed of the heterojunction formed at the interface between the III-V nanowire and the Si substrate, respectively doped n-(or p-) and p-(or n-)type. The maximum efficiencies obtained in these cases were 2.5% for InP/Si and 2.73% for GaN/Si. These results have opened the way towards the direct fabrication of heterojunction solar cells and to the integration of several materials in series along the nanowire axis. In the following section, the possibility of combining semiconductor solar cells in the axial direction of the nanowire and the related advantages is presented.

Table 2: Summary of results obtained so far in *nanowire arrays* under illumination conditions of 1.5 AM.

Material/Substrate	Efficiency (%)	FF	V_{OC} (V)	Reference
Si/metal foil	0.1	0.28	0.13	L. Tsakalakos *et al.* (2007) [29]
Si/Si	0.46	0.33	0.29	E. C. Garnett *et al.* (2008) [30]
Si/glass	4.4	-	0.45	V. Sivakov *et al.* (2009) [39]
Si/electrolyte	2.6†	0.5†	0.41†	S.W. Boettcher *et al.* (2010)
Si/Si	7.9‡	-	-	H. Atwater‡ *et al.* (2010)
GaAs	-	-	-	K. Haraguchi *et al.* (1992) [28]
GaAs	0.83	0.27	0.2	J. A. Czaban *et al.* (2009) [36]
InP	3.37	0.57	0.43	H. Goto *et al.* (2009) [37]
GaN/Si	2.73	0.38	0.95	Y. B. Tian *et al.* (2008) [27]
InP/Si hetero	2.5*	0.52	0.65	W. Wei *et al.* (2009) [38]
CdS-CdTe/Al foil	6	0.43	0.62	Z. Fan *et al.* (2009) [40]

†measured under monochromatic illumination at 808 nm

‡as presented at the 2010 Spring Materials Research Society meeting

*measured at 110K

Solid state solar cell structures (also nanowire based) quite generally rely on the optical generation and separation of electron-hole pairs through an internal electric field generated by a p-n (p-i-n) junction. The electron-hole separation and thereby the carrier collection can be further improved if a type-II heterojunction is used, using the electric field associated with the band offsets of the materials. As an example, p-type CdTe nanowires were coated with n-type CdS layer. The nanowire array exhibited a high degree of order, as it had been synthesized directly in a porous anodic alumina layer fabricated on an aluminum foil. The device was shown to enable a high absorption of light and efficient collection of carriers, with an overall efficiency of 6% [40].

Axial Multijunction Nanowire Solar Cells

A common strategy for the optimization of energy conversion in solar cells is the use of multiple junctions. Multijunction solar cells use multiple sub-cell bandgaps to divide the solar spectrum into smaller sections. By aligning them properly, the utilization of the solar spectrum can be optimized and losses due to thermalization of carriers are minimized. Each layer is optically and electrically in series, with the highest band gap material at the top. The first junction receives the whole spectrum. Photons above the band gap of the first junction are absorbed in the first layer, while those below the band gap of the first layer pass through to the lower layers to be absorbed there.

Typically, the ability to optimize the respective band gaps of the various junctions is hampered by the requirement that each layer must be lattice-matched to all other layers. Working with metamorphic or lattice-mismatched semiconductors would increase the degree of freedom, as it would provide unprecedented flexibility in the bandgap selection. The main challenge of working with lattice-mismatched materials is that above a certain critical thickness, continued crystal growth causes structural dislocations that introduce energy levels in the bandgap which mediate Shockley-Read-Hall recombination [41,42]. As an example, an efficiency of 40.1% was obtained for the lattice-matched triple-junction planar solar cell composed of GaInP/GaInAs/Ge. Moving to a metamorphic design, the bandgaps could be slightly tuned to a better utilization of the solar spectrum and the efficiency was increased to 40.7% [43].

Nanowires and whiskers are especially promising structures for a defect-free stacking of lattice-mismatched materials. Indeed, nanowires containing strongly lattice-mismatched sections with thicknesses well above the 2D critical thicknesses have already been demonstrated [44-47]. The existence of lateral free surfaces in the nanowire geometry allows for lateral strain relaxation. As a consequence, the critical thickness beyond which dislocations appear is larger in the case of nanowires, though it is also radius dependent [6]. Stacking mismatched materials in the form of nanowires, however, is not straightforward in all cases, with other challenges potentially arising. For example, it has been seen that for any pair of materials, it is easier to form a straight nanowire with a particular interface direction. This means that certain material combinations can result in a kinked wire. The presence of such defects could be resolved by the use of surfactants or by tuning the kinetic growth conditions, but it remains a challenge in the synthesis of "free-will" heterostructures [48].

To-date, a double junction nanowire solar cell has been demonstrated [49]. The integration of two cells with a pin(+)-p(+)in structure resulted in an increase in the open circuit voltage of 57% with respect to the single device. This work has shown that the principle of cell stacking can also be applied to nanowires and opens the way for stacking even more cells to form multijunction devices. Still, some fundamental studies have to be realized in order to define the intrinsic performance limits of such a configuration.

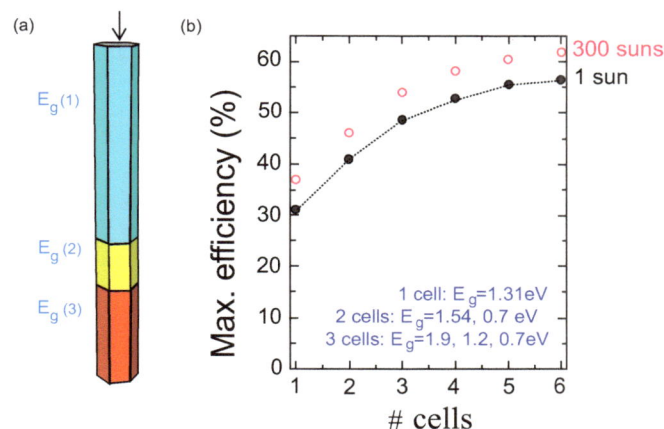

Figure 5: a) Schematic of a nanowire composed by the stacking of semiconductors with different bandgaps, **b)** Maximum efficiency in multijunction solar cells as a function of the number of junctions and under two different illumination conditions, as obtained by detailed balance calculations.

In planar multijunction or axial nanowire solar cells, the subcells are electrically connected in series so that the final cell has two terminals. A major constraint arises from the series connection in that the current through each junction must be the same. If the current for maximum power point of each junction is not the same, as is generally the case, the efficiency degrades. Current-matching of each junction thus represents a critical design consideration in determining what material bandgaps to combine. The sketch of a nanowire composed by the stacking of multiple bandgaps is depicted in Fig. **5(a)**. Detailed balance calculations of the maximum efficiency obtained when a solar cell is composed of multiple junctions are shown in Fig. **5b** [50]. Again, the maximum efficiency that can be achieved with a single junction is slightly above 30% (with E_g=1.31 eV). For a dual junction composed of semiconductors with bandgaps E_g of 1.54 and 0.7 eV, the maximum efficiency is slightly above 40%, which represents an improvement of 33% with respect to the single junction. Stacking three and four subcells results in a maximum efficiency of 48% and 52%, respectively. After four junctions, the gains in efficiency resulting from subsequent stacking of junctions diminish. A slight increase can be further obtained by working under concentrated conditions, as shown by the efficiencies obtained under 300-sun illumination. The detailed balance calculations work both for bulk and nanowire based solar cells. Working with nanowire solar cells in an axial junction geometry only adds the degree of freedom of choosing the materials exhibiting the optimal bandgaps [7].

Another key element for multijunction solar cells is the tunnel junction, which provides a low electrical resistance and optically low-loss connection between two subcells. Without it, the p-doped region of the top cell would be directly connected with the n-doped region of the middle cell. Hence, a p-n junction biased in the reverse direction of the others would appear between the top cell and the middle cell. Consequently, the photovoltage would be lower than if there would be no parasitic diode. The tunnel diode (also known as an Esaki diode) is formed by a p-n junction with the doping so high that the quasi-Fermi level lies within the respective bands [51]. The degeneracy is typically on the order of several $k_B T$ and the depletion layer width is in the range of 10 nm. A schematic of the band alignment and I-V characteristics of such a diode are shown in Fig. **6**.

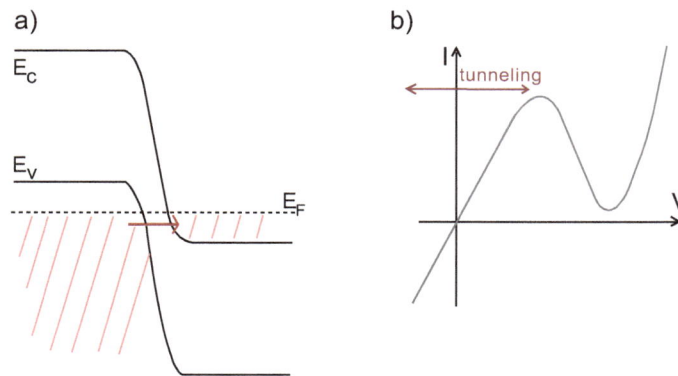

Figure 6: a) Schematic of the band alignment in a p-n junction and of the tunneling of electrons from the valence band of the p[++] side into the conduction band of the n[++] side. **b)** Typical I-V characteristics of a tunnel diode. The bias range in which tunneling is predominant is indicated.

The current-voltage characteristics of such a device present various operating regimes, depending on the respective alignment between the n and p regions. In the particular application of multijunction solar cells, the regime used is the tunneling, occurring in the forward and reverse bias around V=0. As shown in Fig. **6(a)**, for a small reverse bias the electrons tunnel from the valence band on the p[++] side into the conduction band on the n[++] side. For small forward bias, the electrons can tunnel from populated conduction-band states on the n[++] doped side. This is an elastic tunneling process. For small bias, the rectifying behavior of the diode is lost and the tunnel junction allows the current from one cell to be transported to the other cell as if it was a resistance. Please also note in Fig. **6(b)** the regime of negative resistance, which occurs in the regime where the tunneling current ceases. Indeed, at higher forward bias the bands become so separated that the electrons cannot tunnel from the n[++] to the p[++] side.

The design of tunnel junctions is one of the key elements of multijunction solar cells. In the area of nanowires, this is still an area of development. Initial results have recently been obtained by Wallentin *et al.* [52]. In their work, they have used a type II InP(n[++])-GaAs(p[++]) axial heterostructure. A type-II band alignment is commonly used to reduce the tunneling barrier and thereby to increase the interband tunneling current [53]. Single nanowire tunnel diodes with room temperature peak current densities up to 329 A/cm^2 have been obtained, similar to those obtained in state-of-the-art solar cells. These results are extremely promising. In particular, they enable the future fabrication of multijunction solar cells combining subcells formed by $InAs_xP_{1-x}$ and $GaAs_xP_{1-x}$. The bandgap of such materials can be tuned between 0.3 and 2.26 eV, covering an important part of the solar spectrum.

Scattering in Nanowire Forests

One of the primary arguments that make nanowire forests attractive for solar cell and/or photodetector applications is their enhanced light harvesting properties with respect to their planar counterparts. Visually, a bare Si substrate has glossy, reflective surface, while a nanowire sample appears completely matte. However, it remains to be seen whether the nanowire forests themselves are exhibiting enhanced light absorption, *i.e.* whether the reduced surface reflectivity is genuinely a result of light absorption *within* the nanowires or rather a result of the underlying planar substrate. The angular dependence of the absorption provides important information in this regard. Many groups have calculated that nanowire forests should exhibit a reduced reflectivity with respect to planar surfaces, with several existing reports on the calculations of the transmission, absorption and reflection of nanowire arrays for design optimization. For example, Li *et al.* calculate that the ideal ratio between the nanowire diameter and inter-wire distance for a square lattice is about 0.8 for inter-wire spacings between 250 and 1200 nm [54]. Other works suggest that a lattice composed of nanowires with a diameter of 540 nm arranged in a lattice of period 600 nm would result in a silicon solar cell with an efficiency 72.4% higher than the thin film counterpart [55]. Some studies have also been realized for extremely small diameter nanowires in the range between 50 and 80 nm. These calculations demonstrate that nanowire arrays with moderate filling ratio exhibit much lower reflectance compared to thin films [5]. Such calculations represent a good starting point for the design of optimal nanowire arrays. However, the following points are missing: (i) the use of realistic nanowire diameters (we have seen that in the case of silicon they should be in the range of few microns), (ii) a full electromagnetic simulation considering possible waveguiding effects of light coupling into the nanowires which would lead into determining the real absorption in the nanowires, and (iii) the dependence on the angle of incidence.

In this respect, a more realistic approach has been adopted by Kupec *et al.* [56]. They have determined the wave propagations of the direct perpendicular illumination of nanowire arrays by using three dimensional finite element calculations. Based on the field distribution within the nanowires, they calculated the spatially and spectrally dependent power dissipation leading to a spatially resolved optical model. Such modeling predicts highly absorbing modes at low volume fill factors. Combining such calculations with a detailed balance, they find what are the most efficient material combinations and nanowire dimensions/arrangement for single and dual-junction solar cells. They demonstrate that the resulting efficiency limits of the nanowire solar cells are comparable to conventional thin-film single and dual-junction counterparts.

Experimentally, it has been shown that, owing to the vertical orientation of the nanowires, nanowire arrays can show a lower absorption at normal incidence than at other angles [33]. There are multiple reasons for this. First, if the distance between nanowires is in the order or larger than the wavelength, the light absorbed by the nanowires should be proportional to their filling ratio. Second, if the nanowires are closer than this scale, the nanowire layer behaves like a medium with an equivalent effective refractive index corresponding to the filling ratio. The index being an intermediate between the air and the underlying substrate, coupling of light is improved but not necessarily due to an enhancement of absorption at the nanowire level. When the incident light is not normal to the surface, scattering between the nanowires becomes more favorable, thereby increasing the length of the light path in the nanowire layer and fostering the absorption in the nanowire layer. Two methods have to date been shown to be successful in measuring absorption of nanowire arrays: (i) electrolyte junctions and (ii) combined reflectivity/transmission measurements.

In the work of Atwater's group, several nanowire arrangements have been investigated in order to find the ideal nanowire configuration. However, the most effective strategy to enhance light absorption in nanowire forests has been to randomize the light passing through the nanowires. This method increases the length of the light path through the nanowire array. A significant increase in the inter-wire scattering cannot be achieved solely by designing a certain nanowire arrangement (if nanowires are perpendicular to the underlying substrate). Also, back-reflectors and antireflective coatings have only improved the nanowire absorption minimally. To date, the most effective light trapping strategy has been to introduce light scatterers within the nanowire forest, as shown schematically in Fig. **7**, randomizing the direction of propagation and effectively increasing the optical path length. Such scatterers can be as simple as aluminum oxide particles, which have negligible absorption across the wavelengths of interest. One demonstration showed that particles as large as 0.9 μm scatter the light that might otherwise pass between the wires, such that arrays having less than a 5% filling fraction can absorb up to 85% of the above-bandgap normal-incidence sunlight [57].

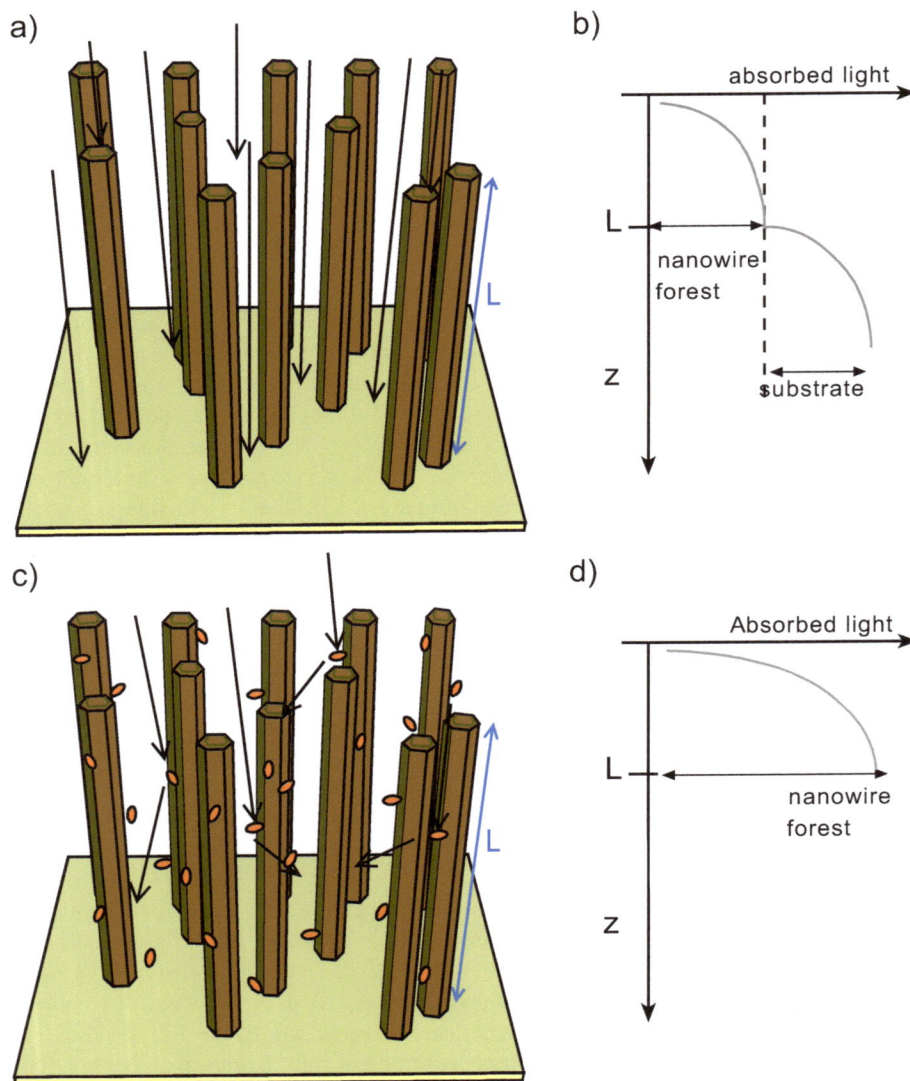

Figure 7: a) Schematic of incident light on a nanowire array without scattering elements. **b)** Schematic of the light absorbed in a structure of type (a). c) Schematic of the light path when the nanowire array includes random scatterers, and d) the resulting increased absorbance in the nanowires.

A second strategy that has been used for enhancing the absorption in solar cells may be thought of as an optical impedance matching *via* refractive index tapering. Indeed, semiconductor materials exhibit a much larger index of refraction than the surrounding environment. As a consequence, it is difficult to achieve a reflection coefficient of zero in any angle and any wavelength of interest. By creating an array of tapered nanostructures (see Fig. **8**), the refractive index is gradually increased from its value of 1 in air to the n_s of the semiconductor. In this way, nanocone arrays serve an impedance-matching function between the semiconductor and the environment at the relevant incident angles and wavelengths. In this case, the nanocone arrangement is not relevant, just the gradual index matching [58], hence acting as both an absorbing and an antireflection layer. Such an approach is extremely promising to enhance the conversion efficiency in solar cells.

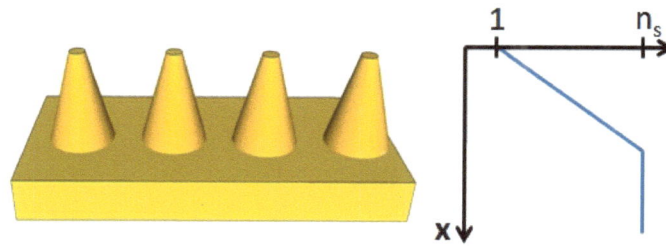

Figure 8: a) Schematic of a nanocone array on a substrate with a similar refractive index, **b)** effective refractive index of the ensemble as a function to the distance to the surface. The refractive index increases from 1 (air) to the n_s of the substrate.

PATENT LANDSCAPE

One interesting means of gauging the direction of nanowire photovoltaic research and development is to follow patent applications on this subject, as these represent concrete directions envisioned for technology transfer from the laboratory to commercial implementation. Patents filed come from both the academic and industrial realms, though of course in either case the essential elements of such a document is identical: background research, claims, and embodiments. The notion of patentability comprises novelty in the form of an inventive step not obvious as a consequence of prior work to others in the field, and as such any idea that has already been disclosed—whether through conference presentations, publications, or the like—is not patentable in principle. In practice, however, the strictness of the implementation of this notion varies by location, with the U.S. patent system, for example, allowing patents to be filed within one year of the initial disclosure. Nonetheless, the aim of the system is to protect the intellectual property of the inventor, and the aim of the inventor is to find a licensee for the intellectual property. With this in mind, the historical development of nanowire-related patent literature takes an unsurprising shape, with early patents describing fairly unspecific nanostructures so as to cover as broad an intellectual property basis as possible. In this section we review some of the most relevant patents for nanowires as related to photovoltaics.

Nanowires as an Active Absorption Medium

The early patent literature on semiconductor nanowires is written at the materials level of the smallest element, in practice then extrapolating in the embodiments section as a means of legal entrenchment. Dating from 1999, for example, one already finds patents granted at the materials level for "metal oxide nanorods", whereby the descriptions essentially include II-VI nanowires [59]. At this individual-element level, semiconductor nanowire patents including III-V materials appear shortly thereafter, for example with axial heterostructures, doping profiles, and radial superlattices all broadly described in filings in 2002 and shortly thereafter [8,9]. In principle any commercialized device using an axial nanowire p-n junction would then have to pay a licensing fee to the holder of the original description. By 2004, one finds patent filings focused on charge separation in nanowires, including radial doping junction geometries, clearly with an eye towards carrier collection and hence germane by implication for photodetector or photovoltaic applications [9].

Also in 2004, a filing by Nanosys patented nanostructures for photovoltaics explicitly, moving away from the individual-nanoelement level instead towards a cell module level with a constituent layer of nano-rods, wires, or crystallites [60]. Even though this particular patent focuses on mixtures of nanostructures with relative type-II band alignments for charge separation, many of the basic elements for the type of nanowire solar cell described in the previous sections are already included. Indeed, one finds several patent applications in the years following concerning nanowire arrays and photovoltaic applications, both from academia and industry: in 2005 by General Electric, with examples for arrayed silicon nanowires [61]; in 2007 by the University of Illinois, Urbana-Champaign, for arrays specifically for photovoltaic applications [62]; in 2008 by QuNano [63] and by Caltech also particularly for nanowire arrays [64]; in 2009 by Bandgap Engineering for nanowire array fabrication processes [65]. Clearly, the growth of the patent landscape from single nanowire elements to arrays to fabrication in the span of less than a decade reflects the combination of economic interest and shifting ecological attitudes of recent years—especially considering that the notion of "semiconductive whiskers" for photovoltaic applications was already described in a patent filing dating to 1977 [66].

Nanowires as Geometrical Structures

As a separate class of device structures, several patents have also appeared exploiting nanowires for their structural geometry as an interdigitating surface enhancer rather than an active medium, for example as part of a hybrid solar cell in combination with a dye sensitizer [67-70]. This idea is particularly well suited for nanowires made from oxides or other wide-gap materials, transparent to a large portion of the solar spectrum, as a conductive channel to speed up transport of one type of charge in the comparatively slow diffusive regime of the sensitizing dye. In another example, a patent filed by Sharp Labs in 2005, nanowires constitute the bottom electrode of a photovoltaic device whose active semiconductor junction area is formed by p- and n-doped layers deposited thereupon [71].

PERSPECTIVES

This chapter has been devoted to explaining the principles of nanowire based solar cells, as well as the advantages and challenges with respect to their bulk counterparts. In the last few years an important progress has been achieved in understanding the main elements leading to higher efficiencies. Still, several other elements are still in the course of development and will certainly bring to a further growth of this exciting research area. In the following, several of these aspects are briefly presented.

Novel Configurations and Materials Combinations

The area of nanowire-based solar cells is currently at a very early stage. Basically, only devices with the most common semiconductors and using relatively traditional processing have been realized. The solar cells of the future will have to combine more materials, either for two- to four-junction solar cells or for creating heterojunction devices that improve the collection. Also, there is the need to find materials that reduce the surface recombination losses, which are particularly important for nanowire-based devices [72].

New Physical Paradigms

Light scattering in single nanowires and/or nanowire arrays is very much a field of intensive research, in which important contributions are continuously arising. As an example, the concept of leaky-mode resonances has been adapted from its classical waveguiding domain, with the resulting claim that high refractive index nanostructures such as nanowires can confine light within the subwavelength dimensions [19]. Such a phenomenon is ideally suited for the enhancement of light absorption, and offers a handle to engineer it. As the understanding of this type of resonance as applied to nanostructure geometries progresses, it is expected that nanowire solar cells with higher efficiencies will be obtained.

Integration of Plasmonic Elements

The scientific discipline of plasmonics studies the unique optical properties of metallic nanostructrures to route and manipulate light at length scales of the nanometer, through collective electron oscillations known as plasmons. In recent years, the field of plasmonics has produced fantastic results ranging from imaging below the diffraction limit to lasers with sub-wavelength mode volumes. In their comparatively limited application to photovoltaics, various effects can be used for improving the light absorption in the photovoltaic absorber layers: i) acting as optically engineered scattering elements, ii) increasing the absorption cross-section by coupling the plasmonic near-field to the semiconductor, and iii) inducing guided modes in a semiconductor slab. Such methods have to date been applied to thin film solar cells, where the thickness of the photovoltaic absorber is usually kept too thin to avoid losses in the carrier collection [73]. However, their implementation in nanowire solar cells will certainly lead to an improvement in light collection and therefore in an increase of the overall efficiency – indeed, the promise of plasmonics integration with photovoltaics has already generated patent applications in this area [74].

Finally, it has been shown that plasmonic structures can enhance a variety of nonlinear processes in materials [75]. Conceptually similar to the plasmonic optical antenna effects under investigation for surface-enhanced Raman spectroscopy, enhancing nonlinear processes such as up- or down-conversion of light using semiconductor nanowires would lead to a more efficient utilization of the solar spectrum and thereby significantly improve the efficiency even of single junction solar cells.

Towards Large Scale Nano-Photovoltaics: Availability of Materials

Materials availability could restrain the size and growth of photovoltaic systems and their global application to an alternative to solid fuel energy technologies. For example, it has been found that devices performing below 10% efficiency deliver the same lifetime energy output as those above 20%, provided 75% of the material used is reduced. As emphasized in Sec. 1, this is exactly one of the motivations of using nanowires in photovoltaics, as they intrinsically reduce the use of material by a greater amount than 75%. This is especially true in the case where III-V semiconductors are obtained on a non III-V substrate. However, it is useful to evaluate the material extraction costs and supply constraints of the chemical elements to understand what semiconductors can be the future photovoltaic materials. Several extended analyses show that such materials are FeS_2, CuO, and Zn_3P_2 [76], although the scarce literature on these materials means that basic properties are known from the theoretical point of view [77] but practical epitaxy and doping procedures are still not well established. In the next years, it will be interesting to develop the processing of these materials both in the form of thin films and of nanostructures.

Transparent Electrodes and the Role of Other Nanomaterials

The most common top electrode used in solar cell devices is Indium Tin Oxide, also known as ITO. ITO is a material fairly transparent to the solar spectrum and it exhibits a relatively low resistivity, both characteristics ideal for the application to solar cell and/or photodetector applications. The low availability of indium in the earth's core has resulted in the search to alternative materials to ITO [78]. Carbon is one of the most available elements in the earth surface and core, which exists in many allotropes such as graphene and nanotubes. Graphene is formed by a layer of carbon atoms arranged in a honeycomb structure, with a transmission coefficient similar to ITO [79-82]. Provided the conductivity can be engineered to acceptable levels and ohmic contacts to the device reliably made, graphene could be a future solution for top electrodes in solar cells [82]. On the same note, transparent conductive thin films have also been achieved by creating thin film composites of carbon nanotubes and polymers. It has been shown that the use of high quality nanotubes can result in films of an optimal low sheet resistance (160 Ω/cm^2) and transparencies up to 87% [83]. An additional advantage of such carbon based electrodes is that in principle they could be deposited in a conformal way by chemistry-based methods to the nanowire arrays, thereby simplifying the fabrication process [84].

ACKNOWLEDGEMENTS

The authors kindly thank financial support of the European Research Council through the starting grant UpCon and the Swiss National Science Foundation through the project number 2000021-121758/1.

REFERENCES

[1]   Green, M. A. *Third Generation Photovoltaics: Advanced Solar Energy Covnersion*; Springer-Verlag: Berlin, **2003**.
[2]   Tsakalakos, L. Nanostructures for photovoltaics. *Materials Science and Engineering: R: Reports* **2008**, *62*, 175-189.
[3]   Tian, B.; Kempa, T.; Lieber, C. Single Nanowire Photovoltaics. *Chem. Soc. Rev.* **2009**, *38*, 16.
[4]   Kayes, B. M.; Atwater, H. A.; Lewis, N. S. Comparison of the device physics principles of planar and radial p-n junction nanorod solar cells. *J. Appl. Phys.* **2005**, *97*, 114302.
[5]   Hu, L.; Chen, G. Analysis of Optical Absorption *in Silico*n Nanowire Arrays for Photovoltaic Applications. *Nano Lett.* **2007**, *7*, 3249-3252.
[6]   Glas, F. Critical dimensions for the plastic relaxation of strained axial heterostructures in free-standing nanowires. *Phys. Rev. B* **2006**, *74*, 121302(R).
[7]   Kandala, A.; Betti, T.; Fontcuberta i Morral, A. General theoretical considerations on nanowire solar cell designs. *Phys. Stat. Sol. (A)* **2009**, *206*, 173-178.
[8]   Majumdar, A.; Shakouri, A.; Sands, T. D.; Yang, P.; Mao, S. S.; Russo, R. E.; Feick, H.; Weber, E. R.; Kind, H.; Huang, M.; Yan, H.; Wu, Y.; Fan, R. Nanowires, nanostructures and devices fabricated therefrom. United States Patent 6,882,051, filed March 29, 2002. See also incorporating patents.
[9]   Samuelson, L. I.; Ohlsson, B. J.; Ledebo, L. Nanowhiskers with pn junctions, doped nanowhiskers, and methods for preparing them. United States Patent 7,432,522, filed April 1, 2004.
[10]   Kelzenberg, M. D.; Turner-Evans, D. B.; Kayes, B. M.; Filler, M. A.; Putnam, M. C.; Lewis, N. S.; Atwater, H. A. Photovoltaic Measurements in Single-Nanowire Silicon Solar Cells. *Nano Lett.* **2008**, *8*, 710-714.

[11]   Nelson, J. *The physics of solar cells*; Imperial College Press, **2003**.

[12]   Shockley, W.; Queisser, H. J. Detailed Balance Limit of Efficiency of p-n Junction Solar Cells. *J. Appl. Phys.* **1961**, *32*, 510.

[13]   Please note that in this model the geometry of the solar cell (e.g. thin film or nanowire) is not considered.

[14]   Aspnes, D. Recombination at semiconductor surfaces and interfaces. *Surface Science* **1983**, *132*, 406-421.

[15]   Reitzenstein, S.; Munch, S.; Hofmann, C.; Forchel, A.; Crankshaw, S.; Chuang, L. C.; Moewe, M.; Chang-Hasnain, C. Time resolved microphotoluminescence studies of single InP nanowires grown by low pressure metal organic chemical vapor deposition. *Appl. Phys. Lett.* **2007**, *91*, 091103-3.

[16]   Spirkoska, D.; Abstreiter, G.; Fontcuberta i Morral, A. GaAs nanowires and related prismatic heterostructures. *Semicond. Sci. Technol.* **2009**, *24*, 113001.

[17]   Tian, B.; Zheng, X.; Kempa, T. J.; Fang, Y.; Yu, N.; Yu, G.; Huang, J.; Lieber, C. M. Coaxial silicon nanowires as solar cells and nanoelectronic power sources. *Nature* **2007**, *449*, 885-889.

[18]   Wagner, R. S.; Ellis, W. C. Vapor-Liquid-Solid Mechanism of Single Crystal Growth. *Appl. Phys. Lett.* **1964**, *4*, 89-90.

[19]   Cao, L.; White, J. S.; Park, J.; Schuller, J. A.; Clemens, B. M.; Brongersma, M. L. Engineering light absorption in semiconductor nanowire devices. *Nat. Mater.* **2009**, *8*, 643-647.

[20]   Colombo, C.; Heiss, M.; Graetzel, M.; Fontcuberta i Morral, A. Gallium arsenide p-i-n radial structures for photovoltaic applications. *Appl. Phys. Lett.* **2009**, *94*, 173108-3.

[21]   Colombo, C.; Spirkoska, D.; Frimmer, M.; Abstreiter, G.; Fontcuberta i Morral, A. Ga-assisted catalyst-free growth mechanism of GaAs nanowires by molecular beam epitaxy. *Phys. Rev. B* **2008**, *77*, 155326.

[22]   Heigoldt, M.; Arbiol, J.; Spirkoska, D.; Rebled, J. M.; Conesa-Boj, S.; Abstreiter, G.; Peiró, F.; Morante, J. R.; Fontcuberta i Morral, A. Long range epitaxial growth of prismatic heterostructures on the facets of catalyst-free GaAs nanowires. *J. Mater. Chem.* **2009**, *19*, 840.

[23]   Fontcuberta i Morral, A.; Spirkoska, D.; Arbiol, J.; Heigoldt, M.; Morante, J. R.; Abstreiter, G. Prismatic Quantum Heterostructures Synthesized on Molecular-Beam Epitaxy GaAs Nanowires. *Small* **2008**, *4*, 899-903.

[24]   Qian, F.; Li, Y.; Gradecak, S.; Park, H.; Dong, Y.; Ding, Y.; Wang, Z. L.; Lieber, C. M. Multi-quantum-well nanowire heterostructures for wavelength-controlled lasers. *Nat. Mater.* **2008**, *7*, 701-706.

[25]   Dong, Y.; Tian, B.; Kempa, T. J.; Lieber, C. M. Coaxial Group III-Nitride Nanowire Photovoltaics. *Nano Lett.* **2009**, *9*, 2183-2187.

[26]   Perraud, S.; Poncet, S.; Noël, S.; Levis, M.; Faucherand, P.; Rouvière, E.; Thony, P.; Jaussaud, C.; Delsol, R. Full process for integrating silicon nanowire arrays into solar cells. *Solar Energy Materials and Solar Cells* **2009**, *93*, 1568-1571.

[27]   Tang, Y. B.; Chen, Z. H.; Song, H. S.; Lee, C. S.; Cong, H. T.; Cheng, H. M.; Zhang, W. J.; Bello, I.; Lee, S. T. Vertically Aligned p-Type Single-Crystalline GaN Nanorod Arrays on n-Type Si for Heterojunction Photovoltaic Cells. *Nano Lett.* **2008**, *8*, 4191-4195.

[28]   Haraguchi, K.; Katsuyama, T.; Hiruma, K.; Ogawa, K. GaAs p-n junction formed in quantum wire crystals. *Appl. Phys. Lett.* **1992**, *60*, 745.

[29]   Tsakalakos, L.; Balch, J.; Fronheiser, J.; Korevaar, B. A.; Sulima, O.; Rand, J. Silicon nanowire solar cells. *Appl. Phys. Lett.* **2007**, *91*, 233117.

[30]   Garnett, E. C.; Yang, P. Silicon Nanowire Radial p–n Junction Solar Cells. *J. Am. Chem. Soc.* **2008**, *130*, 9224-9225.

[31]   Peng, K.; Xu, Y.; Wu, Y.; Yan, Y.; Lee, S.; Zhu, J. Aligned Single-Crystalline Si Nanowire Arrays for Photovoltaic Applications. *Small* **2005**, *1*, 1062-1067.

[32]   Sivakov, V.; Andrä, G.; Gawlik, A.; Berger, A.; Plentz, J.; Falk, F.; Christiansen, S. H. Silicon Nanowire-Based Solar Cells on Glass: Synthesis, Optical Properties, and Cell Parameters. *Nano Lett.* **2009**, *9*, 1549-1554.

[33]   Boettcher, S. W.; Spurgeon, J. M.; Putnam, M. C.; Warren, E. L.; Turner-Evans, D. B.; Kelzenberg, M. D.; Maiolo, J. R.; Atwater, H. A.; Lewis, N. S. Energy-Conversion Properties of Vapor-Liquid-Solid-Grown Silicon Wire-Array Photocathodes. *Science* **2010**, *327*, 185-187.

[34]   Peng, K.; Wang, X.; Wu, X.; Lee, S. Platinum Nanoparticle Decorated Silicon Nanowires for Efficient Solar Energy Conversion. *Nano Lett.* **2009**, *9*, 3704-3709.

[35]   Caram, J.; Sandoval, C.; Tirado, M.; Comedi, D.; Czaban, J.; Thompson, D. A.; LaPierre, R. R. Electrical characteristics of core–shell p–n GaAs nanowire structures with Te as the n-dopant. *Nanotechnology* **2010**, *21*, 134007.

[36]   Czaban, J. A.; Thompson, D. A.; LaPierre, R. R. GaAs Core–Shell Nanowires for Photovoltaic Applications. *Nano Lett.* **2009**, *9*, 148-154.

[37]   Goto, H.; Nosaki, K.; Tomioka, K.; Hara, S.; Hiruma, K.; Motohisa, J.; Fukui, T. Growth of Core–Shell InP Nanowires for Photovoltaic Application by Selective-Area Metal Organic Vapor Phase Epitaxy. *Appl. Phys. Express* **2009**, *2*, 035004.

[38]   Wei, W.; Bao, X.; Soci, C.; Ding, Y.; Wang, Z.; Wang, D. Direct Heteroepitaxy of Vertical InAs Nanowires on Si Substrates for Broad Band Photovoltaics and Photodetection. *Nano Lett.* **2009**, *9*, 2926-2934.

[39]   Sivakov, V.; Andrä, G.; Gawlik, A.; Berger, A.; Plentz, J.; Falk, F.; Christiansen, S. H. Silicon Nanowire-Based Solar Cells on Glass: Synthesis, Optical Properties, and Cell Parameters. *Nano Lett.* **2009**, *9*, 1549-1554.

[40]   Fan, Z.; Razavi, H.; Do, J.; Moriwaki, A.; Ergen, O.; Chueh, Y.; Leu, P. W.; Ho, J. C.; Takahashi, T.; Reichertz, L. A.; Neale, S.; Yu, K.; Wu, M.; Ager, J. W.; Javey, A. Three-dimensional nanopillar-array photovoltaics on low-cost and flexible substrates. *Nat. Mater.* **2009**, *8*, 648-653.

[41]   Shockley, W.; Read, Jr., W. T. Statistics of the Recombinations of Holes and Electrons. *Phys. Rev.* **1952**, *87*, 835-842.

[42]   Hall, R. N. Electron-Hole Recombination in Germanium. *Phys. Rev.* **1952**, *87*, 387.

[43]   King, R. R.; Law, D. C.; Edmondson, K. M.; Fetzer, C. M.; Kinsey, G. S.; Yoon, H.; Sherif, R. A.; Karam, N. H. 40% efficient metamorphic GaInP/GaInAs/Ge multijunction solar cells. *Appl. Phys. Lett.* **2007**, *90*, 183516.

[44]   Björk, M. T.; Ohlsson, B. J.; Sass, T.; Persson, A. I.; Thelander, C.; Magnusson, M. H.; Deppert, K.; Wallenberg, L. R.; Samuelson, L. One-dimensional Steeplechase for Electrons Realized. *Nano Lett.* **2002**, *2*, 87-89.

[45]   Gudiksen, M. S.; Lauhon, L. J.; Wang, J.; Smith, D. C.; Lieber, C. M. Growth of nanowire superlattice structures for nanoscale photonics and electronics. *Nature* **2002**, *415*, 617-620.

[46]   Verheijen, M. A.; Immink, G.; de Smet, T.; Borgström, M. T.; Bakkers, E. P. A. M. Growth Kinetics of Heterostructured GaP−GaAs Nanowires. *J. Am. Chem. Soc.* **2006**, *128*, 1353-1359.

[47]   Chuang, L. C.; Moewe, M.; Chase, C.; Kobayashi, N. P.; Chang-Hasnain, C.; Crankshaw, S. Critical diameter for III-V nanowires grown on lattice-mismatched substrates. *Appl. Phys. Lett.* **2007**, *90*, 043115-3.

[48]   Dick, K. A.; Kodambaka, S.; Reuter, M. C.; Deppert, K.; Samuelson, L.; Seifert, W.; Wallenberg, L. R.; Ross, F. M. The Morphology of Axial and Branched Nanowire Heterostructures. *Nano Lett.* **2007**, *7*, 1817-1822.

[49]   Kempa, T. J.; Tian, B.; Kim, D. R.; Hu, J.; Zheng, X.; Lieber, C. M. Single and Tandem Axial p-i-n Nanowire Photovoltaic Devices. *Nano Lett.* **2008**, *8*, 3456-3460.

[50]   Henry, C. H. Limiting efficiencies of ideal single and multiple energy gap terrestrial solar cells. *J. Appl. Phys.* **1980**, *51*, 4494.

[51]   Esaki, L. New Phenomenon in Narrow Germanium p-n Junctions. *Phys. Rev.* **1958**, *109*, 603-604.

[52]   Wallentin, J.; Persson, J. M.; Wagner, J. B.; Samuelson, L.; Deppert, K.; Borgström, M. T. High-Performance Single Nanowire Tunnel Diodes. *Nano Lett.* **2010**, *10*, 974-979.

[53]   Suzuki, N.; Anan, T.; Hatakeyama, H.; Tsuji, M. Low resistance tunnel junctions with type-II heterostructures. *Appl. Phys. Lett.* **2006**, *88*, 231103.

[54]   Li, J.; Yu, H.; Wong, S. M.; Li, X.; Zhang, G.; Lo, P. G.; Kwong, D. Design guidelines of periodic Si nanowire arrays for solar cell application. *Appl. Phys. Lett.* **2009**, *95*, 243113.

[55]   Lin, C.; Povinelli, M. L. Optical absorption enhancement *in silico*n nanowire arrays with a large lattice constant for photovoltaic applications. *Opt. Express* **2009**, *17*, 19371.

[56]   Kupec, J.; Witzigmann, B. Dispersion, Wave Propagation and Efficiency Analysis of Nanowire Solar Cells. *Opt. Express* **2009**, *17*, 10399.

[57]   Kelzenberg, M. D.; Boettcher, S. W.; Petykiewicz, J. A.; Turner-Evans, D. B.; Putnam, M. C.; Warren, E. L.; Spurgeon, J. M.; Briggs, R. M.; Lewis, N. S.; Atwater, H. A. Enhanced absorption and carrier collection in Si wire arrays for photovoltaic applications. *Nat. Mater.* **2010**, 239-244.

[58]   Zhu, J.; Zongfu, Y.; Burkhard, G. F.; Hsu, C.; Connor, S. T.; Xu, Y.; Wang, Q.; McGehee, M.; Fan, S.; Cui, Y. Optical Absorption Enhancement in Amorphous Silicon Nanowire and Nanocone Arrays. *Nano Lett.* **2009**, *9*, 279-282.

[59]   Lieber, C. M.; Yang, P. Metal oxide nanorods. United States Patent 5,897,945, filed February 26, 1996.

[60]   Scher, E. C.; Buretea, M.; Empedocles, S. A. Nanostructure and nanocomposite based compositions and photovoltaic devices. United States Patent 7,087,833, filed December 9 2004.

[61]   Tsakalakos, L.; Lee, J.; Korman, C. S.; Leboeuf, S. F.; Ebong, A.; Wojnarowski, R.; Srivastava, A. M.; Sulima, O. High efficiency inorganic nanorod-enhanced photovoltaic devices. United States Patent Application 2006/0207647, filed March 16, 2005.

[62]   Bettge, M.; Burdin, S.; MacLaren, S.; Petrov, I.; Sammann, E. Photovoltaic and Photosensing Devices Based on Arrays of Aligned Nanostructures. United States Patent Application 2008/0006319, filed June 5, 2007.

[63]   Samuelson, L.; Magnusson, M.; Capasso, F. Nanowire-based Solar Cell Structure. WIPO 2008/156421, filed June 19, 2008.

[64]   Atwater, H. A.; Kayes, B. M.; Lewis, N. S.; Maiolo, III, J. R.; Spurgeon, J. M. Structures of Ordered Arrays of Semiconductors. United States Patent Application 2009/0020150, filed July 18, 2007.

[65]   Buchine, B. A.; Modawar, F.; Black, M. R. Process for Fabricating Nanowire Arrays. United States Patent Application 2009/0256134, filed April 14, 2009.

[66]   Diepers, H. Solar cell comprising semiconductive whiskers. United States Patent 4,099,986, filed August 23, **1978**.

[67]   Law, M.; Greene, L. E.; Johnson, J. C.; Saykally, R.; Yang, P. Nanowire dye-sensitized solar cells. *Nat. Mater.* **2005**, *4*, 455-459.

[68]   Law, M.; Greene, L. E.; Radenovic, A.; Kuykendall, T.; Liphardt, J.; Yang, P. ZnO–Al2O3 and ZnO–TiO2 Core–Shell Nanowire Dye-Sensitized Solar Cells. *J. Phys. Chem. B* **2006**, *110*, 22652-22663.

[69]   Yang, P.; Greene, L. E.; Law, M. Nanowire array and nanowire solar cells and methods for forming the same. United States Patent 7,545,051, filed September 30, 2007.

[70]   Yang, P.; Greene, L.; Law, M. Nanowire array and nanowire solar cells and methods for forming the same. United States Patent 7,265,037, filed June 14, 2004

[71]   Zhang, F.; Barrowcliff, R. A.; Hsu, S. T. Photovoltaic structure with a conductive nanowire array electrode. United States Patent 7,635,600, filed November 16, 2005.

[72]   Spurgeon, J. M.; Atwater, H. A.; Lewis, N. S. A Comparison Between the Behavior of Nanorod Array and Planar Cd(Se, Te) Photoelectrodes. *J. Phys. Chem. C* **2008**, *112*, 6186-6193.

[73]   Atwater, H. A.; Polman, A. Plasmonics for improved photovoltaic devices. *Nat. Mater.* **2010**, *9*, 205-213.

[74]   Atwater, H. A. Plasmonic Photovoltaics. United States Patent Application 2007/0289623, filed June 7, 2007.

[75]   Schuller, J. A.; Barnard, E. S.; Cai, W.; Jun, Y. C.; White, J. S.; Brongersma, M. L. Plasmonics for extreme light concentration and manipulation. *Nat. Mater.* **2010**, *9*, 193-204.

[76]   Wadia, C.; Alivisatos, A. P.; Kammen, D. M. Materials Availability Expands the Opportunity for Large-Scale Photovoltaics Deployment. *Environ. Sci. Technol.* **2009**, *43*, 2072-2077.

[77]   Andrzejewski, J.; Misiewicz, J. Energy Band Structure of Zn3P2-Type Semiconductors: Analysis of the Crystal Structure Simplifications and Energy Band Calculations. *Phys. Stat. Sol. (B)* **2001**, *227*, 515-540.

[78]   Andersson, B. A. Materials availability for large-scale thin-film photovoltaics. *Prog. Photovolt: Res. Appl.* **2000**, *8*, 61-76.

[79]   Geim, A. K. Graphene: Status and Prospects. *Science* **2009**, *324*, 1530-1534.

[80]   Kim, K. S.; Zhao, Y.; Jang, H.; Lee, S. Y.; Kim, J. M.; Kim, K. S.; Ahn, J.; Kim, P.; Choi, J.; Hong, B. H. Large-scale pattern growth of graphene films for stretchable transparent electrodes. *Nature* **2009**, *457*, 706-710.

[81]   Reina, A.; Jia, X.; Ho, J.; Nezich, D.; Son, H.; Bulovic, V.; Dresselhaus, M. S.; Kong, J. Large Area, Few-Layer Graphene Films on Arbitrary Substrates by Chemical Vapor Deposition. *Nano Lett.* **2009**, *9*, 30-35.

[82]   Wang, X.; Zhi, L.; Mullen, K. Transparent, Conductive Graphene Electrodes for Dye-Sensitized Solar Cells. *Nano Lett.* **2008**, *8*, 323-327.

[83]   Zhang, D.; Ryu, K.; Liu, X.; Polikarpov, E.; Ly, J.; Tompson, M. E.; Zhou, C. Transparent, Conductive, and Flexible Carbon Nanotube Films and Their Application in Organic Light-Emitting Diodes. *Nano Lett.* **2006**, *6*, 1880-1886.

[84]   Hu, L.; Choi, J. W.; Yang, Y.; Jeong, S.; La Mantia, F.; Cui, L.; Cui, Y. Highly conductive paper for energy-storage devices. *Proc. Nat. Acad. Sci.* **2009**, *106*, 21490-21494.

Author Index

A

Rienk E. Algra

B

Erik P.A.M. Bakkers

C

S. Crankshaw

D

Dheeraj L. Dasa

F

Lou-Fé Feiner
Takashi Fukui

G

Frank Glas

H

Shinjiroh Hara
Jean-Christophe Harmand
Kenji Hiruma
Bin Hua
Moira Hocevar

J

Fauzia Jabeen

L

Ray LaPierre
Ludovic Largeau
Jianye Li
Linsheng Liu

M

Faustino Martelli
Zetian Mi
A. Fontcuberta i Morral
Junichi Motohisa

P

Gilles Patriarche
Werner Prost

S

Corinne Sartel

T

Maria Tchernycheva
Franz-Josef Tegude
Katsuhiro Tomioka

V

Kumar S. K. Varadwaj
Marcel A. Verheijen

W

Deli Wang
Helge Weman

Z

Hongmin Zhu

SUBJECT INDEX

www.ingramcontent.com/pod-product-compliance
Lightning Source LLC
Chambersburg PA
CBHW041704210326
41598CB00007B/526